T0202836

Linear Regression
Models

Statistics in the Social and Behavioral Sciences Series

Series Editors
Jeff Gill, Steven Heeringa, Wim J. van der Linden, Tom Snijders

Recently Published Titles

Multivariate Analysis in the Behavioral Sciences, Second Edition
Kimmo Vehkalahti and Brian S. Everitt

Analysis of Integrated Data
Li-Chun Zhang and Raymond L. Chambers

Multilevel Modeling Using R, Second Edition
W. Holmes Finch, Joselyn E. Bolin, and Ken Kelley

Modelling Spatial and Spatial-Temporal Data: A Bayesian Approach
Robert Haining and Guangquan Li

Handbook of Automated Scoring: Theory into Practice
Duanli Yan, André A. Rupp, and Peter W. Foltz

Interviewer Effects from a Total Survey Error Perspective
Kristen Olson, Jolene D. Smyth, Jennifer Dykema, Allyson Holbrook, Frauke Kreuter, and Brady T. West

Measurement Models for Psychological Attributes
Klaas Sijtsma and Andries van der Ark

Big Data and Social Science: Data Science Methods and Tools for Research and Practice, Second Edition
Ian Foster, Rayid Ghani, Ron S. Jarmin, Frauke Kreuter and Julia Lane

Understanding Elections through Statistics: Polling, Prediction, and Testing
Ole J. Forsberg

Analyzing Spatial Models of Choice and Judgment, Second Edition
David A. Armstrong II, Ryan Bakker, Royce Carroll, Christopher Hare, Keith T. Poole and Howard Rosenthal

Introduction to R for Social Scientists: A Tidy Programming Approach
Ryan Kennedy and Philip Waggoner

Linear Regression Models: Applications in R
John P. Hoffmann

Mixed-Mode Surveys: Design and Analysis
Jan van den Brakel, Bart Buelens, Madelon Cremers, Annemieke Luiten, Vivian Meertens, Barry Schouten and Rachel Vis-Visschers

For more information about this series, please visit: https://www.routledge.com/Chapman--HallCRC-Statistics-in-the-Social-and-Behavioral-Sciences/book-series/CHSTSOBESCI

Linear Regression Models

Models

Applications in R

John P. Hoffmann

CRC Press
Taylor & Francis Group
Boca Raton London New York

CRC Press is an imprint of the
Taylor & Francis Group, an **informa** business

A CHAPMAN & HALL BOOK

First edition published 2022
by CRC Press
6000 Broken Sound Parkway NW, Suite 300, Boca Raton, FL 33487-2742
and by CRC Press
2 Park Square, Milton Park, Abingdon, Oxon, OX14 4RN

© 2022 John P. Hoffmann
CRC Press is an imprint of Taylor & Francis Group, LLC

Reasonable efforts have been made to publish reliable data and information, but the author and publisher cannot assume responsibility for the validity of all materials or the consequences of their use. The authors and publishers have attempted to trace the copyright holders of all material reproduced in this publication and apologize to copyright holders if permission to publish in this form has not been obtained. If any copyright material has not been acknowledged please write and let us know so we may rectify in any future reprint.

Except as permitted under U.S. Copyright Law, no part of this book may be reprinted, reproduced, transmitted, or utilized in any form by any electronic, mechanical, or other means, now known or hereafter invented, including photocopying, microfilming, and recording, or in any information storage or retrieval system, without written permission from the publishers.

For permission to photocopy or use material electronically from this work, access www.copyright.com or contact the Copyright Clearance Center, Inc. (CCC), 222 Rosewood Drive, Danvers, MA 01923, 978-750-8400. For works that are not available on CCC please contact mpkbookspermissions@tandf.co.uk

Trademark notice: Product or corporate names may be trademarks or registered trademarks and are used only for identification and explanation without intent to infringe.

ISBN: 9780367753689 (hbk)
ISBN: 9780367753665 (pbk)
ISBN: 9781003162230 (ebk)

Typeset in Palatino
by Deanta Global Publishing Services, Chennai, India

Contents

/

Preface

I wrote this book for researchers, both novice and experienced, whose goal is to investigate statistical associations among quantitative variables. Although several empirical models are available to help researchers meet this goal, I describe one of the oldest and best understood methods: the *method of least squares* or the *linear regression model* (LRM). LRMs are designed to account for or predict the values of a single continuous outcome variable with information from one or more explanatory variables. For example, in the U.S., many jurisdictions, including more than half the states, now allow marijuana—which the federal government continues to classify as having no currently accepted medical use and a high potential for abuse[1]—to be prescribed for medical conditions. A concern of public health officials is that the proliferation of medical marijuana prescriptions will motivate more marijuana use among young people. Suppose a research team wishes to test the notion that use has increased. They might compare the prevalence of marijuana use among youth in states that permit medical marijuana to those that do not. In fact, one research group completed such a comparison and, using an LRM to consider several potential influences in addition to medical marijuana availability, found that allowing physicians to prescribe marijuana had no effect on youth marijuana use.[2]

Some quantitative researchers, unfortunately, dismiss the LRM because more recently developed statistical models are considered more rigorous, complex, or even in vogue. In some situations, it might be necessary to choose another statistical model, but it is unwise to ignore the LRM because it remains a valuable tool for studying associations among variables, making predictions, and even, in certain circumstances, helping to identify causal associations. In addition, it remains popular because (1) it is relatively easy to use and understand; (2) statistical software to estimate LRMs is widely available; and (3) LRMs offer a flexible and powerful method for conducting important types of analyses.

Even though many researchers use LRMs because they offer an effective set of tools for understanding the association between two or more variables, they are sometimes misused by those who fail to take into account the assumptions the underlie the models. For instance, a LRM might provide unstable results when two or more of the explanatory variables (if you don't know this term, you will; just keep reading) have high correlations

---

[1] https://www.dea.gov/drug-scheduling
[2] Arthur Robin Williams et al. (2017). "Loose Regulation of Medical Marijuana Programs Associated with Higher Rates of Adult Marijuana Use but Not Cannabis Use Disorder," *Addiction* 112(11): 1985–1991.

(see Chapter 10). Similarly, if one of the data points is substantially different from the others, then the LRM is forced to compensate, often with questionable results (see Chapter 14). Along with providing a thorough description of LRMs, one of the goals of this book is to provide a relatively painless discussion of the model's assumptions, including how to determine if they are satisfied and what to do if they're not.

Prerequisites

The book is written for researchers, including students just learning about these models and more seasoned scholars who might need some brushing up on LRMs or who wish to learn how to use the statistical software R to estimate the models. It does not assume a background with R, although readers should become acquainted with its more basic functions (Chapter 1 includes advice about resources for learning R). But the book does assume some facility with elementary algebra and the completion of or solid familiarity with material from an introductory (non-calculus) course in statistics. For instance, experience with measures of central tendency (e.g., mean, median), measures of dispersion (e.g., variance, standard deviation), measures of association (e.g., bivariate correlations, t-tests), and statistical inference (e.g., p-values, confidence intervals) provides a suitable background. For those who need brushing up, though, Chapter 2 provides a review of these topics.

Overview of the Chapters

The chapters follow the typical format for books on LRMs. After a general introduction, we review probability and elementary statistics, including important advice about how to interpret statistical analyses. Second, we learn about the *simple LRM*, which includes only one explanatory or predictor variable, and how to understand and interpret its results. Third, we examine how to add additional variables to a model. This transforms it into a *multiple LRM*. Fourth, we learn about goodness-of-fit statistics, model comparison procedures, and indicator variables. Fifth, we embark on an in-depth discussion of the assumptions of LRMs. Several chapters address exciting (okay, too strong of a word) and mystifying topics such as independence, homoscedasticity, collinearity, linearity, model specification issues, measurement error, and influential observations. The goal of these chapters is to help readers understand the model's assumptions, as well as the consequences and some remedies when they are not met.

After addressing assumptions and several derivative issues, we discuss a popular regression approach that is suitable for data that are *nested*. This means that one level of data is an aggregate of the other level (e.g., students nested within schools; people nested within neighborhoods), which are called *multilevel* or *hierarchical* data. We may still use linear regression to analyze this type of data, but the models are treated in their own chapter so we can address some idiosyncrasies. The final substantive chapter is a departure from LRMs because it focuses on situations when the outcome takes on only two categories (e.g., "Do you support the death penalty for murder?" $0 = $ no, $1 = $ yes). The logistic regression model is designed specifically for this situation and, even though it provides a loose fit given the other material, is presented in its own chapter because binary (two-category) outcomes are so common in social, behavioral, and other sciences.

Acknowledgments

I would, first of all, like to thank the many students who have completed courses with me on LRMs and related topics. I have learned much more from them than they will ever learn from me. I have had the privilege of having as research assistants Scott Baldwin, Mandy Workman, Colter Mitchell, Bryan Johnson, Wade Jacobsen, and Liz Warnick. Each has contributed to my courses on statistics in many ways, almost all positive! Karen Spence, Kristina Beacom, Rachel Gates, and Amanda Cooper helped me complete early drafts of the chapters. I am indebted to them for their wonderful assistance. Rob Calver and Vaishali Singh of Taylor & Francis provided excellent editorial guidance and assistance and helped me in myriad ways. The team at Deanta worked tirelessly to save me from mistakes and shepherded my work from amateur musings to a professional book. Finally, this work is dedicated to several remarkable people: my four grandchildren—Rowan, Gideon, Aidan, and August—who have brought me unexpected joy. I am a better person for their presence in my life. And to the late, great, *primus inter pares* statistician, Joseph Hilbe, who wrote some of the finest books on applied statistics and taught generations of researchers—including this one (though please don't blame him for my shortcomings!)—the best statistical practices.

Acknowledgements

Author Biography

John P. Hoffmann is a professor of sociology at Brigham Young University. He holds a PhD in Criminology from the State University of New York at Albany and a Master of Public Health (MPH) from Emory University. He has worked at the U.S. Centers for Disease Control and Prevention (CDC) and the National Opinion Research Center (NORC) of the University of Chicago and taught at Hokkaido University and the University of South Carolina. Hoffmann is the author of more than 100 journal articles and book chapters and 10 books on applied statistics, criminology, and the sociology of religion.

1

Introduction

Think about how often we're exposed to data of some sort. Reports of studies in newspapers, magazines, and online provide data about people, animals, or even abstract entities such as cities, counties, or countries. Life expectancies, crime rates, pollution levels, the prevalence of diseases, unemployment rates, election results, and numerous other phenomena are presented with overwhelming frequency and in painful detail. Understanding statistics—or at least being able to talk intelligently about percentages, means, and margins of error—has become nearly compulsory for the well-informed person. Yet, few people understand enough about statistics to fully grasp not only the strengths but also the weaknesses of the way data are collected and analyzed. What does it mean to say that the life expectancy in the U.S. is 78.7 years? Should we trust exit polls that claim that Wexton will win the election over Comstock by 5% (with a "margin of error" of ± 2%)? When someone claims that "taking calcium supplements is not associated with a significantly lower risk of bone fractures in elderly women," what are they actually saying? These questions, as well as many others, are common in today's world of statistical analysis and numeracy.

For the budding social or behavioral scientist, whether sociologist, psychologist, geographer, political scientist, or economist, avoiding quantitative analyses that move beyond simple statistics such as percentages, means, standard deviations, and t-tests is almost impossible. A large proportion of studies found in professional journals employ statistical models that are designed to predict or explain the occurrence of one variable with information about other variables. The most common type of prediction tool is a *regression model*. Many books and articles describe, for example, how to conduct a *linear regression analysis* (LRA) or estimate an LRM,[1] which, as noted in the Preface, is designed to account for or predict the values of a single outcome variable with information from one or more explanatory variables. Students are usually introduced to this model in a second course on applied statistics, and it is the main focus of this book. Before beginning a detailed description of LRMs, though, let's address some general issues that all researchers and consumers of statistics should bear in mind.

---

[1] The word *linear* is defined as "capable of being represented by a straight line on a graph" (*Oxford English Dictionary*, definition 3, https://www.oed.com). Why the definition specifies a straight line will become clear in Chapter 3.

DOI: 10.1201/9781003162230-1

Our Doubts are Traitors and Make Us Lose the Good We Oft Might Win[2]

A critical issue I hope readers will ponder as they study the material in the following chapters involves perceptions of quantitative research. Statistics has, for better or worse, been maligned by a variety of observers in recent years. For one thing, the so-called "replication crisis" has brought to light the problem that the results of many studies in the social and behavioral sciences cannot be confirmed by subsequent studies.[3] Books with titles such as *How to Lie with Statistics* are also popular[4] and can lend an air of disbelief to many studies that use statistical models. Researchers and statistics educators are often to blame for this disbelief. We frequently fail to impart some important caveats to students and consumers, including:

1. A single study is never the end of the story; multiple studies are needed before we can (or should) reach defensible conclusions about social and behavioral phenomena.

2. Consumers and researchers need to embrace a healthy dose of skepticism when considering the results of research studies.[5] They should ask questions about how data were collected, how variables were measured, and whether the appropriate statistical methods were used. We should also realize that random or sampling "error" (see Chapter 2) affects the results of even the best designed studies.

3. People should be encouraged to use their common sense and reasoning skills when assessing data and the results of analyses. Although it's important to minimize confirmation bias and similar cognitive tendencies that (mis)shape how we process and interpret information, we should still consider whether research findings are based on

---

[2] William Shakespeare, *Measure for Measure*, Act I, Scene IV.

[3] See, for example, Ed Yong (2018), "Psychology's Replication Crisis Is Running Out of Excuses," *The Atlantic*, November 19 (retrieved from https://www.theatlantic.com/science/archive/20 18/11/psychologys-replication-crisis-real/576223).

[4] The book's cover notes that it has sold "over half a million copies" (see Darrell Huff (1993), *How to Lie with Statistics*, New York: W.W. Norton). I suspect this makes it one of the best (if not the best) selling statistics books of all time.

[5] *Healthy* is the operative word here. Unfortunately, I fear that many people have become overly skeptical about the results of scientific studies, even from those that are rigorously designed and executed. In the U.S. there have been increasing skepticism, mistrust of people and institutions, and political polarization that affect worldviews and beliefs. This can motivate some to dismiss important research findings that might otherwise be beneficial (see Esteban Ortiz-Ospina and Max Roser (2019), "Trust," at https://ourworldindata.org/trust; Gleb Tsipursky (2018), "(Dis)trust in Science," *Scientific American*, July 5; and Jan Mewes et al. (2021), "Experiences Matter: A Longitudinal Study of Individual-Level Sources of Declining Social Trust in the United States," *Social Science Research*, retrieved from https://doi.org/ 10.1016/j.ssresearch.2021.102537).

sound premises and follow a logical pattern given what we already know about a phenomenon.

Best Statistical Practices[6]

In the spirit of these three admonitions, it is wise to heed the following advice regarding data analysis in general and regression analysis in particular.

1. Plot your data—early and often.

2. Understand that your dataset is only one of many possible sets of data that could have been observed.

3. Understand the context of your dataset—what is the background science and how were measurements taken (for example, survey questions or direct measures)? What are the limitations of the measurement tools used to collect the data? Are some data missing? Why?

4. Be thoughtful in choosing summary statistics.

5. Decide early which parts of your analysis are exploratory and which parts are confirmatory, and preregister[7] your hypotheses, if not formally then at least in your own mind.

6. If you use *p*-values,[8] which can provide some evidence regarding statistical results, follow these principles:

    a. Report effect sizes and confidence intervals (CIs);

    b. Consider providing graphical evidence of predicted values or effect sizes to display for your audience the magnitude of differences furnished by the analysis;

    c. Report the number of tests you conduct (formal and informal);

    d. Interpret the *p*-value in light of your sample size (and power);

    e. Don't use *p*-values to claim that the null hypothesis of no difference is true; and

---

[6] Adapted from E. Ashley Steel, Martin Liermann, and Peter Guttorp (2019), "Beyond Calculations: A Course in Statistical Thinking," *American Statistician* 73(S1): 392–401.

[7] Preregistration is a growing trend wherein researchers publicly identify the hypotheses that guide their work early in the process. They then restrict the analysis to testing those hypotheses and not others. A common view is that researchers must distinguish the hypothesis generating process from the hypothesis testing process. Preregistration is designed to guard against "fishing expeditions": the tendency to estimate several statistical models and then choose one to report that seems the most interesting or innovative. The point in practice 5 is that we should always preregister hypotheses and the conceptual models or theories that guide them, even if informally, and avoid the temptation to keep estimating statistical models until we "confirm" some attractive, yet post hoc, hypothesis. For additional information, see Brian A. Nosek et al. (2018), "The Preregistration Revolution," *PNAS* 115(11): 2600–2606.

[8] These, as well as effect sizes, CIs, and hypothesis tests, are described in detail in Chapter 2.

   f. Consider the *p*-value as, at best, only *one* source of evidence
      regarding your conclusion rather than the conclusion itself.
7. Consider creating customized, simulation-based statistical tests for
   answering your specific question with your particular dataset.
8. Use simulations to understand the performance of your statistical
   plan on datasets like yours and to test various assumptions.
9. Read results with skepticism, remembering that patterns can easily occur by chance (especially with small samples), and that unexpected results based on small sample sizes are often wrong.
10. Interpret statistical results or patterns in data as being *consistent* or
    inconsistent with a conceptual model or hypothesis instead of claim-
    ing that they reveal or prove some phenomenon or relationship (see
    Chapter 2 for an elaboration of this recommendation).

The material presented in the following chapters is not completely faithful
to these practices. For example, we don't cover how variables are measured,
hypothesis generation, or simulations (but see Appendix B), and we are at
times too willing to trust *p*-values (see Chapter 2). These practices should,
nonetheless, be at the forefront of all researchers' minds as they consider
how to plan, execute, and report their own research.

I hope readers of subsequent chapters will be comfortable thinking about
the results of quantitative studies as they consider this material and as they
embark on their own studies. In fact, I never wish to underemphasize the
importance of careful reasoning among those assessing and using statistical techniques. Nor should we suspend our common sense and knowledge
of the research literature simply because a set of numbers supports some
unusual conclusion. This is not to say that statistical analysis is not valuable or that the results are generally misleading. Numerous findings from
research studies that did not comport with accepted knowledge have been
shown valid in subsequent studies. Statistical analyses have also led to many
noteworthy discoveries in social, behavioral, and health sciences, as well as
informed policy in a productive way. The point I wish to impart is that we
need a combination of tools—including statistical methods, a clear compre-
hension of previous research, and our own ideas and reasoning abilities—to
help us understand social and behavioral issues.

Statistical Software

I have taught courses on regression models for many years. When I first
started out, most social and behavioral scientists used SPSS or SAS to estimate
statistical models. I had used both but became a diehard Stata user. So, after

a few years teaching students to use SPSS for statistical modeling I switched to Stata. But the tide has turned and the statistical software R (www.r-project .org)—a descendant of SPlus—is on the rise in my field. I therefore opted to prepare this book using R for the analytic examples. Since this is not a book on statistical software, however, I strongly urge readers to take the necessary time to learn to use R, which, among its many capabilities, performs all of the analyses presented herein. It has a rather steep learning curve, but once you've mastered the basics of R, estimating univariate, bivariate, and multi-variable statistics, including LRMs, is a straightforward task. Learning to use R is easier if you have experience with another statistical software package such as SAS, SPSS, or Stata; but even a diligent novice can learn enough about it to follow the material in this book in, I suspect, about a week.

Even though R can be challenging to learn, it has several advantages over similar statistical software. First, it has become the go-to software in applied statistics, data science, and research endeavors in many fields of inquiry. Many social and behavioral science departments now teach R rather than SPSS or Stata. Second, its vast user community regularly contributes cutting-edge statistical tools and approaches that are not found among its competitors. Several of its contributors have also written graphical user interfaces (GUIs) for R that make estimating LRMs and similar models easier to implement (see, for example, R Commander at www.rcommander.com and Deducer at www.deducer.org). Third, unlike most other statistical software, R is free for individuals to download and use. Fortunately, despite the precipitous learning curve, many helpful online resources that facilitate learning R are available, including Swirl (swirlstats.com), Learn R (www.youtube .com/user/TheLearnR), and Quick R (www.statmethods.net). Many universities also offer online R instructional tutorials, such as BYU (fhssrsc.byu .edu/Pages/R-Tutorial.aspx), UCLA (stats.idre.ucla.edu/r), the University of Georgia (www.cyclismo.org/tutorial/R), and the University of Denver (math .ucdenver.edu/RTutorial). The main R website (www.r-project.org) has a *Documentation—Other* link (lower left of the page) that includes resources designed to help people learn about the various aspects of R (see, especially, the contributed documentation link: https://cran.r-project.org/other-docs.ht ml). The *R Journal* is also available through the main R website and offers articles published by users about new methods. Finally, numerous books explain how to use R for just about any statistical method imaginable.[9]

---

[9] This includes regression models and related methods. See, for instance, the comprehensive treatments in Julian J. Faraway (2014), *Linear Models with R*, 2nd Ed., Boca Raton, FL: CRC Press; Christopher Hay-Jahans (2012), *An R Companion to Linear Statistical Models*, Boca Raton, FL: CRC Press; and John Fox and Sanford Weisberg (2018), *An R Companion to Applied Regression*, 3rd Ed., Thousand Oaks, CA: Sage. For the economist, a free online book is Christoph Hanck et al.'s (2019) *An Introduction to Econometrics with R* (https://www.econometrics-with-r.org). A helpful introduction to R that covers the basics of analysis through more advanced topics is Robert Kabacoff's (2015) *R in Action: Data Analysis and Graphics with R*, Shelter Island, NY: Manning Publications.

To complement and ease the use of R, the examples in this book were prepared in an integrated development environment (IDE). An IDE is software that combines some basic tools that developers need to write and test software. Most IDEs include a code editor, a compiler or interpreter, and a debugger that users access through a single GUI. RStudio is our IDE of choice for R. It eases many of the routine functions of R and is available to individuals as a free download at www.rstudio.com/products/rstudio/download.

R's help menu is not especially intuitive at assisting users in finding particular code or determining what an error message means, so I recommend becoming familiar with a couple of good online R resources to help you understand code and functions and how they are implemented. In addition to reading the R User Guide[10] and, when needed, documentation that accompanies user-written packages, some helpful resources include *rseek.org*, which is a search engine for R, and Stack Overflow (https://stackoverflow. com/questions/tagged/r) and R-bloggers (www.r-bloggers.com), which are valuable community resources.

In the following chapters, R code and functions, as well as object and variable names, are listed in `Courier New` font to distinguish them from other text. Output from R is presented in a similar manner. For instance, the following includes an R function and the accompanying output.[11]

R code
```
t.test(pincome ~ female, data=GSS2018)
```

R output (abbreviated)
```
t = -7.125, df = 2123.5, p-value = 1.422e-12
```

---

[10] See https://cran.r-project.org/doc/manuals/r-release/R-intro.pdf.

[11] As we'll learn, R is a comprehensive software package because of its more than 14,000 user-written packages. We employ only a small number of these, however. Appendix D lists those used in the following chapters.

2

Review of Elementary Statistical Concepts

You were probably introduced to statistics in a pre-algebra course. But your initial introduction may have occurred in elementary or primary school. When was the first time you heard the word *mean* used to indicate the average of a set of numbers? What about the term *median*? Perhaps in an early math class. Do you recall doing graphing exercises: being given two sets of numbers and plotting them on the *x*- and *y*-axes (also called the *coordinate axes*)? You may remember that the set of numbers for *x* and *y* are called *variables* because they can take on different values (e.g., [12, 14, 17]). Contrast these to *constants*: sets of numbers that contain single values only (e.g., [2,2,2], [17,17,17]).

At some point, you were likely introduced to the concepts of elementary *probability*.[1] Your introduction might have included a motivating question such as, "What is the probability or chance that, if you choose one ball from a closed container that includes one red and five blue balls, it will be the red ball?" You were instructed to first count the number of possible outcomes (since the container includes six balls, any one of which could be chosen, six possible outcomes exist), which served as a denominator. You then counted the particular outcome—choosing the red ball. Only one red ball is in the container, so the count of this outcome is one. This was a numerator. Putting these two counts together in a fraction resulted in a probability of choosing the red ball of 1/6, or about 0.167. The latter number is also called a *proportion*. Probabilities and proportions fall between zero and one; this constitutes the *first rule of probability*.[2] Multiplying them by 100 creates percentages (0.167 × 100 = 16.7%). But what does a probability of 0.167 mean? One interpretation is that we expect to choose a red ball about 16.7% of the time when we pick balls one at a time over and over (don't forget that we need to replace the chosen ball each time—this is called *sampling with replacement*). Each selection is called an *experiment*, so selecting the balls over and over again composes multiple experiments. But is it realistic to expect to choose a red ball 16.7% of the time? We could test this assumption by repeating the experiment again and

---

[1] For an account of the history of probability, see Lorraine Daston (1988), *Classical Probability in the Enlightenment*, Princeton, NJ: Princeton University Press.

[2] The other two rules are (2) the sum of the probabilities of all possible outcomes equals one (e.g., the probability of choosing a red ball (0.167) plus the probability of choosing a blue ball (0.833) equals 1.0), and (3) the probability that an event does not occur equals one minus the probability that it does occur (e.g., the probability that the red ball is not chosen equals the one minus the probability that the red ball is chosen, or $1 - 0.167 = 0.833$).

DOI: 10.1201/9781003162230-2

again and keeping a lengthy record. Statisticians refer to this approach—the theoretical idea rather than the actual tedious counting—as *frequentist* probability or *frequentism* since it examines, but actually assumes, what happens when something countable is repeated over and over.[3]

Probabilities are presented using, not surprisingly, the letter P. One way to symbolize the probability of choosing a red ball is with $P(\text{red})$. We may write $P(\text{red}) = 0.167$ or $P(\text{red}) = 1/6$. You might recall that some statistical tests, such as t-tests or analysis of variance (ANOVA)s, are accompanied by p-values. As we shall learn, p-values are a type of probability used in many statistical tests.

The basic foundations of statistical analysis are established by combining the principles of probability and elementary statistical concepts. Among a variety of descriptions, statistics may be defined as the analysis of data and the use of such data to make decisions in the presence of uncertainty. This two-pronged definition is useful for delineating two general types of statistical analyses: *descriptive* and *inferential*. Researchers use descriptive methods to analyze one or more variables in order to describe or summarize their characteristics, often with *measures of central tendency* and *measures of dispersion*. Descriptive methods are also employed to visualize variables, such as with histograms, density plots, stem-and-leaf plots, and box-and-whisker plots. We'll see examples of some of these later in the chapter.

Inferential statistics are designed to infer or deduce something about a population from a sample and are useful for decision making, policy analysis, and gaining an understanding of patterns of associations among people, states, companies, or other units of interest. But uncertainty is a key issue since inferring something about a population requires acknowledgment that any sample contains limited information about that population. We'll return to the issue of inferential statistics once we've reviewed some tools for describing variables.

Another way to classify statistics is to distinguish methods that address one variable from those that address two or more variables. Many descriptive methods are used to examine a single variable. The bulk of this book, especially in later chapters, addresses a technique designed for analyzing two or more variables simultaneously. A motivating question concerns whether two or more variables are associated in some way. As the values of one variable increase, do the values of the other variable also tend to increase? Or do they decrease? Two methods designed to answer these inquiries are covariances and correlations. But it's important to remember that finding that one variable increases along with another does not mean they have a *causal relationship: correlation does not equal causation*. We'll be cautious about using the term

---

[3] Other probability approaches or theories include Bayesian, classical, propensity, metaphysical, and several others (Philip Dawid (2017), "On Individual Risk," *Sythese* 194(9): 3445–3474). For details about the implications of these for statistical modeling, see Aris Spanos (2019), *Probability Theory and Statistical Inference*, 2nd Ed., New York: Cambridge University Press, chapter 10.

causation in this book because it involves many thorny philosophical issues.[4] One of our main concerns, though, is whether one or more variables are associated with another variable in a systematic way. Determining whether one variable genuinely "causes" another, however, usually requires a carefully constructed experiment, information that falls outside a set of nonexperimental data, or a degree of statistical complexity that is beyond the scope of this presentation.[5] Evaluating the characteristics of associations among a set of variables is a key goal of linear regression, the statistical model addressed in this book. But since the data we use to demonstrate this model are not based on experiments, but rather are from surveys or administrative sources, we shall generally avoid the term causation or making claims about cause-and-effect relationships. This does not imply that linear regression cannot be used with experimental data. This book merely emphasizes the most common research design in quantitative social and behavioral science applications: observational data in which researchers do not have control over the variables and must rely on evaluating natural variation in the data.

Measures of Central Tendency

Now that we have some background information about statistics, let's turn to some statistical measures, including how they are used and computed. We'll begin with measures of central tendency. Suppose we collect data on the weights (in ounces) of several puppies in a litter. We place each puppy on a digital scale, trying to hold them still so we can record their weights. What is your best guess of the average weight of the puppies in the litter? Perhaps

---

[4] For more information about this topic, see Judea Pearl (2009), *Causality: Modeling, Reasoning, and Inference*, 2nd Ed., New York: Cambridge University Press, and Stephen Mumford and Rani Lill Anjum (2013), *Causation*, Oxford: Oxford University Press.

[5] According to the philosopher and polymath John Stuart Mill's adaptation of David Hume's conditions for "temporal regularity," causation is established by satisfying three criteria: (1) the cause must precede the effect, (2) as the cause changes, the effect must also change, and (3) other explanations or factors that account for the cause–effect association must be eliminated. The third criterion is especially difficult to fulfill with a statistical model because researchers rarely have access to *all* the factors that might explain the association between a presumed cause and an effect. Experimental data are useful for establishing causal relations because researchers have control over the potential causes and can eliminate the effects of other factors through randomization. With observational data (such as those obtained by surveys), however, researchers must make strong assumptions about factors that may not be observed (see Chapter 17 for additional discussion about this issue). Nevertheless, several statistical techniques are available that putatively allow causal relationships to be identified with observational data (see Guido W. Imbens and Donald J. Rubin (2015), *Causal Inference for Statistics, Social, and Biomedical Sciences: An Introduction*, New York: Cambridge University Press).

not always the best, but the most common measure is the *arithmetic mean*,[6] which is computed using the formulas in Equation 2.1.

$$E[X] = \mu = \frac{\sum X_i}{N}\left(\text{population}\right) \text{ or } \bar{x} = \frac{\sum x_i}{n}\left(\text{sample}\right) \qquad (2.1)$$

The term on the left-hand side of the first equation, $E[X]$, is a symbolic way of expressing the *expected value* of variable X, which is often used to represent the mean. We could also list this term as $E[$weight in ounces$]$, but, as long as it's clear that $X =$ weight in ounces, using $E[X]$ is satisfactory. The Greek letter μ represents the population mean, whereas $\bar{x}$ in the second part of the equation is the sample mean. The formula for computing the mean is simple. Add all the values of the variable and divide the sum by the number of observations. The cumbersome symbol that looks like an overgrown E in the numerator of Equation 2.1 is the summation sign; it tells us to add whatever is next to it. The symbol X_i or x_i signifies specific values of the variable, or the individual puppy weights we've recorded. The subscript i indicates each observation. The letter N or n is the number of observations. This may be represented as $i \dots n$. If $n = 5$, then five individual observations are in the sample. Uppercase Roman letters represent population values and lowercase Roman letters represent sample values. $E[X] = \bar{x}$ implies that the sample mean is designed to estimate the population expected value or the population mean.

Here's a simple example. Our Siberian Husky, Steppenwolf, sires a litter of puppies. We weigh them and record the following: [48, 52, 58, 62, 70]. The sum of this set is $[48 + 52 + 58 + 62 + 70] = 290$, with a sample mean of $290/5 = 58$ ounces. The mean is also called the *center of gravity*. Suppose we have a plank of wood that is of uniform weight across its span. We order the puppies from lightest to heaviest—trying to space them out proportional to their weights—and place them on the plank of wood. The mean is the point of balance, or the point at which we would place a fulcrum underneath the plank to balance the puppies.

The mean has a couple of interesting features:

1. It is measured in the same units as the observations. If the observations are not all measured in the same unit (e.g., some puppies' weights are in grams, others in ounces), then the mean is not interpretable.

2. The mean provides a suitable measure of central tendency if the variable is measured continuously and is normally distributed.

---

[6] This measure dates back to ancient Greece, Babylonia, and India. Records indicate it was utilized by the Greek astronomer Hipparchus in the 2nd century BCE (R. L. Plackett (1958), "Studies in the History of Probability and Statistics. VII. The Principle of the Arithmetic Mean," *Biometrika* 45(1/2): 130–135).

What do *continuously* and *normally distributed* signify? A variable is measured continuously—or a variable is *continuous*—if it could be any real number, at least within a particular interval, such as between 0 and 100. Measuring things is rarely so precise, so most researchers round measures to whole numbers or integers. Many things must also be measured using positive numbers only; we cannot, for instance, measure a puppy's weight with a negative number. Another type of variable—known as a *discrete* or *categorical*—has a finite number of possible values within a small range. For example, the type of household pet fits into specific categories, such as dog, cat, and potbellied pig, and is therefore a discrete variable. Discrete variables are categorized as ordered and unordered. Ordered variables can be ranked in some reasonable way. If we categorized puppy weights into low, medium, and high, then the variable that included the three categories is ordered (or *ordinal*). Type of pet is an unordered discrete variable since there's not a logical way to say that one type of pet is of higher rank than another (except among cat or dog lovers).

A variable is normally distributed if it follows a bell-shaped curve. How might we determine this? Order the values of the variable from lowest to highest and then plot them by their frequencies or the percentage of observations that have a particular value (assume the variable has many values). We may then view the "shape" of the distribution of the variable. Figure 2.1 illustrates a bell-shaped distribution, which is also called a *normal or Gaussian*

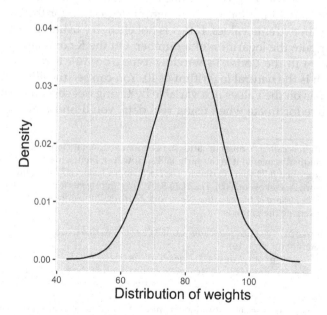

FIGURE 2.1
Simulated normal (Gaussian) distribution of weights.

distribution,[7] using a simulated sample of observations (the graph was generated using R's ggplot function that is part of the ggplot2 package[8]).

We shall return to means and the normal distribution often in this book. To give a flavor of what lies ahead, the LRM is designed, in part, to predict means for particular sets of observations in a sample. For instance, if we have information on the length of our puppies from nose to tail, we may use this information to predict their weights. Our predictions could include calculating the mean weight of puppies who are 20 inches long. An LRM is designed to do this. As discussed later when addressing statistical inference, we might also wish to make reasonable predictions regarding the population of puppies based on information from the puppies in the sample.

Should we use the mean to represent the average value when the variable does not follow a normal distribution? We may, but should do so only if its distribution does not deviate too far from the normal distribution. In many situations in social and behavioral sciences, though, variables do not follow normal distributions. An example of this involves annual income. When information is gathered on individuals' annual incomes, a few people tend to earn a lot more than others. Measures of income are thus usually skewed: when we graph their values they have long right tails. Figure 2.2 furnishes an example of a right-skewed variable.

What are some solutions if we wish to find a good measure of central tendency for a variable such as income? First, we might use the mean if we can make the variable "more normal"—or *normalize* it. A common method, as long as a variable consists of positive values, is to compute its *natural (Naperian*[9]) *logarithm*. A transformation with the natural logarithm (or the base 10 logarithm) pulls in extreme values. If this is not clear, try using the log function in R to compute the logarithm of a number. On the R command line (designated with >) in the Console window, type log(10). R returns the value 2.306, which is the natural logarithm of 10. You can see the effect the natural logarithm has on the values of a variable by trying out several positive numbers. If you're fortunate when using real data, you'll find that transforming

---

[7] The distribution is called "Gaussian" after the German mathematician, Carl Friedrich Gauss, who purportedly discovered it in the early 1800s. However, French mathematician Abraham de Moivre described it 70 years earlier in *The Doctrine of Chances* (1738), calling it "the law of errors" (Clifford A. Pickover (2009), *The Math Book: From Pythagoras to the 57th Dimension—250 Milestones in the History of Mathematics*, New York: Sterling Publishing).

[8] The R code to create the graph is:

```
library(ggplot2) # the package must first be downloaded
a <- rnorm(10000, 80, 10)
wt_density <- ggplot(, aes(a)) + geom_density()
wt_density <- wt_density + scale_x_continuous(name = "Distribution of
            weights") + scale_y_continuous(name = "Density")
wt_density
```

[9] This term is used because the natural logarithm (base *e*) was introduced by Scottish mathematician John Napier (1681) in his book *A Description of the Admirable Table of Logarithmes* (trans. Edward Wright, London: Simon Waterson).

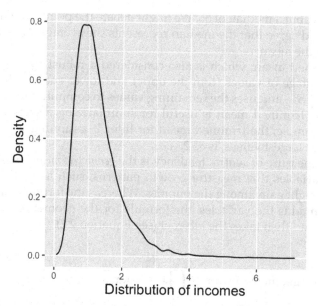

FIGURE 2.2
Simulated right-skewed distribution of annual income.

a variable that has a long right tail using the natural logarithm results in a normal distribution. The square root or cube root may also work to normalize a right-skewed distribution. We'll examine some examples in Chapter 11.

Second, some measures of central tendency are more appropriate than the mean when a variable is not normally distributed or includes extreme values. For example, the median, which is the middle value of a distribution, is not as affected as the mean by extreme values and is thus a *robust statistic*.[10] To determine the median, order the values of the variable from lowest to highest. If the number of observations is odd, choose the middle value; if it's even, take the average of the middle two values. You might recall that the median designates the 50th percentile of a distribution. For example, imagine two puppy weight variables, one that follows a normal distribution (or something close to it) and another that has an extreme value:

Litter 1: [40, 45, 50, 55, 60, 65, 70]

Litter 2: [40, 45, 49, 56, 60, 66, 175]

Litter 1 has a mean of 55 and a median of 55, so the estimate of its central value is the same regardless of which measure is used. In contrast, litter 2 has

---

[10] The earliest work on robust statistics was probably by French mathematician Pierre-Simon Laplace in an 1818 paper on the distribution of the median (Stephen M. Stigler (1973), "Simon Newcomb, Percy Daniell, and the History of Robust Estimation, 1885-1920," *Journal of the American Statistical Association* 68(344): 872–879).

a mean of 70, but a median of 56. We might debate the point, but I think most people would agree that the median represents the average value for litter 2 better than the mean.

The trimmed mean, which is also considered a robust statistic, removes some percentage of values from the upper and lower ends of the distribution, usually 5%, and uses the remaining values to compute a mean. Like the median, the trimmed mean is useful for summarizing a skewed distribution. For example, the trimmed mean for litter 2, assuming we lop off the smallest and largest values, is 55.2.

Another measure of central tendency is the *geometric mean*,[11] which is useful with variables that measure growth patterns, such as if we wished to examine weight gain among the puppies. Whereas the formula for the arithmetic mean adds the variables, the formula for the geometric mean multiplies them and then takes the nth root (see Equation 2.2).

$$\text{geometric mean} = \left(\prod_{i=1}^{n} x_i \right)^{\frac{1}{n}} = \sqrt[n]{x_1 \times x_2 \times x_3 \times \cdots \times x_n} \tag{2.2}$$

The geometric means for the two litters of puppies are 54.1 and 61.7. As indicated by the formula, the geometric mean is appropriate mainly for positive values.

Measures of Dispersion

Knowing a variable's central tendency is just part of the story. As suggested by Figures 2.1 and 2.2, depicting how much a variable fluctuates around the mean is also important. The objective of measures of dispersion is to indicate the spread of the distribution of a variable. You should be familiar with the term *standard deviation*, the most common dispersion measure for continuous variables.[12] Before seeing the formula for this measure, however, let's consider some other measures of dispersion. The most basic for a continuous

---

[11] The astronomer Edmond Halley (of Halley's Comet fame) described this measure in 1695, calling it the *geometrical mean* ("A Most Compendious and Facile Method for Constructing the Logarithms," *Philosophical Transactions of the Royal Society of London* 19(216): 58–67). The geometric mean is often used for computing compound interest on investments. A geometric median—a more robust alternative—also exists.

[12] The first use of this expression was by British mathematician Karl Pearson in 1894 ("Contributions to the Mathematical Theory of Evolution," *Philosophical Transactions of the Royal Society of London, Series A*, 185: 71–110). But the notion of dispersion had been recognized for many years. For instance, Gauss mentioned several measures of variability, such as the *mean error*, in an 1816 paper (H. A. David (1998), "Early Sample Measures of Variability," *Statistical Science* 13(4): 368–377).

variable is the *sum of squares*, or *SS[x]*. Assuming a sample, the formula is supplied in Equation 2.3.

$$SS[x] = \Sigma(x_i - \bar{x})^2 \tag{2.3}$$

We first compute *deviations from the mean* $(x_i - \bar{x})$ for each observation, square each, and then add them. If you've learned about ANOVA models, the sum of squares should be familiar. Perhaps you even recall the various forms of the sum of squares. We'll learn more about these in Chapter 5.

The second measure of dispersion, and one you should recognize, is the *variance*, which is labeled s^2 for samples and σ^2 (*sigma-squared*) for populations. The sample formula is shown in Equation 2.4.

$$var[x] = s^2 = \frac{\Sigma(x_i - \bar{x})^2}{n-1} \tag{2.4}$$

The variance is the sum of squares divided by the sample size minus one and is measured in squared units of the variable. The standard deviation (symbolized as s (sample) or σ (population)), however, is measured in the same units as the variable (see Equation 2.5).

$$sd[x] = s = \sqrt{\frac{\Sigma(x_i - \bar{x})^2}{n-1}} \tag{2.5}$$

A variable's distribution is often represented by its mean and standard deviation (or variance). A variable that follows a normal distribution, for instance, is symbolized as $X \sim N(\mu, \sigma)$ or $x \sim N(\bar{x}, s)$ (the wavy line means "distributed as"). When two variables are measured in the same units and have the same mean, one is less dispersed than the other if its standard deviation is smaller. Recall that the means for the two litters of puppies are 55 and 70. Their standard deviations are 10.8 and 47.1. The weights in the second litter have a much larger standard deviation, which is not surprising given their range. Another useful measure of dispersion is the *coefficient of variation* (CV), which is computed as $s/\bar{x}$ and is usually then multiplied by 100. The CV is valuable for comparing distributions because it shows how much a variable fluctuates about its mean. The CVs for the two litters are 19.6 and 67.1.

Earlier we discussed the median and trimmed mean as robust alternatives to the mean. Robust measures of dispersion are also available, such as the *interquartile range* (75th – 25th quartile) and the *median absolute deviation* (MAD): median $(|x_i - \tilde{x}|)$, or the median of the absolute values of the observed *x*s minus the median. Whereas the standard deviations and CVs of the two puppy litters are far apart, the MADs are the same: 14.8. The one extreme value in litter 2 does not affect this robust measure of dispersion.

As a reminder, different symbols are employed to represent analogous sample and population quantities. Greek or uppercase Roman letters signify population characteristics, which are called *parameters* (e.g., σ, μ), whereas lowercase Roman letters designate statistics (e.g., $\bar{x}$, s^2). This is why one often hears the term *estimator* to refer to a statistic and its formula: an estimator is a method for estimating an unknown parameter (e.g., $\bar{x}$ is an estimate of μ). The next few sections illustrate how this distinction is used in inferential statistics.

Samples and Populations

We learned earlier that one way to classify statistics is to distinguish between descriptive and inferential methods. At the heart of inferential statistics is a question: how do we know that what we find using a sample reflects what occurs in a population? Can we infer what happens in a population with information from a sample? For instance, suppose we're interested in determining who is likely to win the next presidential election in the U.S. Assume only two candidates from whom to choose: Warren and Haley. It would be enormously expensive to ask all people who are likely to vote in the next election their choice of president. But we may take a sample of likely voters and ask them for whom they plan to vote. Can we deduce anything about the population of voters based on this sample? The answer is that it depends on a number of factors. Did we collect a good sample? Were the people who responded honest? Do people change their minds as the election approaches? We don't have space to get into the many issues involved in sampling, so we'll just assume that our sample is a good representation of the population from which it is drawn.[13] Most important for our purposes is this: inferential statistics include a set of methods designed to help researchers answer questions about a population from a sample. Other forms of inferential statistics are not concerned in the same way with a hypothetical population, though. A growing movement is the use of Bayesian inference. We'll refer to this later, but an adequate description is outside the scope of this book.

An aim of many statistical procedures is to infer something about the population from patterns found in samples. Yet, the cynical—but perhaps most honest—answer is that we never know if what we found says anything accurate about a population. Recall that the definition of statistics provided earlier mentioned uncertainty; statistics is occasionally called the *science of*

---

[13] Sampling is a sizeable field, with its own literature. For an introduction, see Robert M. Groves et al. (2009), *Survey Methodology*, 2nd Ed., New York: Wiley. A more in-depth discussion is furnished in Paul S. Levy and Stanley Lemeshow (2013), *Sampling of Populations: Methods and Applications*, 4th Ed., New York: Wiley.

uncertainty. The best we can do given a sample is to offer degrees of confidence that our results reflect characteristics of a population. But what do we mean by *population*? Populations may be divided into *target* populations and *study* populations. Target populations are the group about which we wish to learn something. This might include a group in the future ("I wish to know the average weights of future litters sired by my Siberian Husky") or in the past. Regardless, we typically try to find a population that closely resembles the target population—this is the study population. Many types of populations exist. For instance, we might be interested in the population of seals living on Seal Island off the coast of South Africa; the population of labradoodles in New York City; or the population of voters in Oregon during the 2020 presidential election. Yet some people, when they hear the term population, think it signifies the U.S. population or some other large group.

A sample is a set of items chosen from a population. The best known is the *simple random sample.* Its goal is to select members from the population so that each has an equal chance of being in the sample. Most of the theoretical work on inferential statistics is based on this type of sample. But researchers also use other types, such as clustered samples, stratified samples, and several others.

Sampling Error and Standard Errors

Statistical studies are often deemed valuable because they may be used to deduce something about a population from samples, but keep in mind that researchers usually take only a single sample even though they could conceivably draw many.[14] Any sample statistic we compute or test we run must thus consider the uncertainty involved in sampling—the *sampling error* or the "error" due to using only a portion of a population to estimate a parameter

---

[14] The issue of what it means to have a population and draw a sample from it is one of reasons experts have criticized the typical methods of statistical inference in recent years. Some observers point out that we can never know with any accuracy what occurs in a "population," partly because it tends to be a moving target. Some prefer a different framework, mentioned earlier, known as *Bayesian.* Bayesian inference and analysis are growing in popularity—for good reason, I think—but we remain within a frequentist framework in this book by inferring information about populations from samples. For more on Bayesian and frequentist inference, see James V. Stone (2013), *Bayes' Rule: A Tutorial Introduction to Bayesian Analysis*, Sheffield, UK: Sebtel Press, and a set of papers in *The American Statistician*, 67(1), 2013. For information about using R for Bayesian analysis, see Jean-Michel Marin and Christian P. Robert (2014), *Bayesian Essentials with R*, 2nd Ed., New York: Springer.

from that population.[15] The solution to the problem of uncertainty typically involves using *standard errors* for test statistics, including the mean, the standard deviation, correlations, medians, and, as we shall see, slope coefficients in LRMs. Briefly, a standard error is an estimate of the standard deviation—the variability—of the sampling distribution. The simplest way to understand this is with an example.

Recall that when we compute the variance or the standard deviation, we are concerned with the spread of the distribution of the variable. But imagine drawing many, many samples from a population and computing a mean for each sample. The result is a sample of means from the population ($\bar{x}_i s$) rather than a sample of observations ($x_i s$). We could then compute a mean of these means, or an overall mean, which should reflect pretty accurately—assuming we do a good job of drawing the samples—the actual mean of the population $\left(\dfrac{\sum \bar{x}_{ni}}{n_s} \cong \mu \right)$. Let's expand our examination of puppy litters to help us understand this better.

Litter 1: [40, 45, 50, 55, 60, 65, 70]
Litter 2: [40, 45, 49, 56, 60, 66, 75]
Litter 3: [39, 55, 56, 58, 61, 66, 69]
Litter 4: [42, 44, 48, 55, 57, 60, 66]

The means for the litters are 55, 56, 57, and 53. Their average—the mean of the means—is $(55+56+57+53)/4 = 55.3$. Suppose the samples exhausted the population of puppies. The population mean is thus 55.4. This is close to the mean of the sample means, off by a skosh because of rounding error.

Imagine if we were to take many more samples of puppies. The means from the samples also have a distribution, which is called the *sampling distribution* of the means. We could plot these means to determine if they follow a normal distribution. In fact, an important theorem from mathematical statistics states that, as more and more samples are drawn, their means follow a normal distribution even if they come from a non-normally distributed variable in the population (see Chapter 4). This allows us to make important inferential claims about LRMs. We shall learn about these claims in later chapters.

---

[15] "Error" is a common term in statistics. It may be tempting to think so, yet error is not synonymous with "mistake." For instance, as the definition implies, the sampling error is just the difference in, say, the mean between what a sample shows and what occurs in the population. This type of error is ubiquitous because we don't have all the information from the population of interest. A difference between what we observe in a sample and what exists in the population is almost certain. Other types of errors in research studies, such as errors made in measuring or recording a phenomenon, are closer to meeting the definition of a mistake. An important research goal is to reduce as much error as possible from all sources.

Our concern here is not whether the sample means are normally distributed, at least not directly. Rather, we need to consider a measure of the variability of these means. Statistical theory suggests that a good estimate of their standard deviation is the *standard error* of the mean. The standard error formula in Equation 2.6 includes the sample standard deviation and the sample size.

$$se(\bar{x}) = \frac{s}{\sqrt{n}} \tag{2.6}$$

For example, the standard deviation of weights for litter 4 is 8.8, so the standard error of the mean is $8.8/\sqrt{7} = 3.3$. The actual standard deviation of the sample means using all four litters is 1.7. But if we were to take larger samples from the population, we'd find that the standard error of the mean and the standard deviation of the sample means would get closer. In statistical jargon, we say they *converge*.

Standard errors may also be thought of as quantitative estimates of the uncertainty involved in statistics, uncertainty that arises because we have a sample rather than a population. Several types of standard errors are used in statistical modeling. One type—known as the *standard error of the slope coefficient*—is employed for making inferences about LRMs (see Chapter 3).

Significance Tests

Standard errors are utilized in a couple of ways. First, recall from elementary statistics that when we use, say, a *t*-test, we compare the *t*-value to a table of *p*-values. All else being equal, a larger *t*-value equates to a smaller *p*-value. This approach is known as *significance testing*[16] because we wish to determine

---

[16] The origins of significance testing go back at least to the English physician John Arbuthnot (1711) who noted that, year after year, male births outpaced female births and the chance of this occurring was infinitesimally small. He therefore concluded that divine providence was the cause ("An Argument for Divine Providence, Taken from the Constant Regularity Observ'd in the Births of Both Sexes," *Philosophical Transactions of the Royal Society of London* 27(328): 186–190). Significance testing was developed further by Pierre-Simon Laplace as he also studied male and female births later in the 1700s and developed the framework for CIs; Karl Pearson who introduced the *p*-value in 1900; and Ronald A. Fisher in the 1920s who suggested using 0.05 as the probability threshold at which the results signify that an experiment is worth repeating. For more information, see David Salsburg (2001), *The Lady Tasting Tea*, New York: Owl Books; Eddie Shoesmith (1987), "The Continental Controversy over Arbuthnot's Argument for Divine Providence," *Historia Mathematica* 14(2): 133–146; and Stephen M. Stigler (1986), *The History of Statistics: The Measurement of Uncertainty Before 1900*, Cambridge, MA: Harvard University Press.

if our results are "significantly" different from some other possible result.[17] Significance testing using standard errors is an inferential approach because it is designed to deduce something about a population based on a sample. But the term *significant* does not mean *important*. Rather, it originally meant that the results signified or showed something.[18] A *p*-value is only one piece of evidence that indicates, at best, that a finding is worthy of further consideration; we should not claim that a low *p*-value demonstrates we have found *the* answer or that it reveals the "truth" about some relationship in a population (recall the section on best statistical practices in Chapter 1). A worthwhile adage to remember is "statistical significance is not the same as practical significance." We'll discuss these issues in more detail later in the chapter.

Let's consider an interpretation of a *p*-value and how it's used in a significance test rather than derive its computation. Recall that many statistical exercises are designed to compare two hypotheses: the null and the alternative. The null hypothesis usually claims that the result of some observation or an association in the data is due to chance alone, such as sampling error only, whereas the alternative hypothesis is that the result or association is due to some nonrandom mechanism. Imagine, for instance, we measure weights from the litters of two distinct dog breeds: Siberian Husky and German Shepherd. We compute the two means and find that litter 1's is 5 ounces more than litter 2's. Assuming we treat the two litters as samples from target populations of Siberian Husky and German Shepherd puppies, we wish to determine whether or not the 5-ounce difference suggests a difference in the population means. The null and alternative hypotheses are usually represented as:

Null: H_0: Mean weight, litter 1 (μ_1) = Mean weight, litter 2 (μ_2)

Alternative: H_a: Mean weight, litter 1 (μ_1) ≠ Mean weight, litter 2 (μ_2)

Another way of stating the null hypothesis is that the mean weight of Siberian Husky puppies is actually the same as the mean weight of German Shepherd puppies in the populations of these dog breeds. Because a hypothesis of zero difference is frequently used, though often implicit, some call it the *nil hypothesis*. Recall that the most common way to compare means from two independent groups is with a *t*-test. We'll see a detailed example of this test later. For now, suppose the *t*-test provides a *p*-value of 0.04. One way to interpret this value is with the following garrulous statement:

> If the difference in population means is *zero* $(\mu_1 - \mu_2 = 0)$ and we draw many, many samples from the two populations, we expect to find a

[17] A clear discussion of significance testing is furnished in Amy Gallo (2016), "A Refresher on Statistical Significance," *Harvard Business Review—Analytics*, February 16.
[18] See Salsburg (2001), *op. cit.*

difference in sample means of 5 ounces or more only four times, on average, out of every 100 pairs of samples we examine.

Do you recognize how a *p*-value is a type of probability based on a frequentist inference approach? Researchers are prone to making statements such as "since the *p*-value is below the conventional threshold of 0.05, the *t*-test provides evidence with which to *reject* the null hypothesis" or it "validates the alternative hypothesis."[19] But, as outlined later, such statements should be avoided. The *p*-value provides only one piece of evidence—some argue only a sliver—with which to evaluate hypotheses.

Astute readers may notice the example sets up a *two-tailed* significance test. This assesses the difference in means weights, but it does not judge whether one group's weight is higher or lower than the other. A *one-tailed* test, though, is designed for *directional hypotheses*, or those that examine, say, whether one group's mean weight is larger than the other. The former approach is most common in the social and behavioral sciences, even though directional hypotheses that call for one-tailed tests are widespread (e.g., mean weight, litter 1 > mean weight, litter 2). Yet, some statisticians note that although *p*-values might provide one source of evidence regarding the null hypothesis, they cannot determine the direction of differences. We'll provide additional interpretations of *p*-values later in the chapter.[20]

The second way to use standard errors is to compute CIs.[21] Many statisticians prefer CIs as an inferential tool because they provide a range of values within which parameters are likely to fall. Here we contrast a *point estimate* and an *interval estimate*. The means shown earlier are examples of point

---

[19] As mentioned in footnote 16, Ronald A. Fisher recommended a threshold for the *p*-value of 0.05. But he thought this useful for experiments and did not argue for its use in observational studies that use samples from amorphous populations. It caught on, nonetheless, and is used in many disciplines to signify "statistical significance," which has, as suggested earlier, been regrettably used a synonym for statistical importance.

[20] A curious historical phenomenon is that, although researchers regularly use *p*-values to reach conclusions about hypotheses, the idea of using null and alternative hypotheses was developed by Jerzy Newman and Egon Pearson (Karl's son) as an alternative for Fisher's *p*-values. In fact, aiming a direct criticism at *p*-values, they claimed that "no test based upon a theory of probability can by itself provide any valuable evidence of the truth or falsehood of a hypothesis" (p.291) (Jerzy Neyman and Egon S. Pearson (1933), "On the Problem of the Most Efficient Tests of Statistical Hypotheses." *Philosophical Transactions of the Royal Society of London, Series A* 231(694–706): 289–337). Neyman and Pearson's approach focuses on establishing predetermined type I and type II error levels and then evaluating where a test statistic falls with respect to these levels. Studies rarely use their criteria to evaluate hypotheses, though, which is unfortunate. A comparison of and tutorial about Fisher's, Neyman and Pearson's, and other "hypothesis testing theories" is provided in Jose D. Perezgonzalez (2015), "Fisher, Neyman-Pearson or NHST? A Tutorial for Teaching Data Testing," *Frontiers in Psychology* 6(223).

[21] Jerzy Neyman (1935) is often designated as the inventor of CIs (he also called them *confidence sets*) ("On the Problem of Confidence Intervals," *Annals of Mathematical Statistics* 6(3): 111–116). Pierre-Simon Laplace had developed them for the binomial distribution in the late 1700s, however (E. L. Lehmann (1994), "Jerzy Neyman," *Biographical Memoirs* 34: 395–420).

estimates: single numbers computed from the sample that estimate population parameters ($\bar{x} \cong \mu$). An interval estimate furnishes a range of plausible values that (presumably) contains the actual parameter. Those who prefer CIs, the most prominent type of interval estimate, argue that they provide a better representation of the uncertainty inherent in statistical analyses.[22]

A general formula for a CI is furnished in Equation 2.7.

$$\text{Point estimate} \pm \left[(\text{confidence level}) \times (\text{standard error}) \right] \qquad (2.7)$$

The confidence level represents the percentage of the time, based on a z-statistic or a t-statistic, we wish to be able to claim that the interval includes the parameter. For example, imagine collecting data on household income from a representative sample of 100 counties in the U.S. and attempting to estimate a suitable range of values for the mean income in the "population" of counties. The sample yields a mean of $60,000 with a standard deviation of $1,000. Equation 2.8 demonstrates how to calculate the 95% CI.

$$95\% \text{ CI} = \$60,000 \pm \left[1.96 \times \left(\frac{\$1,000}{\sqrt{100}} \right) \right] = \{\$59,804, \$60,196\} \qquad (2.8)$$

The value of 1.96 for the confidence level is from a table of standard normal values or z-values. It corresponds to a large-sample p-value of 0.05 (two-tailed test). In R, we may compute the large-sample confidence level using the qnorm function (e.g., qnorm(0.975)) or the smaller-sample confidence level with the qt function (e.g., qt(0.975, 19), where 19 is the degrees of freedom). The standard error formula is presented earlier in this chapter.

How should we interpret the interval {$59,804, $60,196}? Two common ways are:

1. Given a sample mean of $60,000, we are 95% confident that the population mean of household income falls in the interval $59,804 and $60,196.
2. If we draw many samples from the population of counties in the U.S. and claim that the population mean falls in the interval $59,804–$60,196, we expect to be accurate about 95% of the time.

The second interpretation offers more precision, but perhaps too much confidence because, not unlike p-values, it suggests that the "true" answer is in

---

[22] I suspect Neyman, who was critical of p-values (see fn. 20), would agree. A compelling argument for using CIs and analogous measures rather than p-values is given in Gerry Hahn, Necip Doganaksoy, and Bill Meeker (2019), "Statistical Intervals, Not Statistical Significance," *Significance* 16(4): 20–22.

the interval.[23] The first interpretation is employed often in research reports but is vague if the term "confident" is not clearly understood. It may also give too much credit to the sample statistic, which, for many who rely on CIs, has no intrinsic meaning. Some experts have called for the term "confidence interval" to be replaced with "uncertainty interval," "compatibility interval," or a similar designation to reflect these ambiguities.[24] But we'll follow tradition and continue to use this term when discussing inferential and evidentiary issues relating to LRMs.

We've alluded to some important concerns about significance testing with *p*-values and CIs. We noted, for instance, that researchers should avoid statements such as "the *t*-test, since it has a statistically significant *p*-value, demonstrates that we may *reject* the null hypothesis." Even CIs, though many researchers prefer them to *p*-values, are frequently used to "reject the null hypothesis": to claim, for example, that since the population mean or some other parameter probably falls in a particular interval, the null hypothesis is invalid or valid. Yet, *p*-values and CIs are only pieces of evidence in what should be a larger evaluation of quantitative associations.

Experts have proposed several supplemental and alternative methods for judging whether an association in a sample tells us something important about a hypothesis or research claim. For instance, comparing effect sizes, such as relative differences in means, and evaluating how they compare to previous studies, is a useful tool that may be combined with a measure of uncertainty such as a CI.[25] Others recommend using Bayesian inference, which, as mentioned earlier, offers an alternative to frequentist approaches that use *p*-values and CIs (see fn. 14). Bayesian models combine prior information—from theories, previous studies, or expert judgments—and information from the sample data to make informed judgments about associations among variables. Chapter 4 includes a discussion of using effect sizes with LRMs, but Bayesian models are beyond the scope of this book.

Another way to think about and interpret statistical models, which might still utilize *p*-values or CIs as sources of evidence, is to consider whether their results are *consistent* or *compatible* with a *conceptual model* or *hypothesis*. A conceptual model—which is often used, formally or informally, to develop hypotheses—provides an abstract representation of anticipated associations among concepts. The model is normally based on previous research and theories about physical phenomena or, in the social and behavioral sciences,

---

[23] See John J. A. Ioannidis (2019), "Options for Publishing Research Without Any *p*-Values." *European Heart Journal* 40(31): 2555–2556.

[24] Andrew Gelman and Sander Greenland (2019), "Are Confidence Intervals Better Termed 'Uncertainty Intervals'?" *BMJ* 366: 15381.

[25] See Ashley Steel et al. (2019), *op cit*. Significance tests are not helpful for determining the direction of an association, so they are usually combined with effect sizes to judge directional hypotheses. This is challenging, however, and has inspired new methods for achieving both objectives (see, for instance, Melinda H. McCann and Joshua D. Habiger (2020), "The Detection of Nonnegligible Directional Effects with Associated Measures of Statistical Significance," *American Statistician* 74(3): 213–217).

human behavior. In judging what the results of a statistical test imply about the conceptual model, researchers use all the available evidence.

Recall the earlier example of the weights of Siberian Husky and German Shepherd puppies. Suppose our study begins with a predetermined, yet simple, conceptual model, based on previous canine research studies, which proposes that the average weight of Husky puppies is not the same as the average weight of German Shepherd puppies. Assessing the samples, we find a mean difference in weights of 5 ounces. The t-test designed to compare the means computes a p-value of 0.04 and a CI of {2, 8}. Rather than claiming that we reject the null hypothesis of zero difference or using the longwinded interpretation of the p-value, we state:

> The evidence from the statistical analysis is *consistent* or *compatible* with the conceptual model of different average puppy weights.

This interpretation may not be as stimulating or offer as much certitude as "rejecting the null hypothesis" but it reflects the spirit of using all the evidence available to reach a reasonable and defensible conclusion.

In fact, discussions about the limitations of p-values and significance testing in general[26] have led to a sea change in statistics education, applied statistics, social science statistics, and various forms of quantitative research. An emergent concern is that many of the statistical guidelines and "wisdom" that have steered research studies for years are imprudent. The emphasis on p-values as a way to gauge the importance of a finding or as guide for whether we should use a statistical result to make decisions about public policy, medical care, and so forth has come under special scrutiny, with calls by some to do away with them.[27]

Much of the statistical community is not so pessimistic about significance testing and p-values, though. For instance, in 2016 the American Statistical Association published six principles regarding p-values, with the goal of tempering, but not abandoning, their use. The principles are:

1. *P*-values can indicate how incompatible the data are with a specified statistical (or conceptual/theoretical) model.

2. *P*-values do not measure the probability that the studied (alternative) hypothesis is true, or the probability that the data were produced by random chance alone.

---

[26] Criticizing hypothesis testing with p-values has become such a cottage industry that it has even spawned its own initialism: NHST (null hypothesis significance testing).

[27] See, for example, Raymond Hubbard and R. Murray Lindsay (2008), "Why *P* Values Are Not a Useful Measure of Evidence in Statistical Significance Testing," *Theory & Psychology* 18(1): 69–88, and Blakeley B. McShane et al. (2019), "Abandon Statistical Significance," *American Statistician* 73(supp. 1): 235–245. Ioannidis (2019), *op. cit.*, discusses some alternatives to p-values.

3. Scientific conclusions and business or policy decisions should not be based only on whether a *p*-value passes a specific threshold.

4. Proper inference requires full reporting and transparency.

5. A *p*-value, or statistical significance, does not measure the size of an effect or the importance of a result.

6. By itself, a *p*-value does not provide a good measure of evidence regarding a model or hypothesis.[28]

Jettisoning *p*-values and the standard decision rules about statistical findings is not likely to happen anytime soon. A clear benefit of critiques of these and related issues, however, is the publication of some excellent recommendations about best statistical practices, such as those listed in Chapter 1, and advice regarding using all available sources of evidence to reach reasonable conclusions.

Unbiasedness and Efficiency

As noted earlier, assembling a good sample is key to obtaining suitable estimates of parameters. This raises the general issue of what makes a good statistical *estimator*, or formula for finding an estimate such as a mean, a median, or, as discussed in later chapters, a regression coefficient. Developing estimates that are accurate and that do not fluctuate too much from one sample to the next is important. Two properties of estimators that are vital for obtaining such estimates are *unbiasedness* and *efficiency*.

Unbiasedness refers to whether the mean of the sampling distribution of a statistic equals the population parameter it estimates. For example, is the arithmetic mean estimated from the sample a good estimate of the corresponding mean in the population? Recall that the formula for the sample standard deviation includes the term $\{n - 1\}$ in the denominator. This is necessary to obtain an unbiased estimate of the sample standard deviation, but it presents a slight degree of bias when estimating the population standard deviation.

Efficiency refers to how stable a statistic is from one sample to the next. A more efficient statistic has less variability across samples and is thus, on average, more precise. The estimators for the mean of the normal distribution and probabilities from binomial distributions are considered efficient. Finally, *consistency* refers to whether the statistic converges to the population

---

[28]Source: https://www.amstat.org/asa/files/pdfs/P-ValueStatement.pdf. See also Daniel J. Benjamin and James O. Berger (2019), "Three Recommendations for Improving the Use of *p*-Values," *American Statistician* 73(S1): 186–191. For a critique of the way *p*-values are employed in applied research, see the set of articles in the journal *Significance* 16(4): 14–22 (2019).

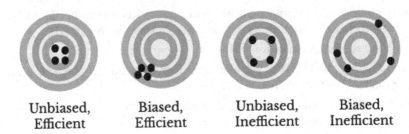

Unbiased,   Biased,   Unbiased,   Biased,
Efficient   Efficient   Inefficient   Inefficient

FIGURE 2.3
Unbiasedness and efficiency of estimators.

parameter as the sample size increases. Thus, it combines characteristics of both unbiasedness and efficiency.[29]

A common way to represent unbiasedness and efficiency is with an archery target. As shown in Figure 2.3, estimators from statistical models can be visualized as trying to "hit" the parameter in the population. Estimators can be unbiased and efficient, biased but efficient, unbiased but inefficient, or neither. The benefits of having unbiased and efficient statistics should be clear.

The Standard Normal Distribution and Z-Scores

Recall that we mentioned z-values in the discussion of CIs. These values are drawn from a standard normal distribution—also called a z-distribution—which has a mean of zero and a standard deviation of one. The standard normal distribution is useful in a couple of situations. First, as discussed earlier, the formula for the large-sample CI utilizes z-values.

Second, they provide a useful transformation for continuous variables that are measured in different units. For instance, suppose we wish to compare the distributions of weights of two litters of puppies, but one is from the U.S. and the weights are measured in ounces and the other is from Germany and the weights are measured in grams. Converting ounces into grams is simple

---

[29] But how do we know that an estimator is unbiased or efficient (at least more so than other estimators)? For many years, statisticians relied on proofs that compared different estimators and determined which ones held these properties. More recent years have seen substantial reliance on simulation studies: experiments using fabricated data to represent populations that are designed to have particular characteristics (e.g., normally distributed or *t*-distributed). Statisticians then determine, through repeated experiments that use multiple "samples" from these "populations," which estimators manifest the least bias and most efficiency or consistency. See Appendix B for some simulations using R.

(1 ounce $=28.35$ grams), but we may also transform the different measurement units using *z-scores*. This results in a comparable measurement scale. A z-score transformation is based on Equation 2.9.

$$z\text{-score} = \frac{(x_i - \bar{x})}{s} \tag{2.9}$$

Each observation of a variable is entered into this formula to yield its z-score, or what are sometimes called *standardized values*. The unit of measurement for z-scores is standard deviations. In R, the scale function computes them for each observation of a variable (the function may also be used to transform variables into other units in addition to z-scores). Let's see how to use it on one of the samples of puppy weights along with a new sample of weights measured in grams.

R code
```
puppy.wt.oz <- c(40, 45, 50, 55, 60, 65, 70)
  # create the first vector of weights in ounces
puppy.wt.gram <- c(1100, 1150, 1200, 1400, 1700, 1725, 1775)
  # create the second vector of weights in grams
z.puppy.wt.oz <- scale(puppy.wt.oz, center = TRUE,
              scale = TRUE)
  # use the scale function to transform to z-scores
z.puppy.wt.gram <- scale(puppy.wt.gram, center = TRUE,
              scale = TRUE)
  # use the scale function to transform to z-scores
z.puppy.wt.oz
z.puppy.wt.gram   # examine the z-scores
```

R output (abbreviated)
```
[-1.39,  -0.93,  -0.46,   0.00, 0.46, 0.93, 1.39]
[-1.14,  -0.97,  -0.80,  -1.21, 0.90, 0.98, 1.15]
```

The two samples now share the same measurement scale. The second sample's observations are less spread out than the first sample's, thus there's less variation in weights. Being able to compare variables that are measured in different units comes in handy as we learn about LRMs.

You may remember that z-scores are also useful for determining the percentage of a distribution that is a certain distance from the mean. For example, 95% of the observations from a normal distribution fall within 1.96 standard deviations of the mean. This translates into 95% of the observations using standardized values falling within 1.96 z-scores of the mean and, as shown in Figure 2.4, 5% of the observations are in the tails. With a slight modification, this phenomenon is helpful if we wish to make inferences from the results of the sample-based LRM to a target population (see Chapter 3).

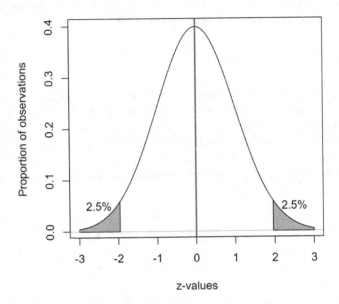

FIGURE 2.4
Standard normal density function: distribution of z-scores.

Covariance and Correlation

We've seen a couple of examples of comparing variables from different sources (e.g., puppy weights from different litters); we now assess whether two variables shift or change together. For instance, is it fair to say that the length and the weight of puppies shift together? Are longer puppies, on average, heavier than shorter puppies? The answer is, on average, most likely yes. In statistical language, we say that length and weight *covary* or are *correlated*. The two measures used most often to assess the association between two continuous variables are, not surprisingly, called the *covariance* and the *correlation*. To be precise, the most common type of correlation is the Pearson's product–moment correlation.[30]

---

[30] Several types of correlations exist, but the best known was developed by Karl Pearson (1895) ("Note on Regression and Inheritance in the Case of Two Parents," *Proceedings of the Royal Society of London* 58(347–352): 240–242). He based it on earlier work by Carl Friedrich Gauss, Auguste Bravais, Francis Galton, and Francis Edgeworth. As they developed an understanding of correlations, Pearsona and Galton also laid the groundwork for the contemporary approach to linear regression (Jeffrey M. Stanton (2001), "Galton, Pearson, and the Peas," *Journal of Statistics Education* 9(3): 1–12).

A covariance is a measure of the joint variation of two continuous variables. Two variables covary when large values of one are accompanied by large or small values of the other. For instance, puppy length and weight covary because large values of one tend to accompany large values of the other in a population or in most samples, though the association is not uniform because of the substantial variation in the lengths and weights of puppies. Equation 2.10 furnishes the formula for the covariance.

$$\text{cov}(x,y) = \frac{\Sigma(x_i - \bar{x})(y_i - \bar{y})}{n-1} \tag{2.10}$$

The covariance formula multiplies deviations from the means of both variables, adds them across the observations, and then divides the sum by the sample size minus one. Don't forget that this implies the xs and ys come from the same unit, whether a puppy, person, place, or thing.

A limitation of the covariance is its dependence on the measurement units of both variables, so its interpretation is not intuitive. It would be helpful to have a measure of association that offered a way to compare various associations of different combinations of variables. The Pearson's product–moment correlation—often shortened to *Pearson's r*—accomplishes this task. Among several formulas for the correlation, Equations 2.11 and 2.12 are the easiest to understand.

$$\text{corr}(x,y) = r = \frac{\text{cov}(x,y)}{\sqrt{\text{var}(x) \times \text{var}(y)}} \tag{2.11}$$

$$\text{corr}(x,y) = r = \frac{\Sigma(z_x)(z_y)}{n-1} \tag{2.12}$$

Equation 2.11 shows that the correlation is the covariance divided by the *pooled standard deviation*. Equation 2.12 displays the relationship between z-scores and correlations. It shows that the correlation may be interpreted as a standardized measure of association. Some characteristics of correlations include:

1. Correlations range from −1 and +1, with positive numbers indicating a positive association and negative numbers indicating a negative association (as one variable increases the other tends to decrease).

2. A correlation of zero implies no statistical association, at least not one that can be measured assuming a straight-line association, between the two variables.

3. The correlation does not change if we add a constant to the values of the variables or if we multiply the values by some constant number. However, these constants must have the same sign, negative or positive.

The following is an example of a covariance and a correlation in R.[31]

R code
```
puppy.length <- c(31, 33, 37, 38, 42, 45, 48)
  # create a vector of the puppies' lengths in centimeters
  # the vector of weights in ounces was created earlier

cov(puppy.wt.oz, puppy.length) # request the covariance
cor(puppy.wt.oz, puppy.length) # request the correlation
```

R output (annotated)
```
66.67 # covariance
0.995 # correlation
```

The interpretation of the covariance is not intuitive, even though 66.7 seems large. The correlation, however, shows a strong statistical association between weight and length. Chapter 4 discusses the relationship between correlations and LRM coefficients.

As mentioned earlier, several other *measures of association* are available in addition to Pearson's *r*. For instance, Spearman's correlation is based on the ranks of the values of variables, rather than the actual values. Similar to the median when compared to the mean, Spearman's is less sensitive to extreme values. In R, the Spearman's correlation of puppy weight and height may be estimated with cor(puppy.wt.oz, puppy.length, method="spearman"). Other measures of association are designed for categorical variables, such as gamma, Cramer's V, lambda, eta, and odds ratios. Odds ratios, in particular, are used often to estimate the association between two binary (two-category) variables. We'll learn more about this measure of association in Chapter 16.

Comparing Means from Two Groups

We referred to a mean-comparison test, the *t*-test, earlier in the chapter when comparing the weights of two litters of puppies. Let's review this test in more detail. First, recall that we may compare many statistics from distributions, including standard deviations and standard errors. A common exercise, however, is to assess, in an inferential sense, whether the mean from one population is likely different from the mean of another population. If we draw samples from two populations, we should consider sampling error. Our samples probably have different means than the true population means,

---

[31] Though not shown here, correlations are often accompanied by *p*-values and CIs. The cor.ci function of R's psych package provides them.

so we should take this likely variation into account. A *t*-test is designed to evaluate whether two means are likely different across populations by, first, taking the difference between the sample means and, second, evaluating the presumed sampling error.

The name *t*-test is used because the *t*-value that is the basis of the test follows a *Student's t-distribution*.[32] This distribution is almost indistinguishable from the normal distribution when the sample size is greater than 50. At smaller sample sizes, the *t*-distribution has fatter tails and is a bit flatter in the middle than the normal distribution.

Equations 2.13 and 2.14 demonstrate how to compute a conventional *t*-test that assumes the means are drawn from two independent populations.

$$t = \frac{\bar{x} - \bar{y}}{s_p \sqrt{\dfrac{1}{n_1} + \dfrac{1}{n_2}}} \tag{2.13}$$

$$\text{where } s_p = \sqrt{\frac{(n_1 - 1)s_1^2 + (n_2 - 1)s_2^2}{(n_1 + n_2) - 2}} \tag{2.14}$$

The s_p in the equations is the pooled standard deviation, which estimates the sampling error. The ns denote the sample sizes and the s^2s are the variances for the two groups. A key assumption of this test is that the variances from the two groups are equal. Some researchers are uncomfortable making this assumption—or it may not be tenable—so they use the estimator shown in Equation 2.15, which is called Welch's *t*-test.[33]

$$t' = \frac{\bar{x} - \bar{y}}{\sqrt{\dfrac{\text{var}(x)}{n_1} + \dfrac{\text{var}(y)}{n_2}}} \tag{2.15}$$

Since the t' does not follow the *t*-distribution, we must rely on special tables to determine the probability of a difference between the two means when using Welch's test. Fortunately, R and other statistical software provide both versions of the *t*-test.

Here's an example of a *t*-test that uses litters 3 and 4 from the earlier puppy samples. As a reminder, the weights in ounces are

---

[32] This distribution and the *t*-test were developed in 1908 by Irish statistician William S. Gosset. He worked for the Guinness Brewing Company and could not include his name on published research, so he used the pseudonym Student see Student [William S. Gosset] (1908), "The Probable Error of a Mean," *Biometrika* 6(1): 1–25; and Stephen E. Fienberg and Nicole Lazar (2001), "William Sealy Gosset," in *Statisticians of the Centuries*, C. C. Heyde et al. (Eds.), New York: Springer).

[33] Introduced in B.L. Welch (1947), "The Generalization of 'Student's' Problem When Several Different Population Variances Are Involved," *Biometrika* 34: 28–35.

Litter 3: [39, 55, 56, 58, 61, 66, 69]

Litter 4: [42, 44, 48, 55, 57, 60, 66]

The means are 57.7 and 53.1, with a difference of 4.6. A *t*-test returns a *t*-value of 0.92 with a *p*-value of 0.37 and a CI of {−6.2, 15.4}. The interpretations of the two inferential measures are:

> *p*-value: if we take many, many samples from the two populations of puppies and the difference in the population means is zero, we expect to find a difference in sample means of 4.6 ounces or more approximately 37 times, on average, out of every 100 pairs of samples we examine.

> 95% CI: given a difference of 4.6 ounces, we are 95% confident that the difference in the population means falls in the interval −6.2 and 15.4.

The interpretations suggest that the chance of getting a difference of 4.6 ounces across the two samples of puppies is pretty high if the difference in the population means is zero. For many researchers, a *p*-value must be much smaller—such as below 0.05 or 0.01—before they conclude that the results are consistent with the presumption that the difference across populations is not zero.[34] The fact that the 95% CI in this example includes zero also suggests that the difference in the population means is small. But is it zero—is the nil hypothesis valid? This seems doubtful, but we cannot be certain without information about the actual means from the populations. Perhaps, as suggested in the section on significance testing, a better interpretation is to claim that the evidence from the *t*-test is *incompatible* with a presumption of a difference in mean weights across the populations.

An important assumption of these mean-comparison procedures is that the variables follow a normal distribution. The *t*-test, for example, does not provide accurate results if the variable from either sample is "non-normal." Other tests are available, such as those designed to compare ranks or medians (e.g., Wilcoxon-Mann-Whitney test), which are appropriate for non-normal variables.

Comparing two means does not exhaust our interest. We may wish to make other types of comparisons. Suppose we're interested in comparing three means, four means, or even ten means. We might, for instance, have samples of puppies from six different breeds. ANOVA procedures are appropriate

---

[34] Look again at Figure 2.4. If we use a two-tailed significance test and the null hypothesis is that the mean difference in the population is zero, then the *p*-value suggests that the difference falls in the non-shaded area of the curve. Researchers often interpret this as meaning that the difference is not "statistically significant." But, of course, a difference in the populations means might still exist.

for comparing means from multiple samples.[35] Multiple comparison procedures are useful for assessing if one of the means is presumed different from one or more of the others in the population. Books that describe ANOVA techniques provide details about these procedures.[36] As discussed in subsequent chapters, LRMs are also useful for computing and comparing means for different groups in the data. In particular, Chapter 7 discusses how to use LRMs to estimate and compare means from different groups based on indicator variables.

Yet another method of comparison is to visualize the joint distribution of two variables with a scatter plot or a similar graph. Along with correlations, this approach helps introduce the LRM (see Chapter 3).

Examples Using R

The file *Nations2018.csv*[37] is a small dataset that contains data from eight nations. The variables are public expenditures (expend), a measure of government expenditures on individual and collective goods and services as a percentage of the nation's gross domestic product, openness to trade with other nations (econopen), and the percentage of the labor force that is unionized (perlabor). Let's use R to compute some of the statistics discussed in this chapter. To begin, after importing the dataset and installing the R package psych,[38] use the following code to obtain descriptive statistics for the public expenditures variable:

---

[35] Scientists had been comparing distributions for many years, but Ronald A. Fisher formalized ANOVA in a 1918 paper ("The Correlation between Relatives on the Supposition of Mendelian Inheritance," *Earth and Environmental Science Transactions of the Royal Society of Edinburgh* 52(2): 399–433).

[36] See, for example, Frank Bretz, Torsten Hothorn, and Peter Westfall (2016), *Multiple Comparisons Using R*, Boca Raton, FL: CRC Press.

[37] The online material on the author's GitHub site [https://github.com/johnhoffmannVA/LinearRegression] includes the datasets used in all the book's examples, as well as codebooks that describe the variables and what they are designed to measure. The datasets are in text files that employ commas to separate each data point. Called *comma-separated values files*, they have the extension *.csv* but are also known as *comma-delimited files*. Such files may be read across a variety of software platforms and are usable in a wide range of settings. You may wish to open the *Nations2018.csv* file in an editor such as Notepad++ or TextEdit to examine its layout. When using RStudio, the File—Import Dataset—From Text (readr) menu sequence provides a quick way to import data files into R. The Environment Window also has an Import Dataset option.

[38] In RStudio, the drop-down menu's Tools—Install Packages is a simple way to find and download user-written packages. The function install.packages may also be placed in an R script file to download packages (e.g., install.packages("psych")). Several functions and packages that provide descriptive statistics are available in R, including the native summary function and the Hmisc and pastecs packages, but the psych package is thorough.

R code
```
library(psych) # to activate the package
describe(Nations2018$expend)
```

R output (abbreviated)
```
vars n   mean   sd   median trimmed mad   min   max
1     8 19.79 2.87   19.8    19.8   1.85 14.1 23.4
range skew kurtosis   se
9.3   -0.6  -0.62    1.01
```

The describe function provides various statistics, including the mean, standard deviation (sd), median, trimmed mean, median absolute deviation (mad), range, skewness, kurtosis, and standard error of the mean (se).

The 95% CI for the mean is simple to calculate using the t.test function (e.g., t.test(Nations2018$expend)). R also has several user-written packages that include CI functions (e.g., Hmisc). For public expenditures, the 95% CI from the t.test function is 17.39 and 22.19. How should we interpret it?

Compute the correlation and covariance between public expenditures and economic openness (hint: see the earlier R function, but you might also wish to review the documentation for the psych package for similar functions). You should find a correlation of 0.64 and a covariance of 36.36.

Let's examine another dataset. Open the data file *GSS2018.csv*.[39] The dataset contains a variable called female, which includes two categories: male and female.[40] We'll use it to compare personal income (labeled pincome) for these two groups using a *t*-test.

R code
```
t.test(GSS2018$pincome ~ GSS2018$female)
```

What does the output show? What is the *t*-value? What is the *p*-value? The 95% CI? How should we interpret the 95% CI? Suppose we wish to test a conceptual model that proposes that males have higher incomes than females in the U.S. Are the results consistent with this model?

Let's practice building some graphs using the variables pincome and sei in the *GSS2018* dataset. What do kernel density plots show about them?[41] Box-and-whisker plots?[42] What measures of central tendency are most

---

[39] The data are an excerpt from the General Social Survey (https://gss.norc.org), a study that collects information every other year from a representative sample of noninstitutionalized adults in the U.S. The data in this file are from just one year of the survey, 2018.

[40] Certain statistical software packages require that variables be numeric in order to execute some analyses. R will conduct many analyses with categorical variables consisting of text (character strings) rather than numbers. Some functions require that these be defined as factor variables, however.

[41] Example: plot(density(GSS2018$pincome))

[42] Example: boxplot(GSS2018$sei)

appropriate for these variables? If one of them is skewed, can you find a transformation that normalizes its distribution?

Chapter Summary

This chapter reviews select information from elementary statistics that is important for understanding LRMs. It shows how to compute several statistics, such as the mean, standard deviation, covariance, and correlation, as well as discussing significance testing, a central part of inferential statistics. Perhaps the most important point about the latter topic is that a *statistically significant* result does not signify a *practically important* result. Remember to use *p*-values and CIs with caution and that statistical results are only a piece of the larger research pie. And don't forget that uncertainty is a fundamental feature of statistical analyses.

Chapter Exercises

The dataset called *American.csv* consists of data from a 2004 national survey of adults in the U.S. Our objective is to examine some variables from this dataset. In addition to an identification variable (id), they include:

- educate: years of formal education.
- american: a continuous measure of what the respondent thinks it means to be "an American" that ranges from believing that being an American means being a Christian, speaking English only, and being born in the U.S. (high end of the scale) to not seeing these as indicators of being an American (low end of the scale).
- group: a binary variable that indicates whether or not the respondent is an immigrant to the U.S.

After importing the dataset into R, complete the following exercises:

1. Compute the means, medians, standard deviations, variances, skewnesses, and standard errors of the means for the variables educate and american.
2. Furnish the number of respondents in each category of the group variable, followed by the percentage of respondents in each category of this variable. What percentage of the sample is in the "Not

immigrant" category? What percentage of the sample is in the "Immigrant" category?

3. Conduct a *t*-test (Welch's version) that compares the means of the variable american for those in the "Not immigrant" group and those in the "Immigrant" group. Report the means for the two groups, the *p*-value from the *t*-test, and the 95% CI from the *t*-test. Interpret the *p*-value and the 95% CI.

4. What is the Pearson's correlation of educate and american? What is the 95% CI of the correlation? Provide a brief interpretation of the Pearson's correlation.

5. Create a kernel density plot and a box plot of the variable american. Describe its distribution.

6. *Challenge*: use R's plot function to create a scatter plot with educate on the *x*-axis and american on the *y*-axis. Describe the pattern shown by the scatter plot. Why is a scatter plot limited in this situation? Search within R or online for the R function jitter. Use this function to modify the scatter plot. Why is the scatter plot still of limited use for understanding the association between the two variables?

3

Simple Linear Regression Models

Chapter 2 describes a conceptual model as an abstract representation of anticipated associations among concepts or ideas designed to represent broader ideas (such as self-esteem, political ideology, or education). Ideally, statistical models are guided by conceptual models, which are used to delineate hypotheses or research questions. Statistical models outline probabilistic relationships among a set of variables, with the goal of estimating whether there are nonrandom patterns among them. Like conceptual models, these models tend to be simplifications of the complexity that occurs in nature but offer enough detail to predict or understand patterns in the data. A useful way of thinking about statistical models is that they assess ways that a set of data may have been produced, or, in statistical parlance, a *data generating process* (DGP).

A regression model is a type of statistical model that aims to estimate the association between one or more *explanatory* variables (*xs*) and a single *outcome* variable (*y*). An outcome variable is presumed to depend on or to be predicted by the explanatory variables. But the explanatory variables are seen as independent predictors of the outcome variable; hence, they are often called *independent* variables. Later chapters discuss why this term can be misleading because these variables may, if the model is set up correctly, relate to one another. Many researchers therefore prefer to call those included in a regression model explanatory and outcome variables (used in this book), *predictor* and *response* variables, *exogenous* and *endogenous* variables, or similar terms. The response or endogenous variable is synonymous with the outcome variable.

An LRM seeks to account for or explain differences in values of the outcome variable with information about values of the explanatory variables. The LRM also seeks, to varying degrees, answers to the following questions:

1. What are the predicted mean levels of the outcome variable for particular values of the explanatory variables?

2. What is the most appropriate equation for representing the association between each explanatory variable and the outcome variable? This includes assessing the direction (positive? negative?) and magnitude of each association.

DOI: 10.1201/9781003162230-3

3. Which explanatory variables are good predictors of the outcome and which are not? The answer is based on several results from the LRM, including the size of the coefficients, differences in predicted means, the *p*-values, and the CIs, but each has limitations.[1]

The first model we'll consider is called the simple linear regression model, or what some designate as a *bivariate* regression model. The term *simple* is used not because a research question is crude or naive, but rather this model includes only one explanatory and one outcome variable. But, you may ask, why not use the Pearson's correlation coefficient to estimate the association between two variables? This is a plausible step because we can learn about its strength and direction by examining their correlation (assuming the variables are continuous). The difference here is conceptual: we think, perhaps because we are testing a theory, that one variable influences the other, with one existing first in a chain of events (e.g., ACT test scores precede and affect college grades). Or we control one variable and wish to assess its influence on another (e.g., we control the amount of fertilizer and determine the consequent height of a set of orchids).

To help you understand the simple LRM, recall those pre-algebra or algebra class exercises when you were asked to plot points on the *x*- and *y*-axes. Let's revisit those days and create a graph. We'll use *y* as a label for the vertical axis and *x* as a label for the horizontal axis. We may then use the well-known equation $y = mx + b$ to either represent the way the points fall on the graph or to decide where to put points on the graph. The equation provides a map that shows where to place objects. The letter *m* represents the slope and *b* denotes the intercept, or the point at which the line crosses the *y*-axis (the *y-intercept*). The simple LRM is easy to understand if you remember these exercises.

Examine Figure 3.1. What is its slope and intercept? Since we don't yet know how *x* and *y* are measured, we can determine only that the slope and the intercept are positive numbers: the line angles upward and it crosses the

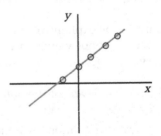

FIGURE 3.1
A positive slope and intercept.

---

[1] These questions are best guided by a conceptual or theoretical model and the criteria for evaluating their answers determined before beginning the data analysis stage of a research project.

y-axis at a positive value of *y*. The figure also implies a high positive correlation because the points fall on the line. Suppose Equation 3.1 represents the association between *x* and *y*.

$$y = 2x + 1.5 \tag{3.1}$$

The slope indicates that as the *x* values increase by one unit the *y* values increase by two units. Its definition as *rise/run* is a shorthand way of saying that as the points shift (*rise*) a certain number of units along the *y*-axis, they also shift (*run*) a certain number of units along the *x*-axis. The shifts could go in either direction, though, since, if the slope (*m*) is a negative number, *y* decreases as *x* increases.

Another way to understand the slope—which is consistent with *rise/run*—is that it represents the change in *y* over the change in *x*. This is often portrayed as $\frac{\Delta y}{\Delta x}$, in which the Greek letter delta (Δ) is used to denote change.

Real data almost never follow a straight-line pattern: rarely do all the points fall on the line. A simple LRM is designed to capture the *average* association between the variables using a straight line. But a straight line—also called a *linear fit line*—may or may not be an accurate representation of their association (see Chapter 11).

Let's examine some real data and picture how well a linear fit line represents the association between two variables. Figure 3.2 is produced in R with the *Nations2018.csv* dataset by using the plot and abline functions.[2] Public

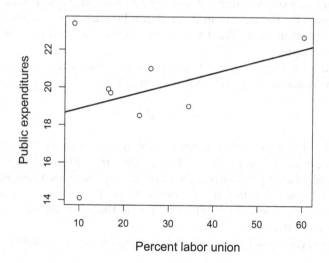

FIGURE 3.2
Scatter plot of public expenditures by percent unionized with linear fit line.

---

[2] Using these two R functions is an example of overlaying one graph on another. In this case, a linear (straight) fit line is overlaid on a scatter plot.

expenditures (expend) is the outcome variable (y) and the percent of the nation's labor force that is unionized (perlabor) is the explanatory variable (x). The figure is built in R using the following two lines of code:

R code for Figure 3.2
```
plot(Nations2018$perlabor, Nations2018$expend,
     xlab="Percent labor union", ylab="Public
     expenditures", pch=1)
       # pch=1 designates the plotting symbol; 1 is the
         default - an open circle
abline(lm(Nations2018$expend ~ Nations2018$perlabor),
           col="red")
  # abline: requests a straight line; lm: linear model;
    ~ symbolizes "distributed as"
```

Once the code is executed in R or RStudio, a graph that looks similar to Figure 3.2 appears in the plot window.[3]

The graph represents a positive association between the two variables, but the points do not fall on the linear fit line. This shows the difference between a mathematical or deterministic relationship, such as in Figure 3.1, and a statistical or probabilistic relationship. Rather than trying to be precise about an association, we claim that, *on average*, public expenditures are higher when percent labor union is higher. But we might also wonder how many more units of percent labor union are, on average, associated with how many more units of public expenditures. This is a core issue addressed with an LRM, but, as shown by Figures 3.1 and 3.2, it can be elusive if we rely only on a graphical representation. We'll see an alternative way to represent the issue later in the chapter.

Rather than using $y = mx + b$ in regression modeling, most researchers employ the notation displayed in Equation 3.2.

$$y_i = \alpha + \beta_1 x_i \qquad (3.2)$$

The Greek letter alpha (α) denotes the intercept, whereas the Greek letter beta (β) signifies the slope. These two terms are statistics in this example since, as suggested by the lowercase y and x, they are from a sample. As suggested in Chapter 2, if based on a population, they are called parameters. The variables are subscripted with i to denote that they are from individual observations in the sample.

Some researchers utilize Equation 3.3 to represent the model.

$$y_i = \beta_0 + \beta_1 x_i \qquad (3.3)$$

---
[3] You may change the dimensions of a graph by changing the size of the window in RStudio or by changing the graph's dimensions when exporting it.

In this depiction, β_0 is the intercept and β_1 the slope, but we'll use alpha (α) to indicate the intercept from now on.

Another way to present the linear regression equation is to use Greek letters but place a "hat" on top of them to indicate they are estimates of population parameters based on sample statistics (see Equation 3.4).[4]

$$\hat{y}_i = \hat{\alpha} + \hat{\beta}_1 x_i \qquad (3.4)$$

You might have noticed a problem with the way the equations are presented thus far. Recall that we use the term *probabilistic* to describe a statistical relationship. If you're not sure what this means, revisit Figure 3.2: the scatter plot between percent labor union and public expenditures. The points are not on the line. Take almost any two variables examined in research studies and they do not fall on a linear fit line in a scatter plot. This may be problematic for those who want their models precise. The equations we've seen so far call for precision, but rarely represent the actual association between variables. The linear fit line is, at best, an approximation of the genuine relationship between X and Y. We should, therefore, revise the linear regression equation to represent the uncertainty of the line, which we achieve by including the *error term*, as shown in Equation 3.5.

$$\hat{y}_i = \hat{\alpha} + \hat{\beta}_1 x_i + \hat{\varepsilon}_i \qquad (3.5)$$

The Greek letter epsilon (ε) (absent the "hat") represents the uncertainty in predicting the outcome variable with the explanatory variable. It indicates how far away the individual y values are from the true mean value of Y for given values of x; hence, ε_i is also known as the *errors of prediction*, which is depicted with the equation: $\varepsilon_i = y_i - Y_i$. This characterization of the error term is loaded with assumptions. We use, for instance, the term *true mean value* of Y. What does this imply? Think about the sample. If we've done a good job collecting a sample, its observations should represent specific groups from the population. For instance, say we've sampled from adults in the U.S. Some of our sample members should then be 25–30-year-olds. We assume that these sample members are a good representation of other 25–30-year-olds in the U.S. Their values on a characteristic such as education should therefore provide a good estimate of mean education among those in this age group of the population. If we wish to use parental education (x) to predict education (y) among 25–30-year-olds then, using an LRM, we assume that the error term represents the distance from the points on the

---

[4] The material that follows does not adhere strictly to one form, although it uses α and β to represent the intercept and slope. We'll also assume in most cases that we have information from a sample with which we wish to infer something about a population, but employ the "hat" notation primarily when discussing inferential issues.

graph to the actual predicted means of Y given particular values of parental education.[5]

Assumptions of Simple LRMs

The LRM rests on several assumptions that dictate how well it operates. Most of these concern characteristics of the population data and focus on the errors of prediction (ε_i). But having access to information from a population is unusual, so we must assess, roughly or indirectly, the assumptions of LRMs with information from a sample. In other words, since we do not have information from the Ys, we cannot compute ε_i directly. The sample includes only the xs and ys, so we must use an estimate of ε_i. This estimate, depicted as the error term $\left(\hat{\varepsilon}_i\right)$ in Equation 3.5, is represented by the *residuals*[6] from the model, which are computed as $\left(y_i - \hat{y}_i\right)$. Rather than distinguishing the errors of prediction from the population and the sample, however, we'll take for granted that the sample provides a good estimate of $\overline{Y}_i$ with $\hat{y}_i$, so that $\left(y_i - \hat{y}_i\right) \cong \left(y_i - \overline{Y}_i\right)$.

Here are the key assumptions of simple LRMs:

1. *Independence*: the errors of prediction (ε_i) are statistically independent of one another. Using the example from the *Nations2018* dataset, we assume that the errors in predicting public expenditures across nations are independent. In practice, this often implies that the observations are independent. One way to (almost) guarantee this is to use simple random sampling. (However, in this example we should ask ourselves: are the economic conditions of these nations likely to be independent?) Chapters 8 and 15 outline additional ways to understand the independence assumption.

2. *Homoscedasticity (constant variance)*: the errors of prediction have equivalent variance for all possible values of X. In other words, the variance of the errors is assumed to be constant across the distribution of X. At this point it may be simpler, yet imprecise, to think about the Y values and ask whether their variability is equivalent at different values of X. Chapter 9 discusses the homoscedasticity assumption.

---

[5] The errors of prediction are due, in the best of circumstances, solely to random error. But remember that, in this context, error is not tantamount to a mistake. To avoid confusion perhaps, economists tend to use the term "disturbance term" rather than "error term," which they may have borrowed from statistician George Udny Yule (1927) who discussed "disturbances" to periodic phenomena, such as the movement of pendulums, that magnify the errors of prediction ("On a Method of Investigating Periodicities Disturbed Series, with Special Reference to Wolfer's Sunspot Numbers," *Philosophical Transactions of the Royal Society of London*, Series A, 226(636–646): 267–298).

[6] The term residual is derived from the Latin term *residuum*, which means "leftover."

3. *Normality*: the errors are a normally distributed random variable. We also assume that the errors have a mean equal to zero in the population, though this is not especially important. Symbolically, the normality assumption is often presented as $\varepsilon_i \sim N(0,\sigma^2)$. As mentioned earlier, the wavy line means "distributed as." The variance portion of the equation (σ^2) has implications for the homoscedasticity assumption.[7]

4. *Linearity*: the mean value of Y is a straight-line function of X. In other words, Y and X have a linear relationship. If they have a nonlinear association, we should modify the regression model. See Chapter 11 for additional information about normality and linearity.

Other derivative assumptions are considered in later chapters, but these four are sufficient for the present discussion.

We should have lots of x values and lots of y values in order to test assumptions about LRMs that use sample data. For example, if we collect data on dozens of nations with, say, 20% of their employees belonging to labor unions, then we expect that the sample provides a good representation of the population of nations at this percent level of labor union. We also expect that those nations with other labor union percentages are good representations of their populations. When using an LRM with percent labor union to predict public expenditures, we furthermore expect the aforementioned assumptions to be satisfied.

Figure 3.3 offers a way to visualize what is meant by some of these statements. We have sets of observations at 20%, 40%, and 60% values of labor union. We assume that the mean public expenditure values for these labor

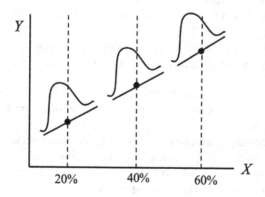

FIGURE 3.3
Visualizing assumptions of a simple linear regression model.

---

[7] Some presentations rearrange the first three assumptions. For example, one often reads that a key assumption is that the errors are *independent and identically distributed* (*iid*). Called the *iid* assumption, it combines suppositions about independence, homoscedasticity, and the distribution of the errors.

union levels are a good representation of the actual means in the population. We also assume that the errors made when predicting public expenditures are normally distributed, with a mean of zero (the underestimation and overestimation of the predictions cancel out) and identical variances regardless of the percent labor union.

An Example of an LRM Using R

You may be confused at this point, though let's hope not. An example using some data should be beneficial. The dataset *StateData2018.csv* includes a number of variables from all 50 states in the U.S. These data include population characteristics, crime rates, substance use rates, and various economic and social factors. We'll treat the data as a sample, even though one might argue that they represent a population. Similar to the code that produces Figure 3.2, the following R code creates a scatter plot and overlays a linear fit line with the number of opioid deaths per 100,000 residents (OpioidODDeathRate) as the outcome (*y*) variable and average life satisfaction (LifeSatis), which is based on state-specific survey data[8] that gauges happiness and satisfaction with one's family life and health among adult residents, as the explanatory (*x*) variable.

R code for Figure 3.4
```
plot(StateData2018$LifeSatis, StateData2018
     $OpioidODDeathRate, xlab="Average life
     satisfaction", ylab="Opioid overdose deaths per
     100,000 population", pch=1)
abline(lm(StateData2018$OpioidODDeathRate ~
     StateData2018$LifeSatis), col="red")
```

Figure 3.4 displays a negative slope. Yet the points diverge from the line; only a few are relatively close to it. Do you see any other patterns in the data relative to the line?

We'll now estimate a simple LRM using these two variables. As you may have already determined given R's abline function that created the linear fit lines in Figures 3.2 and 3.4, an LRM is estimated in R using the lm function. The abbreviation signifies "linear model."

---

[8] The survey data are from the Pew Research Center's Religious Landscape Study, which collected information in 2015 from representative samples of adults in each state of the U.S. The life satisfaction variable is based on the average responses of state residents to the survey questions and is arbitrarily measured on a continuous scale from about 49 to 56 ($\bar{x} = 51.7$), with higher values indicating greater average satisfaction among adults in a state.

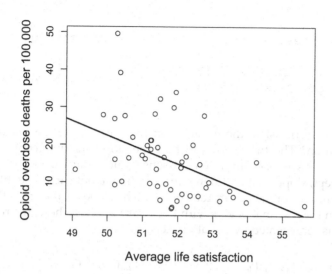

FIGURE 3.4
Scatter plot of opioid overdose deaths by average life satisfaction with linear fit line.

R code
```
LRM3.1 <- lm(OpioidODDeathRate ~ LifeSatis, data =
            StateData2018)
```

After executing this function, R creates an object called LRM3.1 that has information from a model of opioid deaths per 100,000 residents "regressed" on average life satisfaction (see the Global Environment window in RStudio). The summary and confint functions provide some of this information.[9]

R code
```
summary(LRM3.1) # requests default output
confint(LRM3.1) # requests confidence intervals
```

R output (abbreviated)
```
Residuals:
  Min      1Q    Median    3Q      Max
-12.688 -6.952  -1.511  3.408  28.118
```

```
Coefficients:
              Estimate  Std. Error   t value   Pr(>|t|)
(Intercept)   212.056    56.576       3.748    0.000479 ***
LifeSatis      -3.792     1.093       -3.468    0.001116 **
```

---

[9] Executing the function lm(OpioidODDeathRate ~ LifeSatis, data=StateData2018) returns limited information about the model. A better alternative is summary(lm(OpioidOD DeathRate ~ LifeSatis, data = StateData2018)). Saving the R output as an object (e.g., LRM3.x <- lm(...)) followed by the summary function is good practice, however, because it allows an efficient way to use various post-regression options.

```
---
Signif. codes: 0 '***' 0.001 '**' 0.01 '*' 0.05 '.' 0.1

[CIs]          2.5%    97.5%
(Intercept) 98.30    325.81
LifeSatis    -5.98     -1.59
```

How do we translate the numbers in the output into the linear regression equation? The first step is to find the slope and intercept. The term Coefficients is used in a general sense, but in the lm output provides the intercept, slope, and additional information about each. The intercept is the first number in the Estimate column. The slope coefficient is the initial number in the row that begins with LifeSatis. Given these, the regression equation is represented by Equation 3.6.

$$\text{Opioid deaths}\left(\widehat{y_i}\right) = 212.06 + \left\{-3.79 \times \text{Life satisfaction}\left(x_i\right)\right\} + \hat{\varepsilon}_i \quad (3.6)$$

The y and the x are included to remind us that opioid deaths constitute the outcome variable and average life satisfaction the explanatory variable. The y includes a "hat" because we are predicting the outcome with the model. In some representations, the $\hat{\varepsilon}_i$ is omitted because the equation merely predicts the outcome, but it is included here to remind us that LRMs almost always include errors of prediction.

To help us understand what the slope and intercept represent and why they are useful, revisit the scatter plot between these two variables (Figure 3.4). The linear fit line tilts down from left to right. The slope coefficient in the regression equation is negative but supplies more information about the association between the variables. The intercept is positive but is not shown in Figure 3.4 because its location is far off the graph. Imagine expanding the graph and you can visualize the line crossing the y-axis.

The slope coefficient indicates the average number of units the y variable differs for each one-unit difference in the x variable. This leads to an interpretation of the slope in the LRM equation:

> Each one-unit difference in average life satisfaction across states is associated with 3.79 fewer opioid deaths.[10]

Since the unit of measurement for opioid deaths is specified, we should clarify this statement.

---

[10] The term "increase" rather than "difference" is best suited for models in which variables may change over time, such as if we had data on states over several years. But it is also common, though questionable, to see it used in a more general form when interpreting slope coefficients.

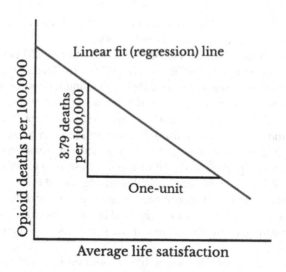

FIGURE 3.5
Linear fit line illustration of regression slope interpretation.

> Each one-unit difference in average life satisfaction is associated with 3.79 fewer opioid deaths per 100,000 residents (see Figure 3.5).[11]

Another way of thinking about the slope coefficient is that it estimates how much the *mean value* of the outcome variable, opioid deaths, is expected to differ (or shift) with each unit difference (or shift) in the explanatory variable, average life satisfaction. In other words, consistent with the first question that motivates LRMs (see p.37), the model is concerned with expected differences in the mean number of opioid deaths based on observed differences in average life satisfaction.

The intercept is also interpretable in terms of expected means. For example, the intercept of 212.2 from LRM3.1 implies:

> The expected number of opioid deaths per 100,000 residents when average life satisfaction is zero is 212.1. An alternative is to state that the average (mean) number of opioid deaths per 100,000 residents when a state's life satisfaction is zero is *expected (or predicted) to be* 212.1.

A problem with this interpretation, though, is that no state has a life satisfaction score of, or even close to, zero (check the data). Estimating means in areas outside the sampling space is not wise and can lead to nonsensical results. Although future predictions based on current values of the variables

---

[11] As noted earlier, we could also use a Pearson's *r* to estimate the association between the two variables. R provides *r* = −0.45. We'll learn in Chapter 4 how correlations and LRM slope coefficients are related.

in LRMs are not uncommon,[12] making claims about what the data show outside the range of the x- and y-variables is risky.

We use the term *associated* in the interpretation of the LRM slope coefficient. Researchers are often uncomfortable with some of the terms that describe the relationship between variables, especially when using regression models in the social and behavioral sciences. *Associated* is a safe term to many; it implies that as one variable shifts the other one also tends to shift in a particular direction. This does not indicate causation, though; for example, each one-unit increase in average life satisfaction does not lead to or produce a decrease in opioid deaths. Several explanations probably account for why these two variables are associated that have little to do with one variable causing the other.[13]

Two additional terms that some researchers use when interpreting slope coefficients are *increase/decrease* and *expected change*. For example, an alternative, though imprecise, interpretation of the previous slope coefficient is

> Each one-unit *increase* in average life satisfaction is associated with a decrease (or expected change) of 3.79 opioid deaths per 100,000 residents.

A problem with the words *increase, decrease,* and *change* is that we don't necessarily observe changes in variables. Researchers often use data that are collected at one time—known as *cross-sectional data.* They don't observe increases or decreases in variables, but rather, like in *the StateData2018* dataset, differences in the values of variables (e.g., Maryland and Missouri have different average life satisfaction values). An alternative interpretation, as implied by the first two interpretations of the slope coefficient, is therefore that the outcome variable is expected to *differ* by a certain amount as the explanatory variable *differs* by one unit.

We may also use the model's information to predict the value of the opioid deaths per 100,000 residents in states with life satisfaction scores at specific values. Equation 3.7 uses 51 as the average life satisfaction value since it's close to the mean and median for this variable.

$$E\left[\text{Opioid deaths}\right] = 212.1 - \{3.79 \times 51\} = 18.8 \text{ per } 100,000 \text{ residents} \quad (3.7)$$

---

[12] A goal of predictive analytics, for example, is to estimate what is expected to happen in the future based on what data suggest happened in the past. This includes how stock markets are expected to behave based on past performance or how many items a store should stock based on previous sales (see, e.g., Ying Liu (2014), "Big Data and Predictive Business Analytics," *Journal of Business Forecasting* 33(4): 40–42). The *Journal of Forecasting* and similar periodicals include many examples of using regression models to predict economic, demographic, electoral, and other outcomes.

[13] Recall the brief discussion of causation in Chapter 2 (see fn. 5). Myriad other factors, many of which are not measured in the dataset, probably affect or even account for the association between average life satisfaction and opioid overdose deaths. Thus, we cannot rule out or "eliminate" explanations of this association without significantly more information. Even though it is probably not causal, however, an intricate connection between life satisfaction and the prevalence of opioid-related deaths likely exists (see Anne Case and Angus Deaton (2020), *Deaths of Despair and the Future of Capitalism,* Princeton, NJ: Princeton University Press).

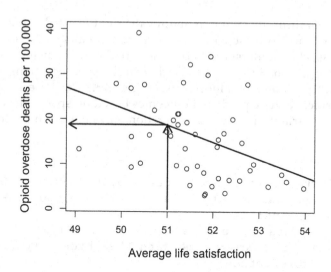

FIGURE 3.6

Scatter plot of opioid overdose deaths by average life satisfaction from LRM3.1: illustration of a predicted value.

On average, states with life satisfaction scores of 51 report approximately 18.8 opioid deaths per 100,000 residents. This is another way of stating that for states with life satisfaction scores of 51, the model estimates that the mean number of opioid deaths is 18.8.

As implied by the earlier discussion about how to interpret the slope coefficient, predicted values are thus *predicted means* of the outcome variable for particular values of the explanatory variable and, as shown by Figure 3.6, fall on the linear fit line.[14] Figure 3.6 provides a visual depiction of the predicted mean based on the model when average life satisfaction is 51. In R, predicted values are also called *fitted values* and, in the example, are part of LRM3.1 object.[15] You may find more information about these by using R's summary function:

R code

```
summary(LRM3.1$fitted.values)
```

---

[14] Sir Francis Galton—whose student was Karl Pearson—helped popularize the notion of linear regression as predicted means in the late 19th century when he collected height data from a sample of parents and their children, finding a positive association. To simplify the association, he computed the average heights of children for small intervals of parental heights. He also used these data to show the phenomenon called *regression toward the mean* (though he initially used the term *reversion*): taller parents tend to have children who are shorter than them and shorter parents tend to have children who are taller than them (Francis Galton (1886), "Regression towards Mediocrity in Hereditary Stature," *Journal of the Anthropological Institute of Great Britain and Ireland* 15: 246–263).

[15] One way to check the fitted or predicted values in RStudio is to click the arrow next to LRM3.1 in the Global Environment window and scroll down. You should see a member of the object called fitted.values with the first few values next to it (17.1, 14.6, 17.2, …).

The mean, minimum, and maximum values of the fitted values are 16.3, 1.1, and 26. How do these compare to the actual values of opioid deaths in the dataset? To understand better the actual and predicted values, re-examine Figure 3.6. Since the data include only 50 observations, you may inspect all the predicted values by executing LRM3.1$fitted.values in the R console. To view predicted outcomes based on specific levels of the explanatory variable, R's predict function may be used in the following manner:

R code
```
 # Predicted opioid deaths per 100,000 based on LRM3.1
   when life satisfaction is at its 25th, 50th, and
   75th percentiles
LRM3.1a <- data.frame(LifeSatis=c(51.1, 51.6, 52.3))
  # create a new data frame with the three levels of
   life satisfaction
predict(LRM3.1, LRM3.1a) # apply the LRM's results to
                             the data frame
```

R output (abbreviated)
```
   1      2      3
 18.30 16.41 13.75
```

Consistent with the LRM results, opioid deaths are expected to be lower in states with higher average levels of life satisfaction.

Let's examine how well the model predicts the outcome in a state that is close to the mean of average life satisfaction. Florida's life satisfaction value is 51.07. Its predicted opioid deaths per 100,000 residents is 18.4, but it's actual number is 16.3. Florida's residual ($\hat{\varepsilon}_i$), which, as noted earlier, estimates the error of prediction, is thus –2.1. Visually, the residual is the vertical distance from the actual value to the predicted value that falls on the linear fit line. Figure 3.7 provides an illustration of the residuals for four states: Nevada, Ohio, Delaware, and Kansas.

We should also ask the following question before and after estimating the LRM: why are opioid deaths negatively associated with states' average life satisfaction? Perhaps places where people tend to be more satisfied in general don't have as much need for pain relieving substances. But another issue is whether more deaths tend to generate lower life satisfaction, so perhaps we have the order of the variables wrong (see Chapter 12). Could there also be a third variable—such as health problems, poor economic conditions, or crime problems—that accounts for the association between the two variables? At this point, good answers are hard to find, but we should nonetheless seek them and be willing to revise our conclusions as new evidence emerges.

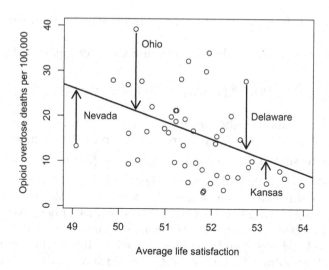

FIGURE 3.7

Scatter plot of opioid overdose deaths by average life satisfaction from LRM3.1: illustration of four predicted values.

Formulas for the Slope Coefficient and Intercept

What is the source of the LRM slope coefficient and intercept? How does R know where to place the linear fit line? Does it plot the data and then try out several lines to see which one does the best job of representing the data? It should come up with a line that is closest, on average, to all the data points in the plot. If we calculate an aggregate measure of distance from the points to the best-fitting line, such as a summary measure of the residuals (see Figure 3.7), it should produce as small a value as possible. Such an exercise is the logic underlying the most common method for fitting the regression line, which R and other software use—*ordinary least squares* (OLS) or the *principle of least squares*. Many researchers refer to the LRM as OLS regression or as an OLS regression model because this estimation technique is used so often.[16] Other estimation routines, such as weighted least squares (WLS) and maximum likelihood (ML), can also estimate regression models. But we'll focus

---

[16] The inventor of the principle of least squares is either Carl Friedrich Gauss—of Gaussian distribution fame—or Adrien-Marie Legendre, a French mathematician. Stephen M. Stigler (1981) proposes that, in all likelihood, Gauss developed the method, but, in 1808, Legendre published the first paper describing it ("Gauss and the Invention of Least Squares." *Annals of Statistics* 9(3): 465–474).

on OLS given its frequent use and because many statistical software routines rely on it.

The goal of OLS is to obtain the minimum value for Equation 3.8.

$$\text{SSE} = \Sigma(y_i - \hat{y}_i)^2 = \Sigma(y_i - \{\alpha + \beta_1 x_i\})^2 \qquad (3.8)$$

SSE is an abbreviation for the *sum of squared errors*.[17] The $(y_i - \hat{y}_i)$ portion represents the residuals, which we learned about in the last section. Thus, the SSE is also the sum of the squared residuals $(\Sigma \hat{\varepsilon}_i^2)$. Think once again about the residuals, such as those depicted in Figure 3.7. If the SSE equals zero, then all the data points fall on the fit line. The Pearson's *r* is also one or negative one (depending on whether the association is positive or negative).

It should be clear by now that residuals are a vital part of LRMs. R places them in the LRM object and provides condensed information in the summary function output. Executing LRM3.1$residuals in the R console or from an R script file, for example, furnishes a list of the residual for each observation. The residuals may also be summarized or plotted by using the methods shown earlier.

The default residuals furnished by R are called *unadjusted* or *raw residuals*. But various other versions useful in regression modeling are available, including *standardized residuals* (residuals transformed into z-scores) and *studentized residuals* (residuals transformed into t-scores). We'll examine these in later chapters.

The straight regression (linear fit) line almost never goes through each data point, so the SSE is always a positive number. The discrepancies from the line exist, in general, for three reasons: (1) data always have variation because of sampling and random error; (2) nonrandom or non-sampling error, such as poor measures of the variables; and (3) a straight-line association is not appropriate (recall the linearity assumption). The first source is a normal part of almost all LRMs. We wish to minimize errors, but natural or random variation always exists because of the intricacies and vicissitudes of behaviors, attitudes, and demographic phenomena. The second reason presents a serious problem: nonrandom errors can lead to erroneous conclusions. For example, imprecise measuring tools usually result in biased LRM estimates (see Chapter 13). If the third situation occurs, we should search for a relationship—referred to as *nonlinear*—that is appropriate. Chapter 11 furnishes a discussion of nonlinear relationships among variables.

---

[17] In this context, the "errors" are understood as the errors of prediction estimated from the sample. The SSE is available for LRM3.1 by using the anova function after executing the lm function (e.g., anova(LRM3.1)). In the printout, the SSE is the number in the Sum Sq column and the Residuals row: 4,144.3. In R and several other statistical environments, the SSE is termed the "Residual Sum of Squares" to avoid confusion about the errors of prediction. Chapter 5 provides additional information about this and related measures.

The SSE in a simple LRM requires a slope and intercept with which to compute the predicted values ($\hat{y}_i$). But what is the source of these estimates? Those who have taken a calculus course might suspect that derivatives provide the answer since they assess differences or changes in variables. This is correct and Cramer's rule and the methods of calculus allow the derivation of some simple formulae. The least squares equation represented in Equation 3.9 for the estimated slope coefficient ($\hat{\beta}$) has been shown as optimal, under certain conditions, at minimizing the SSE. It works particularly well when the assumptions discussed earlier are satisfied.

$$\hat{\beta} = \frac{\Sigma(x_i - \bar{x})(y_i - \bar{y})}{\Sigma(x_i - \bar{x})^2} \tag{3.9}$$

The numerator and denominator in Equation 3.9 should look familiar. The $\bar{x}$ and $\bar{y}$ are the means. The numerator is part of the formula for the covariance and the denominator is the formula for the sum of squares (dispersion) of x (see Chapter 2). Let's think about what happens as the quantities in Equation 3.9 change. What happens to the slope if the covariance increases but the variability of x remains constant? It increases. If other conditions do not change, then as the variation of the explanatory variable (x) increases, the slope decreases. If we presume an association between two variables, then we normally wish to have a large positive or negative slope. But it's important to remember that the magnitude of the slope depends on the variability of x, not just on the size of the covariance.

Once we compute the slope, the formula for the intercept is provided in Equation 3.10.

$$\hat{\alpha} = \bar{y} - \left\{\hat{\beta} \times \bar{x}\right\} \tag{3.10}$$

The formula uses the slope coefficient and the means of the two variables. We don't need to use these formulas to calculate the LRM coefficients, however, since programs such as R compute them. Programs employ matrix routines rather than these equations to speed up the process. But using them to calculate the slope and intercept with a small set of data points can be helpful to get a sense of how they operate.

Hypothesis Tests for the Slope Coefficient

Chapter 2 includes a discussion of hypothesis tests, which are designed to examine whether or not specific statements are valid. In one example, we examined a hypothesis about whether males and females report different average income levels. Similar types of hypotheses are assessed with LRMs. They should be more precise than claiming only that one variable is associated with the other, however. Employ a conceptual model or theory to deduce the expected associations. Or use your imagination, common sense, understanding of the research on the topic, and perhaps even colleagues' ideas as you discuss your research plans to specify the reason there should be an association. Write down the null and alternative hypotheses *before* analyzing the data. If all of these things indicate, for instance, that average life satisfaction should be negatively associated with opioid deaths at the state level, we anticipate a negative slope coefficient in an LRM that assesses these variables. The hypotheses should thus be displayed as in Equation 3.11.

$$H_0: \beta \geq 0$$

$$H_a: \beta < 0 \tag{3.11}$$

Though reasonable, most researchers who use LRMs fail to specify directionality and define the hypotheses as stating, often implicitly, that either the slope coefficient is zero or the slope coefficient is not zero in the population (see Equation 3.12).[18]

$$H_0: \beta = 0 \text{ vs. } H_a: \beta \neq 0 \tag{3.12}$$

We could just look at the sample slope coefficient and determine whether or not it's zero and then assume the same for the population. But don't forget a crucial issue discussed in Chapter 2: the sample we employ is one among many possible samples that might be drawn from a population. Perhaps the sample used in LRM3.1, for example, is the only sample that has a negative slope coefficient, but all others have a positive slope coefficient. How can we be confident that our sample slope does not fall prey to such an event? We can never be absolutely certain, yet significance tests provide some evidence with which to judge whether the results are compatible or incompatible with the hypotheses.[19]

We need to think more about standard errors, which are introduced in Chapter 2, to understand significance tests in LRMs. We already saw how to compute and interpret the standard error of the mean. The *standard error of the slope coefficient* is interpreted in a similar way: it estimates the variability of the estimated slopes that might be computed if we were to draw many, many samples. For instance, imagine we have a population of adults

---

[18] Ask yourself why we should ever expect a slope coefficient to be zero. Assuming a zero vs. non-zero coefficient is a restrictive expectation and, as suggested in Chapter 2, has led experts to criticize the way hypothesis tests are set up in many statistical modeling exercises.

[19] But recall the discussion of significance testing in Chapter 2, especially the admonition that "[b]y itself, a p-value does not provide a good measure of evidence regarding a model or hypothesis" (https://www.amstat.org/asa/files/pdfs/P-ValueStatement.pdf).

in which the correlation between age and alcohol consumption is actually zero. This implies that the population-based slope coefficient in the equation *alcohol use* = α + β(*age*) is zero or the conventional null (nil) hypothesis is valid ($H_0: \beta = 0$). Drawing many samples, can we infer what percentage of the slopes from these samples should fall a certain distance from the true mean slope of zero? We can, if certain assumptions are met, because LRM slope coefficients from samples, if many samples are drawn randomly, follow a *t*-distribution. This suggests that if we have, say, 1,000 samples, and we calculate slopes for each, we expect only about 5% of them to fall more than 1.96 *t*-values from the mean of zero (see Figure 2.4 for an analogy). The occasional sample slope coefficient farther from zero occurs if the null hypothesis is valid, but it should be rare.

Equation 3.13 provides the formula for the standard error of the slope coefficient in a simple LRM.

$$se\left(\hat{\beta}\right) = \sqrt{\frac{\sum\left(y_i - \hat{y}_i\right)^2 / n - 2}{\sum\left(x_i - \bar{x}\right)^2}} = \sqrt{\frac{SSE / n - 2}{SS(x)}} \tag{3.13}$$

The formula includes familiar elements. The numerator contains the SSE and the sample size (n). The denominator includes the sum of squares of x. When the sum of squares of x is larger, the standard error is smaller. When the SSE is larger, the standard error is larger. This should not be surprising if we think about a scatter plot with a fit line (see, e.g., Figure 3.4): larger SSEs indicate more variation—bigger residuals, on average—about the regression line. Our uncertainty about whether we have a good prediction of the population slope should also increase. What happens to the standard error as the sample size increases? It gets smaller. This also makes sense: as the sample gets larger, it gets closer to the population from which it is drawn, and we should gain more confidence that the sample slope reflects the population slope. Yet some analysts criticize studies that use large samples because, even if they are not drawn randomly, researchers can make claims of certainty that are not justifiable.

Once we have computed the standard error of the slope coefficient, the *t*-value is based on Equation 3.14.

$$t\text{-value} = \frac{\hat{\beta}}{se\left(\hat{\beta}\right)} \tag{3.14}$$

Some presentations include one β value minus another in the *t*-value equation ($\hat{\beta} - \beta_p$). This is reasonable, but, as mentioned several times, a common presumption is that the slope implied by the null hypothesis is zero, so there's often no reason to have β_p in the equation (see fn. 18, though).

Each *t*-value is associated with a *p*-value, which depends on the sample size and provides the basis for significance tests used to evaluate the plausibility of the null hypothesis slope ($\beta = 0$). Statistical software furnishes relevant *p*-values.

Recall that LRM3.1 estimates a model with average life satisfaction as the explanatory variable and opioid deaths per 100,000 residents as the outcome variable. Here, once again, is the R output.

R output (abbreviated)
```
Coefficients:
              Estimate  Std. Error  t value   Pr(>|t|)
(Intercept)   212.056      56.576     3.748   0.000479 ***
LifeSatis      -3.792       1.093    -3.468   0.001116 **
---
Signif. codes: 0 '***' 0.001 '**' 0.01 '*' 0.05 '.' 0.1

[CIs]            2.5%      97.5%
(Intercept)    98.30     325.81
LifeSatis      -5.98      -1.59
```

In addition to the slope coefficient and intercept, R furnishes the standard errors (listed under Std. Error), t-values, and p-values (under the column labeled Pr(>|t|)). The t-value for the slope is $-3.79/1.09 = -3.47$. A t-value of this magnitude from a simple LRM with a sample of 50 observations has the small p-value of approximately 0.0011.[20] R supplies significance codes that gauge some standard p-value threshold levels relied on by many researchers. R also provides a p-value for the intercept (0.0005). It compares the intercept to a null hypothesis value of zero.

Rather than focusing on the standard errors or t-values, most researchers emphasize the p-values.[21] As noted earlier, however, p-values can be misused. Research reports often rely on the decision rule that a p-value less than 0.05 implies that the slope coefficient (or other statistic) is *statistically significant*. But, as emphasized in Chapters 1 and 2, this is not synonymous with importance. Yet, the $p < 0.05$ decision rule has, for various reasons, become widely accepted as a marker for a meaningful slope coefficient and as justification for "rejecting" the null hypothesis.[22]

---

[20] One could also use R's pt function to derive the p-value: 2*pt(-abs(-3.468), df=48). R returns 0.001116. However, using this function is rarely necessary since R's summary function following a lm object furnishes precise p-values.

[21] For a lucid discussion, see Gallo (2016), *op. cit.*

[22] See Chapter 2's section on significance testing. Some argue that the threshold for determining "statistical significance" needs to be reduced, such as to $p < 0.005$. A recent report, for example, found that the chance of confirming that a p-value is less than 0.05 is only about 50:50 when studies are replicated, thus casting further doubt on the utility of this approach to significance testing (Leonhard Held et al. (2020), "Replication Power and Regression to the Mean," *Significance* 17(6): 10–11). Alternative measures that serve a similar function as p-values and might obviate some of their limitations are available, such as Bayes factors and D-values (see Eugene Demidenko (2016), "The p-Value You Can't Buy," *American Statistician*, 70(1): 33–38; and Leonhard Held and Manuela Ott (2018), "On p-Values and Bayes Factors," *Annual Review of Statistics and Its Applications*, 5: 393–419). But these are beyond the scope of this book.

Let's attempt to understand *p*-values for slope coefficients better by thinking about their basis and interpretation. We must again make an assumption that the slope coefficient from the sample says something reasonable about the slope coefficient from the population. Suppose, for instance, we find an LRM slope coefficient has a *p*-value of 0.03. Given the presumed distribution of the slope coefficients from multiple samples, the following interpretation is technically precise, even though it's not entirely persuasive:

> If the population slope is zero, we expect to find a sample slope of ($\hat{\beta}$) or one farther from zero about three times out of every 100 samples, on average, if we draw many samples from the population.

This interpretation begins with the conventional null hypothesis ($H_0 : \beta = 0$) and construes the *p*-value as a probability based on many possible samples in a frequentist framework. Another way of interpreting this *p*-value is that it suggests (but does not prove) that, if the null hypothesis ($\beta = 0$) is true in the population, we would "reject" it only about 3% of the time given an LRM slope this far or farther from zero.[23] We are again assuming that we could draw many samples to reach this conclusion.

In the model of opioid deaths and average life satisfaction, the *p*-value suggests that if the population slope is zero, we expect to find a slope coefficient of −3.79 or some value farther from zero less than one time out of every 100 samples drawn (see the two asterisks in the R output). Such a small *p*-value offers some evidence that a non-zero linear statistical association exists between the number of opioid deaths per 100,000 residents and average life satisfaction. But remember that our confidence in the evidence is only as good as the sample and how well it represents the target population.

Consider the phrase "farther from zero" in the *p*-value interpretation. This implies that we are using a *two-tailed significance test*. Examine the *t*-distribution in Figure 3.8. Zero is its middle point (see the vertical line at *t* = 0). Imagine the area under the curve as representing the frequency of slopes from a large number of samples. The mean of zero denotes the null (nil) hypothesis. This curve, therefore, signifies the presumed distribution of slopes from multiple samples given the null hypothesis. The *p*-value is designed to assess how often one would expect sample slope coefficients to fall in a certain part of this distribution. A *p*-value of, say, 0.02 suggests that

---

[23] The term "reject" has been used frequently in this context. But remember, the list of best practices in Chapter 1 includes: don't use *p*-values to claim that the null hypothesis of no difference is true. The interpretation presented here is that, if the null hypothesis is true *in the population*, we expect to get this particular result only a few times with repeated sampling. The caveats are important, though: (a) we don't observe the population and (b) we haven't collected repeated samples. Perhaps inference isn't so inferential after all. These points reinforce the recommendation that *p*-values can be helpful for assessing hypotheses but are most useful as merely one piece of evidence indicating how compatible or incompatible patterns in the data are with the conceptual model.

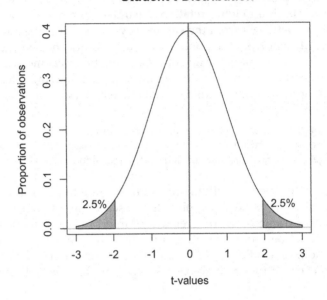

FIGURE 3.8
Student's *t*-distribution.

such a sample slope falls in one of the shaded areas marked 2.5%. Under the null hypothesis, this is an unlikely event, though still feasible.

Programs such as R typically provide two-tailed tests by default, which means that the sample slopes associated with small *p*-values could be in either tail of the distribution (recall that a two-tailed *p*-value of 0.05 corresponds to a *t*-value of |1.96| in large samples). In other words, most programs provide *p*-values that assume that the slope can be in either of the tail areas, above or below the mean (the areas marked 2.5% in Figure 3.8). This implies that a negative slope and a positive slope are equally valid for casting doubt on the null hypothesis as long as their *t*-values are sufficiently large (in absolute value).

But suppose an alternative hypothesis provides direction, such as that the LRM slope is less than zero ($\beta < 0$). In this case, there's justification for using a *one-tailed test*: the conceivable slopes are believed to fall in the lower tail of the *t*-distribution, the 5% area that falls below zero. In large samples this has a threshold at 1.64 *t*-units below zero in the *t*-distribution. How can we convert a two-tailed test, the default option in R and most other statistical software, to a one-tailed test? Take the *p*-value provided by the software and divide it in half (e.g., a *p*-value of 0.04 (two-tailed) is equivalent to a *p*-value of 0.02 (one-tailed)). But, if one wishes to use a one-tailed test, it should be established prior to model estimation. As noted earlier, some critiques have called for more stringent *p*-value thresholds (e.g., $p < 0.005$), whereas others have argued against using *p*-values for directional hypotheses (see Chapter 2), so using one-tailed tests might not be the best approach.

Interpreting p-values can be confusing, especially when thinking about relying on only one sample and then trying to infer something about a population assuming we were able to draw multiple samples. It might be easiest to remember that we want to estimate slopes far from zero if we wish to have evidence that a non-zero linear association exists between an explanatory and an outcome variable. But do this with care: most research studies rely on only one sample out of many conceivable samples from the population.

Given these issues, many researchers, even those who don't wish to eliminate p-values from their data analysis toolkit entirely, argue that significance tests with p-values are misleading because they deceive the reader into thinking that the estimates are precise and fail to account for the uncertainty that is part of any statistical model.[24] An alternative that admits uncertainty is the CI for the slope, which is constructed in a similar manner as the CI for the mean (see Equation 3.15).

$$\text{CI} = \text{Point estimate} \pm \left[(\text{confidence level}) \times (\text{standard error}) \right] \quad (3.15)$$

The confidence level for an LRM is the t-value based on the percentage of time one wishes to be able to claim that the interval includes the population parameter. The most common choice is a t-value that leads to a 95% CI. In a large sample, this value is 1.96. But, for LRM3.1, which uses a sample of 50 with two estimated coefficients ($\hat{\alpha}$ and $\hat{\beta}$), the t-value is 2.01.[25] Based on the slope coefficient and standard error estimated by LRM3.1, the computation of the 95% CI is displayed in Equation 3.16.

$$-3.79 \pm \left[2.01 \times 1.09 \right] = \{-5.98, -1.59\} \quad (3.16)$$

We don't need to go through these calculations to obtain the 95% CI of the slope coefficient. As shown earlier, R's confint function executed after the lm function supplies it: {−5.98, −1.59}. The interpretation of the CI is

> We are 95% confident that the population slope coefficient representing the association between average life satisfaction and opioid deaths per 100,000 residents falls in the interval {−5.98, −1.59}.

The interpretation does not include the sample slope coefficient—even though some researchers include it—because, although it's part of the CI computation, it has no intrinsic interpretation. Rather, the CI focuses on placing the likely population slope within a plausible range of values.

Suppose we wish to obtain a higher degree of confidence, such as 99% confidence about where the population slope falls. Then the interval is wider

---

[24] See, for example, Hahn et al. (2019), *op. cit.*
[25] It can be approximated in R using the following function: abs(qt(0.05/2, 48)). The qt part is short for *quantile of t*, which is pth quantile of the Student's t distribution.

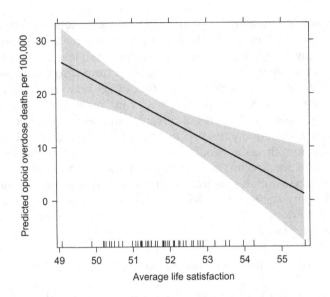

FIGURE 3.9
Predictor effects plot of opioid overdose deaths by average life satisfaction from LRM3.1.

because more uncertainty exists. The *t*-value used in the equation is thus larger. For example, a *t*-value corresponding to the 99% confidence level for LRM3.1 is about 2.68, so the 99% CI is {−6.72, −0.86}. Try to figure out how to obtain a 99% CI for an LRM slope coefficient in R.

A helpful way to represent visually what LRM3.1 implies, including its inherent uncertainty, is with a *predictor effects plot*, which graphs the predicted values from the model, along with 95% confidence bands for the linear fit line.[26] This plot is available in R's `effects` package. The confidence bands or regions reflect the uncertainty about the regression (linear fit) line. According to Figure 3.9, the negative association is characterized by relatively low uncertainty in the middle of the life satisfaction distribution but more uncertainty in its tails. One reason we see this varying uncertainty is because most of the observations fall in the middle of the joint distribution of the two variables, so there is relatively less uncertainty.

R code for Figure 3.9
```
library(effects)
plot(predictorEffects(LRM3.1))
```

An interesting relationship between *p*-values and CIs occurs when using the same *t*-value for determining the threshold and the confidence level. Suppose, in a large sample, we use as a threshold a *p*-value of 0.05 (two-tailed test) and a *t*-value of 1.96 for the confidence level. We should, under

---

[26] See Fox and Weisberg (2018), *op. cit.*, chapter 4.

conventional significance testing guidelines, interpret the evidence provided by each as supporting the same conclusion regarding the null hypothesis. This is because the p-value of less than 0.05 corresponds to a 95% CI that does not include zero. Our conclusions about statistical significance are similar whether we use p-values or CIs. As mentioned in Chapter 2, a key advantage of CIs is that they offer a better reflection of the uncertainty that is a fundamental part of statistical modeling exercises. Some researchers prefer, therefore, to present just the CIs and avoid p-values altogether.[27] But rather than making claims about the null hypothesis, even CIs are best viewed as one piece of evidence with which to judge whether patterns in the data are consistent with the conceptual model driving the research.[28] Taking into account all the evidence, including the p-value, the CI, the sign and the magnitude of the slope coefficient, and the predictor effects plot, the most reasonable interpretation of the model in LRM3.1 might therefore be

> The evidence from the linear regression model is consistent with a conceptual model that presumes that higher average life satisfaction is negatively associated with opioid deaths at the state-level in the U.S.

Chapter Summary

This chapter introduces the simple LRM that involves a single explanatory variable (x) designed to predict or account for a single outcome variable (y). The key issues covered in this chapter include: (a) the purpose of LRMs; (b) how to estimate and interpret LRM coefficients; (c) predicted (fitted) values and residuals that result from the model; (d) assumptions of the LRM; (e) the formulas used to estimate the LRM coefficients; and (f) significance tests for the slope coefficient.

---

[27] But this doesn't avoid some of the criticisms of significance tests: they rely on samples that may not be representative of a target population (if such a thing exists) and fail to account for nonrandom sources of error, such as poor measures of variables. In an amusing but helpful recommendation, David Spiegelhalter (2019) writes, "My personal, rather skeptical heuristic is that any quoted margin of error [CI] in a poll [or survey] should be doubled to allow for systematic errors made in the polling" (*The Art of Statistics: Learning from Data*, London: Penguin, pp. 245, 247).

[28] As mentioned earlier, a number of researchers prefer Bayesian inference to frequentist (or classical) inference. Bayesian inference is concerned with estimating *credible intervals* for some parameter: one estimates the probability of various values of a parameter given the data. For more information, see David L. Weakliem (2016), *Hypothesis Testing and Model Selection in the Social Sciences*, New York: Guilford. Marin and Robert (2014), *op. cit.*, discuss Bayesian analysis in R. An excellent treatment of regression models using Bayesian and classical inference is provided in Andrew Gelman, Jennifer Hill, and Aki Vehtari (2020), *Regression and Other Stories*, New York: Cambridge University Press.

Chapter Exercises

The dataset called *HighSchool.csv* consists of data from a 2000 national survey of high school students in the U.S. Our objective is to examine the associations among some variables from this dataset. Several of them are coded strangely, so just focus on what the higher and lower values imply. In addition to identification variables (Row, IDNumber), the variables include:

- SportsParticipation   participation in school athletics, with higher values indicating playing more high school sports

- AcademicClubs   involvement in school academic clubs and honor societies, with higher values indicating being a member of more clubs and societies

- AlcoholUse   frequency of alcohol use—higher values indicate more frequent drinking

- GPA   grade point average in high school on a standard four point scale (0–4), with higher values indicating better grades

After importing the dataset into R, complete the following exercises.

1. Compute the mean, median, standard deviation, skewness, and kurtosis of the AlcoholUse variable. Based on this information, comment on its likely distribution.

2. Create a kernel density plot in R of AlcoholUse. Describe the distribution of this variable.

3. Create a scatter plot in R that specifies AlcoholUse as the *y*-axis. On the *x*-axis, use the substantive variable (Not the row or ID variable) that has the highest Pearson's correlation (farthest from zero) with AlcoholUse. Include a red linear fit line in the plot. Include a blue horizontal line in the plot that represents the mean of AlcoholUse. Describe the linear association between the two variables.

4. Estimate a LRM that utilizes AlcoholUse as the outcome variable and, as the explanatory variable, the variable you used on the *x*-axis in exercise 3.

   a. Interpret the intercept and the slope coefficient associated with the explanatory variable.

   b. Interpret the *p*-value and 95% CI associated with the slope coefficient.

5. Use the LRM in exercise 4 to predict the level of `AlcoholUse` when the explanatory variable is at its minimum, maximum, and mean levels.

6. What is the Pearson's correlation of the predicted values (R labels them `fitted.values`) from the model estimated in exercise 4 and `AlcoholUse`?

7. *Challenge*: create a kernel density plot of the residuals from the LRM estimated in exercise 4. Describe the distribution of the residuals. What does the distribution indicate about whether the model satisfies the normality assumption?

4

Multiple Linear Regression Models

We've learned about several aspects of simple LRMs, in particular how to interpret slopes and intercepts, their source equations and assumptions, and some inferential issues. The model introduced in the last chapter considers only the LRM with a single explanatory variable. We'll now extend the model to include more than one explanatory variable, the *multiple LRM*. Many of the same issues exist, but a key difference between simple and multiple LRMs involves the interpretation of the coefficients.

Most simple LRMs are not interesting. In the last chapter, for instance, we assessed the association between average life satisfaction and opioid deaths among states in the U.S. and discovered a negative association. But you don't need to be a sociologist or epidemiologist to realize that other factors also relate to opioid deaths at an aggregate level. Some potential factors include economic conditions, crime rates, substance use rates, the age structure; we could go on. The point is that many potential explanatory variables might be useful for predicting an outcome variable. The selection of these variables should be guided by theoretical and conceptual models, yet limiting LRMs to one explanatory variable is often incomplete and unsatisfactory.

One reason to include other variables in a model—in addition to theoretical concerns—is that they may account for the association between one of the explanatory variables and the outcome variable. This is known as *confounding*, since we say that if variable x_2 accounts for the association between x_1 and y, x_2 is a *confounder* of their association.[1] For example, suppose we're interested in the association between the number of lighters purchased and the rate of lung disease across a sample of U.S. cities. Is there a positive association between these two variables? Probably, but is it fair to conclude that purchasing lighters "causes" or produces lung disease? No, because cigarette smoking is related to both purchasing lighters and lung disease. Cigarette smoking is thus a confounding variable, and the association between the number of lighters purchased and the rate of lung disease is called *spurious*. Smoking should be included in a regression model that predicts lung disease, especially if the model also includes the frequency of lighter purchases. (Smoking should always be in models predicting lung disease.)

---

[1] Chapter 6 provides a detailed discussion of confounding. Recall, moreover, that one of Mill's criteria for causation is that one should eliminate other explanations or factors that might account for the association between a presumed cause and effect (see Chapter 2, fn. 5). If our goal is to identify causal relationships among variables then surely we must consider a number of factors.

DOI: 10.1201/9781003162230-4

An Example of a Multiple LRM

Like simple LRMs, multiple LRMs are estimated with ordinary least squares (OLS) (see Chapter 3), but adding explanatory variables makes the coefficients and other features of the model more challenging to interpret. Before examining some of these features, let's estimate a multiple LRM in R and examine its output. We'll then learn how to interpret its slope coefficients. The *StateData2018.csv* dataset used in the previous chapter includes measures of the states' number of violent crimes per 100,000 residents (ViolentCrimeRate), which we shall treat as the outcome variable, percent of children living in poverty (PerChildPoverty), and median household income (MedHHIncome). Before estimating an LRM, let's ask R to create a subset of the *StateData2018* dataset that consists of these three variables. We then use this subset to examine the correlations among the variables to see, in a rough sense, the direction and strength of their associations. R has several functions that compute correlations; we'll use the corr.test function in the psych package since it provides correlations and their *p*-values.

R code
```
library(psych) # activate the psych package
sub.corr <- StateData2018[c("ViolentCrimeRate",
          "PerChildPoverty", "MedHHIncome")]
  # create a subset of the data

corr.test(sub.corr)  # request the correlation matrix
                     using psych's corr.test function
```

R output (annotated and abbreviated)
```
              # Pearson's r #
          Violent PerChildPov MedHHInc
Violent    1.00       0.49      -0.21
ChildPov   0.49       1.00      -0.76
MedHHInc  -0.21      -0.76       1.00
               # p-values² #
          Violent PerChildPov MedHHInc
Violent               0.0003    0.1471
ChildPov   0.0003               0.0000
MedHHInc   0.1471    0.0000
```

Violent crimes have a positive correlation with child poverty and a negative correlation with median household income. The latter correlation is not

---

[2] Two of the *p*-values are listed as zero in the R output. Yet, *p*-values are never zero; R has rounded down a small number. When a *p*-value is listed with several digits but appears as, say, 0.0000, it indicates that it is less than 0.0001. When presenting small *p*-values, they should be listed as $p < 0.001$ or with some similar designation.

below the standard *p*-value threshold of 0.05, though ($p = 0.15$). Child poverty and median household income also have a substantial negative correlation ($r = -0.76, p < 0.001$). If we hypothesize that child poverty and median household income are explanatory variables that predict violent crime rates, we have tentative evidence that the key variables are associated.

What does an LRM indicate about these associations? We'll begin with a simple LRM using the percent of children in poverty as an explanatory variable (see LRM4.1). The interpretation of the slope coefficient should be straightforward:

> Each 1% difference (or increase) in children below the poverty level across states is associated with 13.34 more violent crimes per 100,000 residents.

The *p*-value suggests that, if the population slope is zero, we'd expect to find a slope coefficient of 13.34 or one farther from zero less than one time out of every 1,000 samples (what are your thoughts on this?). The CI denotes that we may be 95% confident that the population slope representing the association between child poverty and violent crimes falls in the interval {6.49, 20.18}. The intercept indicates that the expected number (mean) of violent crimes per 100,000 is 122 when the percent of children in poverty is zero (is this a reasonable number?). In general, then, the evidence from the statistical model is compatible with a conceptual model asserting a positive association between child poverty and violent crimes.

R code
```
LRM4.1 <- lm(ViolentCrimeRate ~ PerChildPoverty,
             data=StateData2018)
summary(LRM4.1)
confint(LRM4.1))
```

R output (abbreviated)
```
Coefficients:
                 Estimate  Std. Error  t value  Pr(>|t|)
(Intercept)       122.244      59.556    2.053  0.045587 *
PerChildPoverty    13.335       3.406    3.915  0.000285 ***
---
Signif. codes: 0 '***' 0.001 '**' 0.01 '*' 0.05 '.' 0.1

Residual standard error: 113.3 on 48 degrees of freedom
Multiple R-squared: 0.242, Adjusted R-squared: 0.2262
F-statistic: 15.33 on 1 and 48 DF, p-value: 0.000285

[CIs]                 2.5%      97.5%
(Intercept)          2.498    241.990
PerChildPoverty      6.486     20.184
```

Let's now estimate a multiple LRM by adding median household income to the lm function (see LRM4.2). The interpretation of the slope coefficient is simpler if we transform the measurement scale of this variable to $1,000s.

R code
```
StateData2018$MedHHInc <- StateData2018$MedHHIncome/1000
 # transform its measurement scale to $1,000s
LRM4.2 <- lm(ViolentCrimeRate ~ PerChildPoverty +
                MedHHInc, data=StateData2018)
summary(LRM4.2)
confint(LRM4.2)
```

R output (abbreviated)
```
Coefficients:
                 Estimate Std. Error t value Pr(>|t|)
(Intercept)      -310.566    217.913  -1.425 0.160712
PerChildPoverty    21.180      5.038   4.204 0.000116 ***
MedHHInc            4.991      2.423   2.060 0.045008 *
---
Signif. codes: 0 '***' 0.001 '**' 0.01 '*' 0.05 '.' 0.1

Residual standard error: 109.7 on 47 degrees of freedom
Multiple R-squared: 0.3048, Adjusted R-squared: 0.2752
F-statistic: 10.3 on 2 and 47 DF, p-value: 0.000195

[CIs]                  2.5%     97.5%
(Intercept)        -748.950   127.819
PerChildPoverty      11.046    31.314
MedHHInc              0.116     9.866
```

The output lists two slope coefficients, one for each explanatory variable. What does a predictor effects plot (recall Figure 3.9) indicate about the model? Figure 4.1 represents the positive linear association between percent child poverty and predicted violent crimes after statistically *adjusting for* the effects of median household income in $1,000s. Recall that the confidence bands reflect uncertainty in the association.

R code for Figure 4.1
```
library(effects)
plot(predictorEffect("PerChildPoverty", LRM4.2,
     xlevels=60), main="", xlab="Percent child poverty",
     ylab="Predicted violent crimes per 100,000")
  # set median household income at its approximate mean
    of 60 (xlevels=60)
```

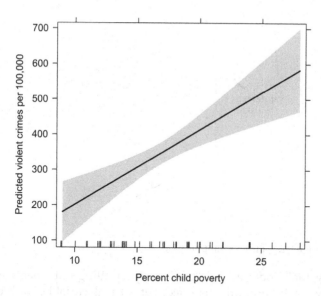

FIGURE 4.1
Predictor effects plot of violent crimes by percent child poverty from LRM4.2.

But what does it mean to claim that the association *adjusts for* the effects of the other explanatory variable? Let's first learn how to interpret the intercept and slope coefficients in the R output and then try to understand how a multiple LRM produces them.

We now have three variables in the model, so we may envision the associations in three dimensions, with the intercept represented as the point on the *y*-axis when the other two variables' axes are at zero (see Figure 4.2). The intercept is thus the expected value of the outcome variable when both of the explanatory variables are zero. If an imaginary state has no child poverty and zero median income, we expect its mean number of violent crimes to be −310.6 per 100,000. The intercept is meaningless, however, since there cannot be a negative number of violent crimes nor states that have no child poverty or zero median income.

The slope coefficients are interpreted in a familiar manner, with a phrase added to each statement. The child poverty slope coefficient is interpreted as:

> Statistically adjusting for the effects of median household income, each 1% difference (or increase) in children living below the poverty level across states is associated with 21.2 more violent crimes per 100,000 residents.

Similar to the interpretation for Figure 4.1, we use the phrase "statistically adjusting for." Some researchers employ alternative phrases such as

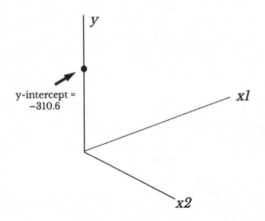

FIGURE 4.2
Illustration of an intercept from a multiple LRM.

"controlling for," "holding constant," or "partialling out." Because we are, presumably, partialling out the effects of a third variable, multiple linear regression coefficients are also called *partial regression coefficients, partial slope coefficients,* or *partial slopes.* The terms "controlling for" or "holding constant" may be misleading, however, because they imply that researchers have control over a variable. But "controlling" variables is rare in social and behavioral sciences, especially with observational data. Experimental designs are typically needed if researchers are to control the level of or exposure to a variable (see Chapter 2). For example, in the *StateData2018* dataset, we have no control over—and cannot change—a state's median household income. But, as we'll learn later in the chapter, claiming that we are holding another variable "constant" as we assess the linear association between an explanatory variable and the outcome variable offers a useful way to understand how to interpret an LRM. We'll typically use the phrase "adjusting for," though, when interpreting a model's slope coefficients.

Statistical adjustment can be a difficult concept, but one way to understand it is that we are estimating the slope of one explanatory variable on the outcome regardless of the level of the other explanatory variable (or by setting it at a specific value, such as in Figure 4.1—median household income is set at its mean). For instance, if we claim that each 1% difference in child poverty across states is associated with 21.2 more violent crimes, we're assuming that this occurs for any value of median household income, whether $50,000 or $70,000. Since statistical adjustment is such an essential topic, the following discussion provides four ways to understand it.

The first is designed for those with good spatial perception skills. It utilizes a three-dimensional graph to visualize the relationship among the three variables. R has several options for creating these graphs, including functions available in the packages plotly and rgl. The following function from the plotly package creates a dynamic three-dimensional graph that

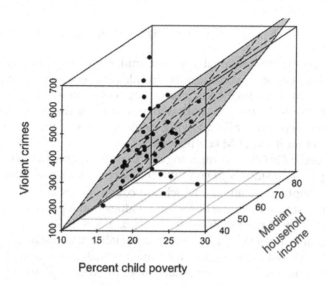

FIGURE 4.3
Three-dimensional representation of a multiple LRM slope.

may be rotated to examine the associations among the three variables from a variety of angles.

R code
```
library(plotly)
plot_ly(StateData2018, x=~PerChildPoverty, y=~MedHHInc,
        z=~ViolentCrimeRate, type="scatter3d",
        mode="markers")
```

The dataset is not particularly large ($n = 50$), but recognizing patterns in the associations is difficult. If the angle is just right, though, the positive association between child poverty and violent crimes is perceptible. Figure 4.3 offers an alternative depiction of the associations. The plane in gray represents the association between percent child poverty and violent crimes per 100,000 residents, adjusting for the effects of median household income. There's a slight tilt to the plane, but it shows that the regression surface is flat regardless of the level of median household income. A flat surface suggests that the association between percent child poverty and violent crimes does not vary by median household income. The association is also implied by Figure 4.1, though it is not as evident.

A second way to understand statistical adjustment assumes experience with calculus. Suppose y is a function of two variables, x and w, that we wish to include in a regression model. If w is held constant (e.g., $w = w_0$), then y is a function of a single variable x. Its derivative at a particular value of x is called the *partial derivative* of y concerning x, which is represented by

$\dfrac{\partial y}{\partial x}$ or $\dfrac{\partial f(x,w)}{\partial x}$ where $y = f(x, w)$. Partial derivatives, which reflect partial slope coefficients, offer a valuable way to understand statistical adjustment. But this book assumes no background in calculus, so we won't go into more detail. Many calculus textbooks and online tools include graphical depictions of partial derivatives that allow one to explore the notion of holding one variable constant while allowing another to vary, thus providing an illustration of multiple LRM coefficients.[3]

A third method involves computing residuals from two distinct LRMs and then using these residuals to compute the partial slope coefficient. As discussed in Chapter 3, residuals, which are computed as $\hat{\varepsilon}_i = (y_i - \hat{y}_i)$, gauge the vertical distance from the observed y values to the predicted values ($\hat{y}_i$) represented by the linear fit line (see Figure 3.7). Residuals also measure the variation that is left over in an outcome variable after accounting for the systematic part that is associated with an explanatory variable. Part of the remaining variation may be associated with another explanatory variable. This part is represented by the partial slope coefficient.

To illustrate how residuals can help us understand statistical adjustment, consider the following steps in R:

1. Estimate a simple LRM with violent crimes as the outcome variable and median household income as the explanatory variable (leave child poverty out of the model). Call this model `resid1`:

   ```
   resid1 <- lm(ViolentCrimeRate ~ MedHHInc, data =
               StateData2018)
   ```

   Save the residuals from this model in a new R object called `resid1a`:

   ```
   resid1a <- resid1$residuals
   ```

   The residuals (`resid1a`) measure the variability in violent crimes not accounted for by median household income. Review them using RStudio's `View` option or in the `Global Environment` window.

2. Estimate a second LRM with child poverty as the outcome variable and median household income as the explanatory variable. Call this model `resid2`:

   ```
   resid2 <- lm(PerChildPoverty ~ MedHHInc, data =
               StateData2018)
   ```

   Save the residuals from this model in a new object called `resid2a`:

   ```
   resid2a <- resid2$residuals
   ```

---

[3] See, for example, Robert P. Gilbert et al. (2020), *Multivariable Calculus with Mathematica*, Boca Raton, FL: Chapman and Hall/CRC Press.

These residuals (resid2a) measure the variability in child poverty that is not accounted for by median household income.

3. Estimate an LRM that uses the residuals from step 1 (resid1a) as the outcome variable and the residuals from step 2 (resid2a) as the explanatory variable:

   ```
   summary(lm(resid1a ~ resid2a))
   ```

   There's no reason to examine the residuals from this model.

The slope coefficient for resid2a from the third model is 21.18, which is the same number as the partial slope coefficient for child poverty from LRM4.2.[4] The slopes are identical because they represent the same thing: the shared variability between child poverty and violent crimes that does not involve median household income.

The final method that helps illustrate statistical adjustment is similar to the previous one, except it's visual. Figure 4.4 shows three overlapping circles labeled y, x_1, and x_2 because they represent three variables in an LRM. The total area of each circle represents the variable's dispersion, such as its sum of squares or variance. The overlapping areas symbolize their joint variability or covariance. The cross-hatched area is the overlap between y and x_1 that does not include the circle representing x_2. This area represents the joint

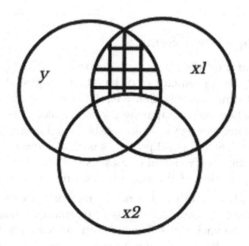

FIGURE 4.4
Overlapping variability of three variables to represent statistical adjustment.

---

[4] Don't get confused by R's use of scientific notation to represent the slope coefficient in the LRM that uses the residuals. The value 2.118e + 01 is 2.118 × 10 = 21.18. R uses scientific notation when a number has a large number of digits before or after the decimal place. If you don't care for scientific notation, you may turn it off using the function options(scipen = 999).

variability of y and x_1 that is not accounted for by the variable x_2. It thus signifies the partial slope coefficient for x_1 from a multiple LRM.

As Figure 4.4 demonstrates, the explanatory variables, x_1 and x_2, do not need to be completely independent; they may covary. As mentioned in Chapter 3, some researchers use the term *independent variable* to describe explanatory variables in LRMs, even though this can be misleading. The explanatory variables are not independent of one another, but, in a multiple LRM, they are interpreted as *independently* predicting the outcome variable (though in some models we may examine their joint association with the outcome—see Chapter 11's discussion of interaction terms).

Multiple LRMs with more than two explanatory variables require no deeper level of understanding than what we now possess. We should continue to mention the additional explanatory variables when interpreting slope coefficients; for instance, "statistically adjusting for the effects of the other variables in the model, each one-unit increase/difference in variable x_1 is associated with a [partial slope, $\hat{\beta}_1$] unit increase/difference in variable y." But remember that a more precise interpretation uses the units of the variables, such as percentages, dollars, pounds, centimeters, and so forth (e.g., "each $1,000 difference in median household income is associated with ...").

Comparing Slope Coefficients

One of the aims of multiple LRMs is to determine which explanatory variable is the best (linear) predictor of or has the strongest statistical association with the outcome. Because many explanatory variables are measured in different units, however, comparing the size of slope coefficients directly is rarely appropriate.[5] For example, in LRM4.2, median household income's slope coefficient is 4.99 and child poverty's is 21.18. Does child poverty have a stronger association with violent crimes? Is it a better predictor? Consider the generic interpretation of the slope coefficient: "a *one-unit* difference in x is associated with a $\hat{\beta}$-unit difference or increase/decrease in y." A one-unit difference is not the same for the two explanatory variables in LRM4.2, though, because one variable is measured in percentages and the other is measured in $1,000s. We should not attempt to compare coefficients based

---

[5] Some researchers also use p-values to compare the relative predictive strength of explanatory variables. If one variable's p-value is, say, below 0.05 and another's is above 0.05, the temptation is to consider the latter as "more significant" or more strongly related than the other to the outcome. Avoid this—it is not appropriate (see Andrew Gelman and Hal Stern (2006), "The Difference Between 'Significant' and 'Not Significant' Is Not Itself Statistically Significant," *American Statistician* 60(4): 328–331, and Adeline Lo et al. (2015), "Why Significant Variables Aren't Automatically Good Predictors," *Proceedings of the National Academy of Sciences* 112(45): 13892–13897).

on what R provides in its summary function unless the x variables are measured in the same manner (e.g., dollars, percentages).[6]

Researchers have developed several techniques to compare the predictive strength of explanatory variables that are measured in different units. Two common methods fall under the classification of *effect sizes*. An effect size measures the strength of the statistical association between two variables on a numeric scale, typically a standardized scale. Effect sizes for LRMs include correlations and standardized slope coefficients.[7]

The simplest effect size measure for LRMs utilizes a bivariate correlation matrix and identifies the largest Pearson's correlation between an explanatory variable and the outcome variable. Inspecting the correlation matrix at the beginning of the chapter, for instance, it appears that child poverty has a larger correlation with violent crimes than median household income ($r = 0.49$ vs. -0.21), so we might view it as a stronger predictor. Using bivariate correlations can be misleading, though, because they don't account for the associations among the explanatory variables (see Figure 4.4). Once these are considered, the association between the x and y variables might change. For instance, the Pearson's correlation between median household income and violent crimes is -0.21 with a p-value of 0.15. But the partial regression slope coefficient for this variable is 4.99 with a p-value of 0.045. Not only does the 95% CI for the slope coefficient not include zero {0.12, 9.87} and the p-value (0.045) fall below the threshold of 0.05, but, in contrast to the correlation, the slope is positive.

Some researchers prefer to use *standardized slope coefficients* as effect sizes to compare associations in LRMs. Their relationship to unstandardized slope coefficients—those provided by R's summary function—is shown in Equation 4.1.

$$\text{Standardized slope}\left(\hat{\beta}_k^*\right) = \hat{\beta}_k\left(\frac{s_{x_k}}{s_y}\right) \tag{4.1}$$

The term s_{x_k} denotes the standard deviation of x for variable k. The s_y term is the standard deviation of y. Based on LRM4.2, the standardized slope coefficient for child poverty is computed in Equation 4.2.

$$21.18 \times \left(\frac{4.75}{128.82}\right) = 0.78 \tag{4.2}$$

---

[6] Chapter 5 provides an example of comparing slope coefficients from variables measured in the same way (see LRM5.3).

[7] The best-known effect size metric is called *Cohen's d*, which measures standardized differences in means (see Jacob Cohen (1988), *Statistical Power Analysis for the Behavioral Sciences*, 2nd Ed., Lawrence Erlbaum). Cohen's d is calculated by taking the difference between the means from two groups and dividing by the pooled standard deviation. A d value of $|0.5|$ or larger is considered a substantial effect size. This metric has been adapted for use with LRMs but is usually based on comparing the explained variance (R^2s) (see Chapter 5) from separate models.

R provides standardized slope coefficients with the lm.beta package. After re-estimating the original multiple LRM, the R code for LRM4.3.beta demonstrates how to request these coefficients.

R code
```
library(lm.beta)
LRM4.3 <- lm(ViolentCrimeRate ~ PerChildPoverty +
             MedHHInc, data=StateData2018)
LRM4.3.beta <- lm.beta(LRM4.3) # use the lm.beta
                                         function
print(LRM4.3.beta) # print the standardized coefficients
```

R output (abbreviated)
```
Standardized Coefficients:
  (Intercept) ChildPoverty  MedHHInc
     0.000        0.781       0.383
```

Whether computed by hand, calculator, or with the lm.beta function, the standardized slope coefficients are 0.78 (child poverty) and 0.38 (median household income).[8]

These coefficients, which are also called *beta weights*, are interpreted using standard deviations. Another way to understand them is to imagine transforming the explanatory variables and the outcome variable into z-scores, estimating the LRM, and inspecting the slope coefficients. They thus represent the association between variables in standard deviation units and are identical to the Standardized Coefficients produced by R's lm.test function. For example, the interpretation of median household income's beta weight in LRM4.3.beta is

> Statistically adjusting for the effects of child poverty, each one standard deviation difference (increase) in median household income is associated with 0.38 standard deviation-unit additional violent crimes per 100,000 residents.

Some researchers prefer to use and report beta weights because they argue that this allows direct comparisons of the slopes' magnitudes within the same LRM. They maintain that if one explanatory variable has a beta weight farther from zero than another (e.g., $\left|\hat{\beta}^*\right|(child\,poverty) > \left|\hat{\beta}^*\right|(median\,income)$), it also has a stronger association with the outcome variable. This assumes, though, that the distributions of the explanatory variables are similar. But one variable might be more skewed than another, so standard deviation

---

[8] Recall from Chapter 2 that the correlation is a standardized version of the covariance. In a simple LRM, the standardized slope coefficient is the Pearson's correlation between x and y. In a multiple LRM, they are not equivalent, however, at least not in simple form, but instead are called *partial correlations* $(cor(x, y \mid z))$. R's ppcor package computes partial correlations.

shifts in the two variables are not equivalent. (Are the two explanatory variables in LRM4.3 distributed similarly?) Beta weights also have no clear interpretation for indicator (binary) variables (see Chapter 7) and are therefore a limited way to compare the strength of the associations between particular explanatory variables and the outcome variable.[9]

Slope coefficients from LRMs may be compared with several other metrics. The R package relaimpo (*relative importance*) furnishes five different measures designed to identify the strongest predictor in an LRM.[10] The following example extends LRM4.3 by adding two explanatory variables—the state's unemployment rate and the percent of residents without health insurance—and then uses the package's calc.relimp function to compute the relative importance metrics (see LRM4.4 and the R code that follows).

R code
```
library(relaimpo)
LRM4.4 <- lm(ViolentCrimeRate ~ PerChildPoverty +
            MedHHInc + UnemployRate + PercentUninsured,
            data=StateData2018)
summary(LRM4.4)
X.compare <- calc.relimp(LRM4.4, type = c("lmg",
                        "first", "last", "betasq",
                        "pratt")) # requests five
                        metrics
X.compare      # outputs the five metrics
plot(X.compare) # provides bar graphs of the results
                (results not shown)
```

R output (abbreviated)

	Estimate	Std. Error	t value	Pr(>|t|)	
(Intercept)	-398.488	219.041	-1.819	0.0755	.
PerChildPoverty	15.418	5.913	2.608	0.0123	*
MedHHInc	4.688	2.579	1.818	0.0758	.
UnemployRate	27.633	23.150	1.194	0.2389	
PercentUninsured	10.313	4.722	2.184	0.0342	*

```
---
Signif. codes:  0 '***' 0.001 '**' 0.01 '*' 0.05 '.' 0.1

Response variable: ViolentCrimeRate
Total response variance: 16594.39
```

---

[9] Standardized slope coefficients are helpful when the explanatory variables one wishes to compare are (a) continuous and (b) normally distributed (or distributed in a similar manner). In other situations, they should be used with caution.

[10] For information about this R package and the metrics it estimates, see Ulrike Gromping (2006), "Relative Importance for Linear Regression in R: The Package relaimpo," *Journal of Statistical Software* 17(1): 1–27.

```
Analysis based on 50 observations

4 Regressors:
PerChildPoverty MedHHInc UnemployRate PercentUninsured
Proportion of variance explained by model: 39.44%
Metrics are not normalized (rela=FALSE).

Relative importance metrics:
              lmg    last   first  betasq  pratt
PerChildPov 0.155  0.092  0.242   0.324   0.280
MedHHInc    0.039  0.044  0.043   0.130  -0.075
Unemploy    0.106  0.019  0.200   0.029   0.077
Uninsured   0.094  0.064  0.158   0.080   0.113
```

The metrics are based on decomposing how much of the variance in the outcome variable is attributable to the explanatory variables. The metric first is the least rigorous since it uses simple LRMs to compare the strength of the associations, similar to comparing Pearson's correlations. The metric last examines how much more of the variance in y is accounted for by each explanatory variable after the others are in an LRM. The metrics lmg and pratt are based on more complex approaches, whereas betasq is the beta weight squared (see LRM4.3.beta).[11]

The results favor child poverty as the strongest predictor of violent crimes per 100,000 residents, with the largest value in each of the relative importance metrics. The unemployment rate is the second largest value in two of the metrics (lmg and first), whereas percent uninsured is the second largest in two others (last and pratt). Median household income is the second largest value in the beta-squared (betasq) column, but since it's the squared value of its beta weight, it has limitations as a comparison tool.

We should now ask why child poverty is the strongest predictor of violent crimes. Is this what we hypothesized? If we assume a sample, have we taken into consideration sampling error when determining the strongest predictor? Examine the documentation for the relaimpo package to see if it provides CIs for its measures. CIs might be useful for making a more exhaustive determination about the explanatory variables' predictive power.

Another approach that helps us understand the practical—in addition to the statistical—significance of LRM results is to compare predicted means of the outcome for different levels of the explanatory variables. For example, what is the difference in the predicted means of violent crimes based on the results of LRM4.3? One tactic is to compare differences based on quantiles of the explanatory variables. We might choose, for instance, to compute predicted means at the 25th and 75th percentiles of the explanatory variables

---

[11] Gromping (2006), *op. cit.*, and Johnson and LeBreton (2004) recommend using lmg to compare slope coefficients ("History and Use of Relative Importance Indices in Organizational Research," *Organizational Research Methods* 7(3): 238–257).

TABLE 4.1

Predicted violent crimes per 100,000 population

	25th percentile	75th percentile	Raw difference	Percentage difference (%)
Percent child poverty	260	408	148	57
Median household income	304	377	73	24

and determine how many violent crimes are expected for each percentile. The 25th and 75th percentiles are 13% and 30% for child poverty and 53.1% and 67.7% for median household income (in $1,000s). Based on the following code that utilizes R's predict function, the predicted numbers of violent crimes for these percentiles are provided in Table 4.1.[12]

R code

```
LRM4.3a <- data.frame(PerChildPoverty=c(13, 30),
          MedHHInc=59.2)
predict(LRM4.3, LRM4.3a)
LRM4.3b <- data.frame(PerChildPoverty=16.5,
          MedHHInc=c(53.1, 67.7))
predict(LRM4.3, LRM4.3b)
```

The fourth and fifth columns suggest that differences in percent childhood poverty are associated with larger differences in predicted violent crimes. Violent crimes per 100,000 residents are expected to be 57% higher in states at the 75th percentile of child poverty compared to those at the 25th percentile. The corresponding difference based on median household income is 24%. This result reinforces the notion that median household income's association with violent crimes is weaker. But compare the results to what Figure 4.1 estimates for child poverty relative to what a similar graph indicates for median household income. Is there more uncertainty at lower or higher levels of the explanatory variables? Does this affect your interpretations?

Although comparing predicted means is helpful, the best approach for understanding the results of an LRM and its explanatory capabilities is to begin before estimation with a conceptual model based on previous research and theoretical concerns. After estimating the LRM, combine evidence from the coefficients (magnitudes, CIs, predictor effects plots), relative importance metrics, and percentage differences in predicted outcomes to reach sensible conclusions about the associations. These careful steps will allow a

---

[12] The R code sets one explanatory variable at its median—median household income → 59.2; percent child poverty → 16.5—and estimates the two predicted means based on the other. Chapter 7 provides another example of comparing predicted means.

reasonable determination of whether or not an LRM's results are compatible with a predetermined conceptual model or hypothesis.

Assumptions of Multiple LRMs

We now have a basic understanding of simple and multiple LRMs, including how to interpret their slope coefficients, intercepts, *p*-values, and CIs, as well as some of their practical implications. Let's next consider the assumptions of the model. We learned about some of these assumptions in Chapter 3. We'll revisit them, introduce a new one, and mention some special situations when they might not be satisfied. The following is a brief overview because several of the subsequent chapters examine the assumptions in detail.

1. *Independence*: the errors of prediction are independent of one another. This affects the bias and efficiency of the estimates. We can understand this assumption better now that we've examined a couple of LRMs. For example, when analyzing state-level data, we assume the errors of prediction are independent across states. But is this true? States that share borders are similar in many ways relative to states that are far apart. The errors in prediction are likely to be similar in adjacent states but different in states that are far away from one another. When we collect data over time, errors of prediction from those time points closer together are usually more alike than those farther apart. A simple way to predict whether the errors of prediction are not independent is when the units of observation in a dataset are also not independent. In this situation, the careful researcher will take steps to address the likely dependence of the errors. Chapter 8 provides more information about the independence assumption.

2. *Homoscedasticity (constant variance)*: the variance of the errors is constant for all combinations of Xs. Homoscedasticity means "same scatter." Its antonym is *heteroscedasticity* ("different scatter"). This important assumption, when not satisfied, has implications for the efficiency of the LRM slope coefficients. We'll learn more about this critical issue in Chapter 9.

3. *Collinearity*: no combination of the Xs has a perfect association—they are not *perfectly collinear*. This assumption is not listed in Chapter 3 since the simple LRM includes only one explanatory variable. Revisit the overlapping circles in Figure 4.4 and you can visualize what collinearity implies. Suppose circles x_1 and x_2 overlap completely. Is it possible to estimate the covariance between x_1 and y exclusive of x_2? No, because no variability is left over in x_1 once we consider its

association with x_2. We'll discover in Chapter 10 that even a higher degree of collinearity among explanatory variables can lead to unusual results in LRMs.

4. *Normality*: the errors are a normally distributed random variable with a mean of zero. This statement includes two assumptions—normality and mean of zero—that are combined for convenience. The first part is considered a weak assumption since, even when contravened, the model performs fairly well, especially in large samples.[13] The second part—that the mean is zero—is important for estimating the correct intercept. As we learned earlier, though, the intercept in many models is of little use since its value often falls outside the range of the explanatory and outcome variables. Chapter 11 provides a detailed discussion of the normality assumption.[14]

5. *Linearity*: the mean value of Y for each specific combination of the Xs is a linear function of the Xs. In other words, the regression surface is assumed flat in three dimensions (see Figure 4.3). In simple LRMs we assume straight-line relationships in two dimensions, but multiple LRMs include two or more explanatory variables so we must move to higher dimensions. One way to understand this is to imagine three-dimensional space and then visualize the difference between a flat surface and a curved surface. We assume that the relationship between the Xs and Y is not curved. In Chapter 11, we'll learn about some tools for analyzing relationships that are not linear.

These assumptions may not be satisfied in several situations that are not uncommon in research applications, including the following:

1. *Specification error*: we assume that the covariance (or correlation) between each explanatory variable and the errors of prediction is zero, which is symbolized $\text{cov}(X, \varepsilon_i) = 0$. A nonzero covariance suggests we've left something important out of the model and might reach the wrong conclusions. In Chapter 12, we'll learn that this problem involves whether we have *specified* the correct model, hence the term specification error. Its occurrence contravenes the independence assumption. Since this issue has such important implications for the way empirical models are developed and tested, it warrants

---

[13] A large sample for LRMs has about 100 or more observations, though the implications for the normality assumption also depend on the degree of non-normality and the number of explanatory variables (see Thomas Lumley et al. (2002), "The Importance of the Normality Assumption in Large Public Health Data Sets," *Annual Review of Public Health* 23(1): 151–169).

[14] As mentioned in Chapter 3, fn. 7, some researchers rearrange and combine aspects of some assumptions with the *iid* assumption. In this presentation, we'll adhere to the distinctions listed here.

its own chapter. Another type of specification error arises when non-linear associations exist among the Xs and Y, so we have not satisfied the linearity assumption (see Chapter 11).

2. *Measurement errors*: we presume that the xs and y are measured without error. Lowercase letters designate the explanatory and outcome variables because we are concerned primarily with sample measures (even though the same concerns apply to population measures). As the name implies, knowing whether we have accurate measures of the variables is crucial. When we ask people to record their family incomes, do they provide correct information? When we ask people whether they are happy, might some interpret this question differently than others? As we'll learn in Chapter 13, measurement error is a special problem since it involves the independence assumption in general and is a type of specification error in particular. Claiming that this problem is a common and substantial nuisance in social and behavioral sciences is not an exaggeration.

3. *Influential observations*: we should evaluate discrepant or extreme values and whether they affect the model. They can affect whether the model satisfies several of the assumptions, including linearity, normality, and homoscedasticity. Suppose we measure annual income and find the following values: $25,000, $35,000, $50,000, $75,000, and $33,000,000. As we'll see in Chapter 14, the last entry is labeled either a *high leverage point* (if income is an explanatory variable) or an *outlier* (if income is the outcome variable). High leverage points and outliers—which are known collectively as influential observations—can affect, sometimes in untoward ways, LRM results. The question when confronted with such values is why they have occurred. Did someone record a wrong number, perhaps by placing a decimal place in the wrong spot? Or is the value accurate? Does someone in the sample earn that much money per year? If the value is a coding error, there's an easy fix. But extreme values that are measured accurately tend to have a disproportionate effect on LRMs, so they require our attention.

Evaluating assumptions of LRMs and their derivative issues involves two concerns: (1) how to test whether the assumptions are satisfied and (2) what to do if they are not. Diagnostic tests, which are known collectively as *regression diagnostics*, are available for each of them. Methods to adjust the model are available if one or more assumptions are contravened. If these don't work, alternative regression models exist. Subsequent chapters discuss diagnostic tests and various solutions.

Some of these assumptions are stringent, but, one argument goes, the models are often saved by statistical theory's notable *Central Limit Theorem* (CLT). This occurs, for example, when the normality of errors assumption is not met, yet we have a large random sample. As you might recall from an introductory statistics book or course, the CLT states that, for relatively large samples, the sampling distribution of the mean of a variable is approximately normally distributed even if the distribution of the underlying variable is not normally distributed.[15] More formally, the CLT states:

> For random variables with finite sample variance, the sampling distribution of the standardized sample means approaches the standard normal distribution as the sample size approaches infinity.

The theorem concerns the sampling distribution of the mean, which, as we saw in Chapter 2, assumes taking many samples from the population. If these samples are drawn randomly, the distribution of means tends to approximate the normal distribution after about 30 samples. But if the underlying variable's distribution is highly skewed, it may take more samples to approximate the normal. Since intercepts and slope coefficients are related to means, they also follow particular normal-like distributions (such as the *t*-distribution). Given a large enough sample, we can thus infer that even if the errors in predicting the outcome variable are not normally distributed, the results of an LRM estimated with OLS tend to be unbiased. Learning how to use techniques appropriate for situations where assumptions like normality are not met is important, though. And this discussion should also reinforce the idea that using random samples or having control over explanatory variables is valuable.[16]

Let's assume, though, that the assumptions are met. In particular, if the assumptions regarding independence, homoscedasticity, collinearity, and linearity are satisfied, then, according to the Gauss–Markov theorem,[17] the OLS estimator offers the best linear unbiased estimator (BLUE) among the class of linear estimators: no other linear estimator hits the population target as often, on average, as the OLS estimator.[18]

---

[15] See Neil A. Weiss (1999), *Introductory Statistics*, 5th Ed., Reading, MA: Addison-Wesley, p.427, for a helpful review. Bernard W. Lindgren (1993), *Statistical Theory*, 4th Ed., New York: Chapman & Hall, p.140, provides a standard proof.

[16] Appendix B provides statistical simulations that examine several of the assumptions.

[17] See Lindgren (1993), *op. cit.*, p.510. Carl Friedrich Gauss proved an early version of the theorem in his 1823 work *Theoria combinationis Observationum Erroribus Minimis Obnoxiae* (*The Theory of the Combination of Errors in Observations*) (Göttingen, DE: Apud Henricum Dieterich). Russian mathematician Andrei Andreevich Markov provided another application and proof of the theorem in 1900.

[18] When the normality assumption is also satisfied, the OLS estimator is the most *efficient* unbiased estimator (see John Fox (2016), *Applied Regression Analysis and Generalized Linear Models*, 3rd Ed., Los Angeles: Sage). Its sampling variability is smaller than other estimators. This is also called the *minimum variance unbiased* property (S. D. Silvey (1975), *Statistical Inference*, Boca Raton, FL: CRC Press).

Some Important Characteristics of Multiple LRMs

As mentioned earlier, OLS is the main technique used to estimate LRMs. We've now learned that OLS has some nice features that make it especially useful in regression analysis. The OLS regression equation also provides the *linear combination* of the xs (recall the summation signs in the linear regression equation) that has the largest possible correlation with the y variable $(cor(y_i, \hat{y}_i))$. Since we hope the model explains as much of the variability in the outcome variable as possible with the explanatory variables, this is a beneficial property. We'll learn how to estimate this correlation in Chapter 5.

Chapter 3 notes that statistical software uses matrix routines to compute slope coefficients, standard errors, and other features of LRMs. For those familiar with vectors and matrices, think of the y values as a vector of observations and the x values as a matrix of observations, as illustrated in Equation 4.3.[19]

$$\mathbf{Y} = \begin{bmatrix} y_1 \\ y_2 \\ \vdots \\ y_n \end{bmatrix} \quad \mathbf{X} = \begin{bmatrix} 1 & x_{11} & \cdots & x_{1k} \\ 1 & x_{21} & \cdots & x_{2k} \\ \vdots & \vdots & \cdots & \vdots \\ 1 & x_{n1} & \cdots & x_{nk} \end{bmatrix} \tag{4.3}$$

Expressing the explanatory and outcome variables this way leads to an abbreviated depiction of the multiple linear regression equation: $\hat{\mathbf{Y}} = \mathbf{X}\hat{\boldsymbol{\beta}}$. $\hat{\boldsymbol{\beta}}$ is a vector of the intercept (denoted in the X matrix with 1s) and slope coefficients for the explanatory variables. The X matrix is listed first in this equation because it is *postmultiplied* by the vector of slope coefficients.[20]

The matrix formula in Equation 4.4 estimates the vector of slope coefficients with X and Y.

$$\hat{\boldsymbol{\beta}} = \left(\mathbf{X'X}\right)^{-1}\mathbf{X'Y} \tag{4.4}$$

---

[19] Translate the vector and matrix into spreadsheet format, such as with R's View function, if it's not clear how this works, and you will understand the utility of representing data in matrix form.

[20] Readers who have experience with linear algebra may recognize the key role that matrix routines play in solving systems of linear equations. They should not be surprised, therefore, that estimation with OLS uses matrices. But other routines for solving systems of linear equations also exist. Perhaps the most common in regression modeling is maximum likelihood estimation (MLE), which iterates through a set of reasonable estimates and determines which set is most likely given the data, though it tends to be slower than OLS and may lead to biases with small samples or sparse data matrices. Recent advances that use "coordinated guesses" of sets of estimates may improve the efficiency and speed of solving these equations even beyond matrix manipulations, however (Richard Peng and Santosh Vempala (2021), "Solving Sparse Linear Systems Faster than Matrix Multiplication," retrieved from https://arxiv.org/abs/2007.10254).

The accent next to the matrix X denotes its *transpose* and the superscript -1 indicates the inverse of the product in parentheses. Using R with matrix routines and a small dataset facilitates a deeper understanding of LRMs.[21]

Matrix algebra is also useful for estimating several other features of multiple LRMs. For example, the standard errors of the coefficients are estimated by taking the square roots of the diagonal elements of the matrix shown in Equation 4.5.

$$V = \left(X'X\right)^{-1} \hat{\sigma}^2 \quad \text{where} \quad \hat{\sigma}^2 = \frac{\hat{\varepsilon}'\hat{\varepsilon}}{n-k} \tag{4.5}$$

The n refers to the sample size and k denotes the number of explanatory variables in the model. We'll learn more about $\hat{\sigma}^2$ in Chapter 5 when discussing goodness-of-fit statistics; in brief it is a measure of dispersion: it measures the amount of variability of the residuals around the regression line—the *residual variance* or the *mean square error* (MSE).[22]

Equation 4.6 also estimates the standard errors of multiple LRM slope coefficients.

$$se\left(\hat{\beta}_i\right) = \sqrt{\frac{\Sigma\left(y_i - \hat{y}_i\right)^2}{\Sigma\left(x_i - \bar{x}\right)^2 \left(1 - R_i^2\right)\left(n - k - 1\right)}} \tag{4.6}$$

The R^2 in the equation is from an LRM—called an *auxiliary regression model*—with x_i as the outcome variable and all the other explanatory variables as predictors (e.g., $x_{1i} = \alpha + \beta_1 x_{2i} + \ldots + \beta_k x_{ki}$). $\left(1 - R_i^2\right)$ is the *tolerance*. Continuing with the discussion surrounding Equation 3.13 in Chapter 3, the standard error increases as (a) the R^2 from the auxiliary regression model increases, (b) the variability of x decreases, (c) the variability of y increases, or (d) the sample size decreases. Think about the practical implications these factors have for significance testing with LRMs.

Chapter Summary

This chapter provides an introduction to multiple LRMs—models with two or more explanatory variables. A key difference between simple and

---

[21] For more information about matrix routines in R, see Nick Fieller (2016), *Basics of Matrix Algebra for Statistics with R*, Boca Raton, FL: CRC Press.

[22] Faraway (2014), *op. cit.*, and David G. Kleinbaum et al. (1998), *Applied Regression Analysis and Other Multivariable Methods*, 3rd Ed., Pacific Grove, CA: Duxbury Press (Appendix B), provide perspicuous overviews of matrix routines used in LRMs. A more formidable treatment, but one worth the effort for understanding the role of matrix routines, is in James R. Schott (2016), *Matrix Analysis for Statistics*, 3rd Ed., Hoboken, NJ: Wiley.

multiple LRMs is in how we interpret the slope coefficients. The multiple LRM's slopes are "statistically adjusted" for the effects of the other explanatory variables. The same assumptions apply to both models, but multiple LRMs are also concerned with collinearity—statistical associations among explanatory variables. Understanding the assumptions is so important that several of the following chapters provide details about each. Before getting to this, though, we'll learn about some other features of LRMs in Chapters 5–7.

Chapter Exercises

The dataset called *TeenBirths.csv* consists of data from almost 3,000 counties in the U.S. Our objective is to estimate a multiple LRM with a set of variables and provide interpretations of the results. The variables in the dataset include

- `state`   State name
- `county`   County name
- `teen_birth_rate`   percentage of births to teenage mothers
- `per_uninsured`   percentage of residents with no health insurance
- `per_hsgrads`   percentage of adult residents who are high school graduates
- `per_child_poverty`   percentage of children living in poverty
- `per_singleparent`   percentage of children living in single parent households

After importing the dataset into R, complete the following exercises.

1. Compute the Pearson's correlations of the following variables: percentage of births to teenage mothers, percentage of children living in poverty, percentage of children living in single parent households, and the percentage of residents with no health insurance. Which variable appears to be the strongest predictor of the percentage of births to teenage mothers? Why?

2. Estimate a simple LRM with the percentage of births to teenage mothers as the outcome variable and percent of children living in poverty as the explanatory variable.

   a. Interpret the slope coefficient.

   b. Interpret the *p*-value associated with the slope coefficient. What are some of its limitations for judging the results of the model?

   c. Interpret the 95% CI associated with the slope coefficient.

3. Estimate a multiple LRM with the percentage of births to teenage mothers as the outcome variable and the following explanatory variables: the percentage of children living in poverty, the percentage of children living in single parent households, and the percentage of residents with no health insurance.

   a. Interpret the slope coefficient associated with the percentage of children living in poverty. How does this slope coefficient compare with the slope coefficient in exercise 2 (a)?

   b. Interpret the slope coefficient associated with the percentage of residents with no health insurance.

4. Compute the standardized regression coefficients (beta weights) from the LRM in exercise 3. What do they suggest about the relative strength of the associations between each explanatory variable and percentage of births to teenage mothers? What are some possible limitations of using these coefficients for judging the strengths of the associations?

5. Compute the relative importance measures from the LRM in exercise 3. What do the results suggest about the best predictor(s) of the percentage of births to teenage mothers?

6. Compute the predicted means of the percentage of births to teenage mothers for the following groups (set the other variables at their means):

   a. Counties at the 25th percentile of the percentage of children living in poverty.

   b. Counties at the 75th percentile of the percentage of children living in poverty.

   c. Counties at the 25th percentile of the percentage of residents with no health insurance.

   d. Counties at the 75th percentile of the percentage of residents with no health insurance.

   What do these predicted means suggest about the associations of the explanatory variables with the outcome variable?

7. *Challenge*: save the predicted values (R labels them `fitted.values`) from the LRM in exercise 3. Compute the Pearson's correlation between the predicted values and each explanatory variable. What do the correlations suggest about the model and its predictive capabilities? Does this information strengthen your conclusions in exercises 5 and 6? Why or why not?

5

The ANOVA Table and Goodness-of-Fit Statistics

We are now adept at fitting LRMs to predict or account for relationships among a single continuous outcome variable and one or more explanatory variables. The slope coefficients represent one way of assessing the direction and strength of associations, and the standard errors, *t*-values, *p*-values, and CIs are normally used to make inferences from the sample to the target population. As a reminder, though, we should use all the information available from the model to assess whether its results are consistent with a conceptual model or hypothesis. To do this it would be helpful to have additional information about how well the model, in a more general sense, functions to predict or explain patterns in the outcome variable. As described in this chapter, we may use information from R's summary output, as well as the *ANOVA table*, to meet this objective.

LRMs and ANOVA models share some characteristics. Both are useful, for example, if we wish to estimate the means of specific groups represented by a set of explanatory variables. Both models provide significance tests if we want to infer something about the target population, such as whether the results suggest that an outcome differs for two or more groups defined by an explanatory variable. You may recall, however, that a fundamental purpose of ANOVA is to partition the variance of a variable into component parts: one, due to its systematic association with another variable and, the second, due to random error. LRMs serve a similar purpose but are useful for a few additional reasons.

Statistical software packages often include ANOVA tables as part of their standard LRM output. R is an exception, though, and we must request this table with a function. Let's first re-estimate LRM4.1 from Chapter 4 that includes percent child poverty as the explanatory variable and violent crimes per 100,000 residents as the outcome variable and then request the ANOVA table. Information from the lm output also includes some statistics of interest (see LRM5.1).

R code
```
LRM5.1 <- lm(ViolentCrimeRate ~ PerChildPoverty,
             data=StateData2018)
summary(LRM5.1)
anova(LRM5.1) # requests the ANOVA table
```

DOI: 10.1201/9781003162230-5

R output (residuals and coefficient output omitted)
Residual standard error: 113.3 on 48 degrees of freedom
Multiple R-squared: 0.242, Adjusted R-squared: 0.2262
F-statistic: 15.33 on 1 and 48 DF, p-value: 0.000285

Analysis of Variance Table
Response: ViolentCrimeRate

	Df	Sum Sq	Mean Sq	F value	Pr(>F)
PerChildPoverty	1	196800	196800	15.327	0.000285 ***
Residuals	48	616325	12840		

Signif. codes: 0 '***' 0.001 '**' 0.01 '*' 0.05 '.' 0.1

You may now be wondering why this is called an ANOVA table. Let's examine its attributes to answer this inquiry. As in a standard ANOVA, a set of numbers is called the sum of squares (Sum Sq in the output). Recall from Chapter 2 that we discussed a measure of dispersion called the sum of squares of x, abbreviated $SS[x]$. The formula for this measure is $SS[x] = \Sigma(x_i - \bar{x})^2$. But we now have an outcome variable, labeled y, and, as seen in the last chapter, we wish to assess something about the overlap of the explanatory and the outcome variables. It makes sense, therefore, to try to come up with a measure of this overlap. If the sum of squares of the outcome variable, $SS[y]$, measures its total area, we might also wish for a measure of the area that overlaps with the explanatory variable (for a visual representation, revisit Figure 4.4). This is what the ANOVA table provides. Under the Sum Sq column, we may see how $SS[y]$, which, in this example, concerns the violent crimes variable, is partitioned into two components: the overlap with the variation in the percent child poverty (x), which is called the *regression sum of squares* (RSS), and the portion that is left over—the sum of squares due to the residuals. The latter measure is often termed the *residual sum of squares* but is more frequently called the sum of squared errors (SSE), which we learned about in Chapter 3. In the ANOVA table, the RSS and the SSE are 196,800 and 616,325. Hence, the *total sum of squares* (TSS)—which is simply $SS[y]$—is computed by adding these two numbers: 196,800 + 616,325 = 813,125.[1] This may be thought of as the total area of a circle representing the variation in violent crimes.

Another way of understanding the RSS is that it measures the variation from the mean of y of the values predicted from the regression equation. Recall that the predicted values are designated as $\hat{y}_i$ and are calculated from the LRM. The RSS computation is furnished in Equation 5.1.

$$RSS = \Sigma(\hat{y}_i - \bar{y})^2 \tag{5.1}$$

---

[1] A simple way to find this quantity for violent crimes is by using the var function and multiplying by $n - 1$: var(StateData2018$ViolentCrimeRate)*49. R returns 813,125.

As already noted, the RSS represents the total area of overlap between the variability of x and the variability of y (since the values of y are predicted based on the x values). It may also be thought of as the improvement in prediction over the mean with information from the explanatory variable(s).

The SSE is the variability in y leftover after accounting for its overlap with the explanatory variable. As noted in Chapter 4, when the assumptions are met, OLS minimizes this quantity. As a reminder, the SSE is computed using Equation 5.2.

$$SSE = \Sigma(y_i - \hat{y}_i)^2 = \Sigma \hat{\varepsilon}_i^2 \qquad (5.2)$$

The right-hand side of Equation 5.2 is a reminder that the SSE also represents the sum of the squared residuals (see Chapter 3).

But why is the sum of the RSS and SSE equal to the TSS of the outcome variable? Because we assume we have accounted for all of the variation in y with either the variability in the explanatory variable (x) or random variability (this does not, however, rule out that there might be other variables that account for portions of the variability of y).

Another way of representing the equations inherent in the ANOVA table is to combine the sums of squares equations into one general equation. Such an equation should represent the partitioning of the area representing the TSS of violent crimes into its two component parts. Figure 5.1 and Equation 5.3 that follows it illustrate this concept.

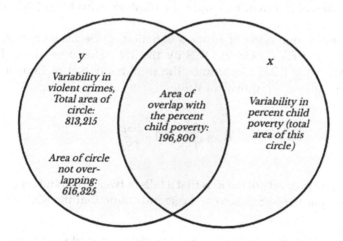

FIGURE 5.1
Illustration of disaggregating the sum of squares of y into the regression sum of squares and the residual sum of squares.

$$\Sigma(y_i - \bar{y})^2 = \Sigma(\hat{y}_i - \bar{y})^2 + \Sigma(y_i - \hat{y})^2 \qquad (5.3)$$

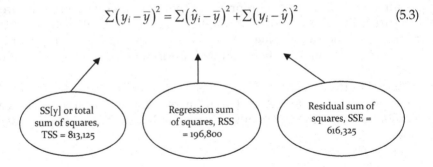

The next issue is whether we can use information on overlapping or joint variability to come up with a measure of the overall fit of the model. In most types of regression modeling exercises researchers are concerned with what are known as *goodness-of-fit* (GOF) statistics, which assess what portion of the variability in the outcome variable is accounted for by the LRM.

One assessment is to use the sums of squares to estimate the amount of variability shared and not shared among the variables. Since we have measures of "area," we may estimate the proportion or percentage of variation in the outcome variable that is accounted for by, or overlaps with, the explanatory variable(s). In the R output, one finds a number labeled `Multiple R-squared`; in many research presentations it is labeled R^2. Known as the *coefficient of determination* or the *squared multiple correlation coefficient*, it indicates the proportion of variability in the outcome variable that is accounted for by (some use the phrase "explained by") the explanatory variable(s). A simple transformation changes it to a percentage. In LRM5.1, the R^2 is 0.242, which implies 24.2%. In other words, the LRM accounts for 24.2% of the variability in violent crimes.[2]

Returning to our sums of squares equation, it should be apparent how to compute the R^2: divide the RSS by the TSS. Therefore, using LRM5.1: 196,800/813,125 = 0.242. Sometimes the right-hand side of Equation 5.4 is used but results in an equivalent value.

$$R^2 = \frac{RSS}{TSS} = 1 - \frac{SSE}{TSS} \qquad (5.4)$$

An important property of the R^2 is that it falls between zero and one, with zero indicating that the RSS is zero and one indicating that the RSS = TSS, with

---

[2] What constitutes a large or sufficient R^2 is open to debate. In social and behavioral sciences, for instance, an R^2 of 0.4 might be considered substantial for individual-level data, but moderate for aggregate-level data. R^2 values also tend to be larger with experimental data than with observational data.

the SSE equal to zero.[3] An R^2 of one denotes that the explanatory variable(s) perfectly predicts the outcome variable. Returning to the graphs representing the LRM (see, for example, Figure 3.4), perfect prediction denotes that the observations fall directly on the linear fit line.

When we add explanatory variables to the model, the R^2 increases whether or not the explanatory variables add anything of substance to the model's ability to predict y. Many researchers are understandably bothered by this since they do not wish to be misled into thinking that their model provides good predictive power when it does not. Most analysts therefore use a measure known as the *adjusted* R^2. The term *adjusted* indicates that the R^2 is adjusted for the number of explanatory variables in the model. If we simply add extraneous explanatory variables, the adjusted R^2 can actually decrease from one model to the next. The formula for this measure of model fit is shown in Equation 5.5.

$$\bar{R}^2 = \left(R^2 - \frac{k}{n-1} \right)\left(\frac{n-1}{n-k-1} \right) \tag{5.5}$$

As before, n denotes the sample size and k indicates the number of explanatory variables in the model. The summary output of LRM5.1 displays an adjusted R^2 of 0.226.

The R output that furnishes the ANOVA table, R^2, adjusted R^2, and other statistics was shown a few pages ago, so we'll exhibit it again.

R output (LRM5.1)
```
Residual standard error: 113.3 on 48 degrees of freedom
Multiple R-squared: 0.242, Adjusted R-squared: 0.2262
F-statistic: 15.33 on 1 and 48 DF, p-value: 0.000285

Analysis of Variance Table

Response: ViolentCrimeRate
                 Df  Sum Sq  Mean Sq  F value   Pr(>F)
PerChildPoverty   1  196800   196800   15.327  0.000285 ***
Residuals        48  616325    12840
---
Signif. codes: 0 '***' 0.001 '**' 0.01 '*' 0.05 '.' 0.1
```

The two remaining statistics in the output are the Residual standard error and the F-statistic. Since we've just discussed the R^2 statistic, we'll first consider the F-statistic—which is frequently called the F-value—since it has important implications for how we evaluate the R^2.

---

[3] Another interesting nuance is how one interprets R, the square root of R^2 (not the software). R represents the correlation between the observed y values and the predicted y values from the LRM: $corr(y_i, \hat{y}_i)$. This property is the source of the term "squared multiple correlation coefficient."

Consider that, as with other statistics derived from a sample, such as means and slope coefficients, the R^2 has a sampling distribution. If we were to take many samples from the population, we could compute R^2s based on an LRM for each sample. We may therefore consider significance tests if we wish to infer something about the population parameter. Before seeing such a test, though, we have to determine how R^2 values are distributed. We've thus far considered the normal distribution, the standard normal distribution (z-distribution), and the t-distribution. R^2 values follow an F-distribution. Perhaps you remember the F-distribution from the ANOVA model. If so, you'll recall that it has two degrees of freedom, known as the numerator and denominator degrees of freedom. In R's summary output, two numbers precede DF (degrees of freedom). The first, representing the numerator, is the number of explanatory variables in the model, of which only one appears in LRM5.1. The second, representing the denominator, is the sample size minus the number of explanatory variables minus one ($n - k - 1$). For LRM5.1, this is $50 - 1 - 1 = 48$. Thus, the degrees of freedom for the F-test, which relies on the F-value, is {1, 48}.

The output furnishes an F-value of 15.33. This is computed from the values in the Mean Sq column of the ANOVA table. Take the first value in the column (196,800) and divide it by the second value in the column (12,840) and this provides the F-value. A more general formula for the F-value is shown in Equation 5.6, with an alternative representation in Equation 5.7.

$$F \text{ value} = \frac{\text{regression sum of squares} \, (\text{RSS}) / \text{number of explanatory variables} \, (k)}{\text{sum of squared errors} \, (\text{SSE}) / (n - k - 1)}$$

(5.6)

or

$$F \text{ value} = \frac{\text{Mean square due to regression} \, (\text{MSR})}{\text{Mean square error} \, (\text{MSE})} \qquad (5.7)$$

Similar to the way we compute and utilize t-values for the slope coefficients, the F-value is an inferential measure that helps us deduce something about the R^2 in the population from which the sample was drawn.

As with t-values, p-values are associated with F-values and their degrees of freedom. The R output shows that the p-value associated with the F-value of 15.33 (df = 1, 48) is 0.0003. What does this mean, though? Since R^2 values follow an F-distribution, the p-value implies something about the population R^2. The null hypothesis (H_0) used by the summary function from an LRM is that the R^2 is equal to zero in the population.[4] The alternative hypothesis (H_a)

---

[4] Not to be overly skeptical, but perhaps we should ask whether we genuinely expect any R^2 value to be precisely zero (cf. Chapter 3, fn. 18).

is that the R^2 is greater than zero in the population. A relatively large F-value and small p-value provides tentative evidence that the R^2 in the population is greater than zero. The p-value associated with the F-value in LRM5.1 may be interpreted as:

> If the R^2 in the population is zero, we would expect to find a sample R^2 of 0.24 or one larger than zero only about three times out of every 10,000 samples, if we were to draw many samples from the population.

We might also compute a CI for the estimated R^2. The lm function does not provide a CI for the R^2, but it can be computed with R's boot package.[5]

Most researchers from social and behavioral sciences use the R^2 and accompanying F-test to determine something about the importance of the model. You will often find statements such as "the model accounts for 42% of the variability in the tendency to twitch, and this percentage is significantly greater than zero." However, you should be cautious about such conclusions. Some statisticians argue that we should rely on either the substantive conclusions recommended by the partial regression slopes or the mean square error (MSE, often denoted as $\hat{\sigma}^2$, was introduced in Chapter 4 and is part of the formula for the F-value [see Equation 5.7]) to judge how well the set of explanatory variables predicts the outcome variable.[6] Of course, we should also not forget the importance of the conceptual model driving the analysis—are the LRM results, including the fit measures, consistent with the pre-conceived conceptual model?

The MSE, which is found in the R output in the Mean Sq column and the Residuals row, results from Equation 5.8.

$$MSE = \hat{\sigma}^2 = \frac{SSE}{n-k-1} \quad \text{or} \quad \frac{\hat{\varepsilon}'\hat{\varepsilon}}{n-k} \left(\text{see Chapter 4}\right) \tag{5.8}$$

The SSE is divided by its degrees of freedom, which in this case is $(n - k - 1)$. The MSE in LRM5.1 is 616,325 / 48 = 12,840 (see the R summary output).

The MSE is also called the *residual variance* because, as implied by $\hat{\sigma}^2$, it quantifies the variance of the residuals. But is there an analogous measure of the standard deviation of the residuals? Yes, see the Residual standard error listed in the LRM5.1 output, which is often termed the *root mean square error* (RMSE). Recall that the standard deviation is the square root of the variance (see Chapter 2). Hence, whereas the MSE is the variance, the RMSE denotes the standard deviation of the residuals. The RMSE is also

---

[5] A bootstrap based on R's boot package reveals a CI for the R^2 in LRM5.1 of {0.04, 0.52}. How should it be interpreted? For more information about the bootstrap, see Chapter 9.

[6] See, in particular, Franklin A. Graybill and Hariharan K. Iyer (1994), *Regression Analysis: Concepts and Applications*, Belmont, CA: Duxbury Press (chapter 4), for a discussion of this issue.

called the *standard error of the estimate* or the *standard error of the regression* (S_E or $S_{y|x}$). In the LRM5.1 output the RMSE is $\sqrt{12,840} = 113.3$.

The MSE and RMSE are employed as GOF measures since smaller values indicate less variation in the residuals of the model. The predicted values are closer, on average, to the actual values and thus the model is doing a better job of predicting the outcome variable.

The RMSE is also useful for estimating *prediction intervals*, which, similar to CIs, are inferential measures. The prediction intervals estimate the proportion of time we expect future observations to fall within a range of values of the outcome variable given a single value of an explanatory variable. For example, say we wish to determine the interval of predicted violent crimes in states with a specific percent of children in poverty. Equation 5.9 provides the formula for a prediction interval.

$$PI = \hat{y} \pm (t_{n-2}) \times (\text{RMSE}) \times \sqrt{1 + \frac{1}{n} + \frac{(x_0 - \bar{x})^2}{(n-1)s_x}} \tag{5.9}$$

Suppose the specific percent is 16.8, which is represented as x_0 in the formula. The *t*-value is the confidence level we wish to use with degrees of freedom equal to $n - 2$. For example, LRM5.1 has a sample size of 50, so we use a *t*-value of 2.011 for a confidence level of 95%, the RMSE (S_E) (113.3), the mean of x ($\bar{x} = 16.8$), and the standard deviation of x ($s_x = 4.8$) in Equation 5.9.

R makes it easy to compute a prediction interval. The steps furnished in the next example begins with a simple LRM (LRM5.2) and requests a prediction interval for violent crimes that assumes the percent of children living in poverty is 16.5%, its median.

R code
```
LRM5.2 <- lm(ViolentCrimeRate ~ PerChildPoverty,
             data=StateData2018)
predLRM5.2 = data.frame(PerChildPoverty=16.5)
 # Create a new data frame with a single child poverty
   value of 16.5; then use it in the next function
 # LRM5.2 contains the results of the model for which we
   wish to have the prediction interval
predict(LRM5.2, newdata=predLRM5.2, interval =
        "prediction")
```
R output (abbreviated)
```
    fit     lwr     upr
342.272 112.160 572.384
```

The R output provides the estimated number of violent crimes for this type of state: 342.3, along with the upper and lower values of the 95% prediction interval (112.2 and 572.4). If we were to hypothetically collect many future

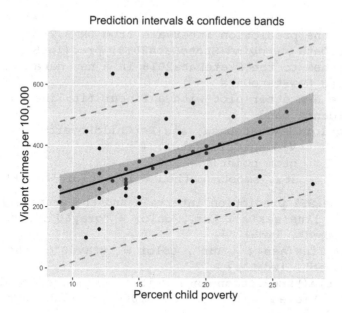

FIGURE 5.2
Scatter plot of predicted violent crimes by percent child poverty with 95% confidence bands and prediction intervals from LRM5.2.

observations of states with 16.5% of their children living in poverty, we'd expect the number of violent crimes to be between 112 and 572 about 95% of the time (does this seem rather uncertain to you?).

We may also plot 95% prediction intervals as well as 95% confidence bands using various R tools. The next example uses the ggplot2 package's ggplot function in conjunction with the predict function to create Figure 5.2. The scatter plot includes a gray-shaded area that denotes the confidence band for the slope (linear fit line) and dashed lines that designate the prediction intervals. Since the prediction intervals suggest the uncertainty around single values, they are wider than the estimated confidence band. Consider, for instance, when a state has 15% of its children living in poverty. The prediction interval—the expected future range of the outcome—is about 100–550 violent crimes per 100,000 residents. The confidence band, which reflects uncertainty about the slope (linear fit line) and is thus interpreted differently, is much narrower.[7]

R code for Figure 5.2
```
library(ggplot2)
```

---

[7] We may also use the DescTools package's lines function to graph prediction intervals:

```
library(DescTools)
plot(ViolentCrimeRate ~ PerChildPoverty, data=StateData2018)
lines(LRM5.2, pred.level=0.95, col="gray10")
```

```
pred.int5.2 <- predict(LRM5.2, interval = "prediction")
 # save the prediction intervals from LRM5.2
NewStateData <- cbind(StateData2018, pred.int5.2)
 # add them to the StateData2018 in a new data frame
   called NewStateData
# create a scatter plot with a linear fit line and 95%
  confidence band
p <- ggplot(NewStateData, aes(PerChildPoverty,
          ViolentCrimeRate)) +
          geom_point() +
          stat_smooth(method = lm, color="black")
```

```
# add the 95% prediction intervals and the graph's labels
p + geom_line(aes(y = lwr), color = "gray60", linetype =
   "dashed", lwd=1) +
   geom_line(aes(y = upr), color = "gray60", linetype =
   "dashed", lwd=1) +
   ggtitle("Prediction intervals & confidence bands") +
   ylab("Violent crimes per 100,000") + xlab("Percent
   child poverty")
```

Another Example of a Multiple LRM

Before completing our discussion of the ANOVA table and GOF statistics, let's see another example of a multiple LRM. The dataset *GPA.csv* has information on first-year college grade point average (gpa), scholastic aptitude test (SAT) scores (SATMath, SATReadWrite), and high school grades in English and mathematics (HSMath, HSEnglish) for 20 students. Consider LRM5.3 that uses GPA as the outcome variable and the following explanatory variables: SATMath, SATReadWrite, and HSMath.

R code
```
# First, rescale the SAT scores into 100s - this makes
  their slope coefficients simpler to interpret
GPA$SATMath <- GPA$SATMath/100
GPA$SATReadWrite <- GPA$SATReadWrite/100
LRM5.3 <- lm(gpa ~ SATMath + SATReadWrite + HSMath,
             data=GPA)
summary(LRM5.3)
confint(LRM5.3)
anova(LRM5.3)
```

R output (abbreviated)
Coefficients:

	Estimate	Std. Error	t value	Pr(>|t|)	
(Intercept)	0.29828	0.26498	1.126	0.276911	
SATMath	0.21849	0.04553	4.799	0.000197	***
SATReadWrite	0.13123	0.05252	2.499	0.023738	*
HSMath	0.17987	0.08768	2.051	0.056964	.

Signif. codes: 0 `***` 0.001 `**` 0.01 `*` 0.05 `.` 0.1

Residual standard error: 0.2621 on 16 degrees of freedom
Multiple R-squared: 0.8504, Adjusted R-squared: 0.8223
F-statistic: 30.31 on 3 and 16 DF, p-value: 7.816e-07

[CIs]	2.5%	97.5%
(Intercept)	-0.263	0.860
SATMath	0.122	0.315
SATReadWrite	0.020	0.243
HSMath	-0.006	0.366

Analysis of Variance Table
Response: gpa

	Df	Sum Sq	Mean Sq	F value	Pr(>F)	
SATMath	1	5.3015	5.3015	77.1663	1.614e-07	***
SATReadWrite	1	0.6559	0.6559	9.5468	0.007029	**
HSMath	1	0.2891	0.2891	4.2085	0.056964	.
Residuals	16	1.0992	0.0687			

Signif. codes: 0 `***` 0.001 `**` 0.01 `*` 0.05 `.` 0.1

First, examine the model fit information. The R^2 is 0.85, the adjusted R^2 is 0.82, and the RMSE (S_E) is 0.26. We may claim that, using the three explanatory variables, the model accounts for 85% of the variability in first-year college GPA. As suggested by the adjusted R^2, this percentage explained is not affected much by the number of explanatory variables in the model. The results are also inconsistent with a null hypothesis that the R^2 equals zero in the population—the F-value of 30.3 ($df = 3, 16$) is accompanied by a p-value of less than 0.0001.

The slope coefficients provide several pieces of information. We'll ignore the intercept in this model because, first, SAT scores do not have a zero value (the minimum is 200) and, second, it is unlikely that a student who had a grade of zero in high school math would be attending college! The slope coefficient for SAT math scores suggests:

> Statistically adjusting for the effects of SAT reading/writing scores and high school math grades, students who score one-unit higher on SAT math tests tend to have, on average, first-year college GPAs 0.22 units higher than others.

This seems fairly substantial. But what is a one-unit increase? Recall that we rescaled the SAT scores prior to estimating the model so they ranged from 2 to 8. A 100-point increase in SAT math scores is thus associated with a 0.22 higher first-year college GPA. We might predict, for example, that among students who are similar on SAT reading/writing scores and high school math grades, those who scored 600 had a GPA 0.22 units higher than those who scored 500 on the SAT math portion. As an exercise, create a predictor effects plot of the predicted GPA scores by the SAT math scores to see how these differences play out. Don't forget to consider the confidence band.

How do the model's coefficients compare? Since SAT scores for math and reading/writing abilities are measured on a similar scale, we may consider which type of score has a stronger association with first-year GPA. It appears that SAT math scores ($\beta = 0.22$) matter more than SAT reading/writing scores ($\beta = 0.13$). Remember the issue of sampling error, however: we do not know based only on the point estimates whether the two SAT coefficients are distinct in an inferential manner. Methods for determining this are available,[8] though a crude way is to examine the CIs for the coefficients and see whether there's any overlap. For instance, the 95% CI is {0.122, 0.315} for the `SATMath` coefficient and {0.020, 0.243} for the `SATReadWrite` coefficient. Since these overlap by a substantial margin, we should not be confident that math scores are a stronger predictor than reading/writing scores of first GPAs.[9] This conclusion should not be surprising, however, because the sample size is small ($n = 20$).

An alternative method is to use an *F*-test to compare the coefficients, but it should be used only if the variables are measured on the same metric. The null hypothesis of the test is that the slope coefficients are equal. The following R code shows how to use the `car` package to execute the *F*-test.

R code
```
library(car)
linearHypothesis(LRM5.3, "SATMath = SATReadWrite")
```

R output
```
Linear hypothesis test
Hypothesis:
SATMath - SATReadWrite = 0
Model 1: restricted model
Model 2: gpa ~ SATMath + SATReadWrite + HSMath
```

---

[8] See also Samprit Chatterjee and Ali S. Hadi (2012), *Regression Analysis by Example*, 5th Ed., New York: Wiley, §3.10, and Graybill and Iyer (1994), *op. cit.*, chapter 3.

[9] Using CIs to determine if slope coefficients are distinct results in a conservative test and are also influenced by whether or not their standard errors are similar (Mark E. Payton, Matthew H. Greenstone, and Nathaniel Schenker (2003), "Overlapping Confidence Intervals or Standard Error Intervals: What Do They Mean in Terms of Statistical Significance?" *Journal of Insect Science* 3: 34–39).

```
   Res.Df  RSS    Df Sum of Sq   F      Pr(>F)
1    17   1.1837
2    16   1.0993 1  0.084409  1.2286  0.2841
```

Similar to the overlapping CIs, we see that the difference between the two regression coefficients has a p-value (0.28) that exceeds conventional thresholds such as $p < 0.05$. It would therefore be unwise to judge SAT math scores as more consequential than SAT reading/writing scores for first-year college grades, at least with information from this particular sample.[10]

The coefficient for high school math grades offers an interesting problem. The p-value for this coefficient is 0.057 (95% CI: {−0.01, 0.37}), just a bit above the usual threshold ($p < 0.05$) for making claims about statistical significance. Yet, if we are comfortable judging coefficients based on p-values, we might argue that the result is statistically significant. How? Well, it seems reasonable to claim, *a priori*, that students with better high school math grades should also have higher first-year college GPAs. If this is the case—although you might contend "it's too late for that!"[11]—then the underlying hypothesis is directional and we might feel justified using a one-tailed significance test or CI. As mentioned earlier, to switch from a two-tailed test to a one-tailed test, divide the p-value in half. The one-tailed p-value is thus 0.057/2 = 0.0285. Voilà! We now have a "statistically significant" result. This demonstrates the uncertainty involved in statistical endeavors, one that should be recognized consistently. But some researchers might see this statistical legerdemain—or is it prestidigitation?—as defensible, whereas most experts would argue that it is disingenuous. We should be careful to set up the hypotheses or conceptual model and describe the types of significance or other decision rules we plan to use before estimating the LRM. This is good research practice and is consistent with the spirit of preregistration (see the section on best statistical practices in Chapter 1). Moreover, we should consider evidence from the model in addition to p-values and CIs before concluding that the results are meaningful or compatible with a theory or conceptual model.

Chapter Summary

This chapter provides additional information about the LRM estimated with OLS. The main focus is on how to determine whether the model provides a good fit—and good predictions—given the observed patterns

---

[10] For additional information on the F-test, and similar tests, for comparing regression coefficients, see John Fox, Michael Friendly, and Sanford Weisberg (2013), "Hypothesis Tests for Multivariate Linear Models Using the car Package," *The R Journal* 5(1): 39–52.

[11] And you would be correct to make this argument.

in the data. Perhaps the most common measure of model fit is the R^2. It provides an estimate of the proportion of variability in the outcome variable that is accounted for by the explanatory variables. The MSE and the RMSE measure the variability of the residuals and furnish alternative measures of fit. Smaller values suggest less variation in the errors of prediction, thus indicating better correspondence between what the model predicts and patterns in the data.

Chapter Exercises

The dataset called *QBPassers2019.csv* includes passing statistics from 50 quarterbacks who played in the National Football League (NFL) over the past several decades. A common way to judge the efficiency of an NFL quarterback is with a passer rating, which combines data from a number of passing statistics. Our goal is to examine a model that predicts the rating variable with several passing variables and determine the fit of the model.

The variables in the dataset include:

- name   name of quarterback
- firstyear   first year in NFL
- lastyear   last year in NFL (missing: still active)
- games   number of games played in career
- attempts   career pass attempts
- completions   career passes completed
- compperc   career completion percentage
- avercomp   career average yards per completion
- yards   career passing yards
- interceptions   career passes intercepted
- interatt   career interceptions per 100 pass attempts
- touchdowns   career number of touchdowns thrown
- tdsatt   career touchdowns per 100 pass attempts
- longestPass   longest pass thrown in a regular season game
- rating   career quarterback rating

After importing the dataset into R, complete the following exercises.

1. Estimate an LRM with `rating` as the outcome variable and the following explanatory variables: completion percentage and interceptions per 100 pass attempts.

    a. Interpret the two slope coefficients from the model.

    b. Interpret the 95% CIs associated with the two slope coefficients.

2. Interpret the Multiple R-squared provided in the R `summary` output. How does the Multiple R-squared differ from the Adjusted R-squared?

3. Interpret the *p*-value associated with the *F*-value that R provides in its `summary` output. What are some limitations of this inferential exercise?

4. Use R's anova function to obtain additional information from the model. Consider Equation 5.3 that follows Figure 5.1. Create a similar equation for the LRM using the output provided by the anova function.

5. Estimate a prediction interval for `rating` based on the LRM in exercise 1 by setting the explanatory variables at their median values (hint: `predLRM = data.frame(compperc = [place its median here], interatt = [place its median here]))`. Interpret the prediction interval.

6. Estimate another LRM that includes the same variables as the LRM in exercise 1 but adds the following explanatory variables: average yards per completion and touchdowns per 100 pass attempts.

    a. Interpret the R^2 from this model.

    b. Which LRM do you think does a better job of predicting `rating`? Explain your answer.

7. *Challenge*: use R to compute the variance and the standard deviation of the residuals from the LRM estimated in exercise 1. Why might these be valuable as measures of model fit? Examine R's `summary` and anova output from the LRM in exercise 1. What statistics from the output are similar to the variance and standard deviation of the residuals? Why do you think these similarities exist?

6

Comparing Linear Regression Models

Now that we've learned about ANOVA tables and goodness-of-fit statistics, the next step is to use this information to compare different LRMs. The motivating issue is whether one LRM does a better job than another of accounting for or predicting an outcome variable. For instance, suppose we extend the model in Chapter 5 designed to predict first-year college GPA and add another explanatory variable. Can we determine if the expanded model, to use a common phrase, "fits the data better" than the initial model?[1] Another way of considering this is to ask whether the second model is more accurate at predicting GPA than the first model? On its face, the answer is simple: we may determine whether the additional variable is a statistically significant predictor of GPA. Or, we might assess whether the R^2 increased from one model to the next. These approaches might work well. For instance, if we decide, using significance tests, a reasonable conceptual framework, and other functional criteria, that all three variables in the first model are important predictors of GPA, then determining whether a fourth variable is suitable and adds to the explanatory or predictive power of the model is usually a straightforward task. Thus, comparing models is not difficult.

Regression models can be peculiar things, however, and you may find that various differences arise when you begin to add and remove variables from a model. We discuss finding an appropriate set of predictors in greater detail in Chapter 12. For now, though, we address model fitting procedures, mainly by comparing models with different variables. We first need to introduce a couple of terms. A *nested model* is one composed of a smaller subset of explanatory variables than another model. Suppose we estimate, for example, two regression models designed to predict serum cholesterol levels among a sample of adults. The first model includes the following explanatory variables:

Model 1: resting heart rate, exercise frequency, age

The second model adds a couple of explanatory variables:

Model 2: resting heart rate, exercise frequency, age, triglyceride levels, smoking status

---

[1] In an LRM estimated with OLS, this phrase is a brief way of denoting whether an increase in the explained variance occurs from one model to the next.

DOI: 10.1201/9781003162230-6

We say that Model 1 is *nested* within Model 2 because it includes a subset of the variables. The model with the entire set of explanatory variables is usually called the *full* or *unconstrained* model; the nested model is also known as the *restricted* or *constrained* model.

Another issue involves *non-nested models* in which two (or more) regression models do not include the same subsets of variables. For example, if the first model includes resting heart rate, exercise frequency, and age as explanatory variables, and the second includes triglyceride levels, smoking status, and time spent each day watching television, we cannot say that one model is nested within another. Since comparing models in the non-nested situation is challenging,[2] we will not pursue this issue.

The Partial *F*-Test and Multiple Partial *F*-Test

Suppose we wish to compare two LRMs, one nested within another. Let's begin with the simplest scenario: we add one more explanatory variable to the model and wish to ascertain if it improves our ability to predict the outcome variable. Two general tests we may use are a *t*-test and a *partial F-test* (also called a *nested F-test*), both of which are inferential. As shown in Equation 6.1, the partial *F*-test relates the extra sum of squares attributable to adding one explanatory variable to the model to the MSE from the full model.

$$F\text{-value} = \frac{\text{RSS(full)} - \text{RSS(nested)}}{\text{MSE(full)}} \quad df = 1, n - k - 1 (\text{full model}) \quad (6.1)$$

We wish to learn if the regression sum of squares has increased from the nested to the full model. If it has, then we may infer that the errors of prediction have decreased and the second model fits the data better than the first.[3] Another way to view this is that it tests whether the R^2 has increased to a statistically significant degree.

One of the advantages of this test is that, as we shall see, it generalizes well when we wish to add more than one explanatory variable to the model. It may be inefficient in the current situation, however, because we need to check a table of *F*-values to determine the *p*-value (or ask an R function to

---

[2] See Marno Verbeek (2017), *A Guide to Modern Econometrics*, 5th Ed., Hoboken, NJ: Wiley, chapter 3.
[3] The *F*-test discussed in this section is not the same as subtracting the *F*-values from two LRMs. This latter procedure is not appropriate as a model comparison tool.

identify the *p*-value).[4] As an alternative, most researchers use the *t*-test. In the present situation, the *t*-value from a two-tailed *t*-test is the square root of the *F*-value from the partial *F*-test ($t = \sqrt{F}$). If the *t*-test for the additional variable's coefficient is associated with a low *p*-value (e.g., $p < 0.05$ or 0.01) or CI that does not include zero, we have evidence that the full model fits the data better than the nested model.

But imagine that we wish to compare two models, one of which is nested in the other, but the full model contains two or more additional explanatory variables. In this situation, we depend on the *multiple partial F-test* (often, just like the partial *F*-test, called a *nested F-test*). Equation 6.2 furnishes the formula for this test and shows it is a logical extension of the partial *F*-test.

$$F\text{-value} = \frac{\left[\text{RSS}(\text{full}) - \text{RSS}(\text{nested})\right]/q}{\text{MSE}(\text{full})} \quad df = q, \ n - k - 1 \,(\text{full model}) \quad (6.2)$$

The formula computes an *F*-value based on the difference between the regression sums of squares from the two models and divides this difference by *q*, the number of explanatory variables added to the full model. As with the partial *F*-test, this quantity is then divided by the MSE of the full model. The degrees of freedom for the resultant *F*-value are *q* and ($n - k - 1$), the degrees of freedom associated with the residual sum of squares (SSE) for the full model.

To see an example, let's return to the *StateData2018.csv* dataset. The two LRMs we compare use the number of opioid deaths per 100,000 residents as the outcome variable (OpioidODDeathRate). The explanatory variables in the nested model (LRM6.1) are average life satisfaction (LifeSatis) and percent of the state's population 18 years old and younger (PerAge0_18). The full model (LRM6.2) includes these two variables plus the percent of the state's population that is White (PerWhite) and the unemployment rate (UnemployRate). The lm output and ANOVA information from the nested LRM are in the following R output (the R code is omitted):

R output: Nested model (LRM6.1)
```
Coefficients:
               Estimate  Std. Error  t value   Pr(>|t|)
(Intercept)    190.1200   50.8221      3.741   0.000498 ***
LifeSatis       -2.4025    1.0468      -2.295   0.026237 *
PerAge0_18      -2.1050    0.5766      -3.651   0.000655 ***
---
Signif. codes:  0 '***' 0.001 '**' 0.01 '*' 0.05 '.' 0.1
```

---

[4] The R function drop1 (for example, drop1(LRM6.2), test="F") provides partial *F*-tests for each explanatory variable and furnishes the SSE (labeled RSS in the drop1 output) for each model without a specific variable. The *p*-values, however, are identical to those associated with the slope coefficients in the lm summary output.

```
Residual standard error: 8.288 on 47 degrees of freedom
Multiple R-squared: 0.377, Adjusted R-squared: 0.3505
F-statistic: 14.22 on 2 and 47 DF, p-value: 1.479e-05

Analysis of Variance Table
Response: OpioidODDeathRate
            Df Sum Sq  Mean Sq  F value    Pr(>F)
LifeSatis    1 1038.5  1038.50   15.117  0.0003162 ***
PerAge0_18   1  915.5   915.54   13.327  0.0006555 ***
Residuals   47 3228.7    68.70
---
Signif. codes: 0 '***' 0.001 '**' 0.01 '*' 0.05 '.' 0.1
```

The results indicate that the two explanatory variables account for about 38% of the variability in opioid deaths per 100,000 residents. Moreover, the slope coefficients indicate a negative association between opioid deaths and each of the explanatory variables: states whose residents report higher life satisfaction or with a greater proportion of the population 18 years old or younger tend to experience fewer opioid deaths.

Let's examine the full model that adds the unemployment rate and the percent White (LRM6.2). The following output provides the results (again, the R code is omitted):

```
R output: Full model (LRM6.2)
Coefficients:
               Estimate  Std. Error t value  Pr(>|t|)
(Intercept)   138.84305    57.02128   2.435   0.01892 *
LifeSatis      -2.01291     1.15444  -1.744   0.08805 .
PerAge0_18     -1.96410     0.56083  -3.502   0.00105 **
PerWhite        0.20303     0.07358   2.759   0.00835 **
UnemployRate    3.69623     1.58331   2.335   0.02409 *
---
Signif. codes: 0 '***' 0.001 '**' 0.01 '*' 0.05 '.' 0.1

Residual standard error: 7.525 on 45 degrees of freedom
Multiple R-squared: 0.5083, Adjusted R-squared: 0.4646
F-statistic: 11.63 on 4 and 45 DF, p-value: 1.436e-06

Analysis of Variance Table
Response: OpioidODDeathRate
              Df  Sum Sq  Mean Sq  F value   Pr(>F)
LifeSatis      1 1038.50  1038.50  18.3395 9.579e-05 ***
PerAge0_18     1  915.54   915.54  16.1682 0.0002185 ***
PerWhite       1  371.93   371.93   6.5682 0.0137968 *
UnemployRate   1  308.61   308.61   5.4499 0.0240916 *
Residuals     45 2548.18    56.63
---
Signif. codes: 0 '***' 0.001 '**' 0.01 '*' 0.05 '.' 0.1
```

It appears we've added some important information to the model. The two additional explanatory variables have what appear to be substantial associations with opioid deaths per 100,000 after adjusting for the effects of the other variables. In fact, if the beta weights (not shown in the output) can be trusted, the percent of the state's population ages 0–18 has the strongest association with opioid deaths among the four variables[5] (confirm this for yourself; also consider estimating the relative importance measures described in Chapter 4).

But does the addition of the two explanatory variables improve the predictive power of the model? Perhaps it would be useful to compare the R^2 values. After all, they measure the proportion of variability in the outcome variable that is accounted for by the model. The R^2 for the nested model is 0.38 and for the full model 0.51. Unfortunately, simply comparing these statistics is not sufficient. In Chapter 5 we learned that the R^2 increases when we add explanatory variables, whether or not they are good predictors. An alternative is to compare the adjusted R^2 values since they are not as affected by the addition of explanatory variables. The values are 0.35 for the nested model and 0.46 for the full model, which appears to be a substantial increase: the adjusted R^2 increases by about ([0.465 − 0.351] / 0.351) × 100 = 32%.[6] But suppose we wish to use a significance test to judge whether the increase in explained variance is statistically significant—in other words, is the increase in the R^2 statistically significant? We may use the multiple partial F-test to execute this task (see Equation 6.3).[7]

$$F = \frac{[2634.5 - 1,954]/2}{56.6} = 6.01 \left(df = 2, 45\right) \tag{6.3}$$

The p-value for an F-value of 6.01 with $df = (2, 45)$ may be computed in R using the following function: 1 - pf(6.01, 2, 45). R returns a p-value of 0.005, which is lower than the threshold recommended by most criteria. In this situation, the multiple partial F-test offers evidence that the addition of percent White and the unemployment rate increases the explained variance.

---

[5] Recall from Chapter 4 that, when comparing beta weights, we should examine the distributions of the relevant explanatory variables to determine if they are comparable.

[6] Some researchers also compare the RMSEs (labeled Residual standard error in the R output; see Chapter 5), noting that a reduction from one model to the next demonstrates that the errors of prediction in the sample are diminished and this indicates a better fitting model. For instance, the RMSE drops from 8.3 in the nested model to 7.5 in the full model.

[7] The ANOVA tables do not provide the regression sum of squares directly. We must add them for each explanatory variable first and then plug this information into the equation. For instance, the RSS from the nested model is 1,038.5 + 915.5 = 1,954.

R furnishes a simple way to compute a multiple partial F-test with the anova function. Recall that we labeled the nested model as object LRM6.1 and the full model as object LRM6.2.

R code
```
anova(LRM6.1, LRM6.2)
```

R output
```
Analysis of Variance Table
Model 1: OpioidODDeathRate ~ LifeSatis + PerAge0_18
Model 2: OpioidODDeathRate ~ LifeSatis + PerAge0_18 +
        PerWhite + UnemployRate
 Res.Df    RSS    Df Sum of Sq    F      Pr(>F)
1    47  3228.7
2    45  2548.2   2    680.54   6.009   0.004864 **
---
Signif. codes: 0 '***' 0.001 '**' 0.01 '*' 0.05 '.' 0.1
```

The R output confirms what we learned from the earlier assessment: an F-value of 6.01 with a p-value of about 0.005.[8] In addition to signifying something about the increase in explained variance and the R^2, another way of understanding this test is that it evaluates whether or not the slope coefficients from the two variables added to the full model are jointly equal to zero, which is a key purpose of these types of F-tests. You should now understand why the first model is called a *constrained model*: by leaving the explanatory variables from the full model out of the initial model, their slope coefficients are constrained to be equal to zero.

But is this all we would wish to do with the model? We added two variables, both of which appear to be modestly associated with opioid deaths per 100,000. Yet, we now see that the coefficient for average life satisfaction is no longer below the threshold of $p < 0.05$.[9] Perhaps we should compute a partial F-test to determine whether the presence of average life satisfaction adds much to the model's explained variance. However, at some point—many argue it should be very early in the analysis process—the researcher must decide which variables are important and which

---

[8] As with the R^2 (see Chapter 5), obtaining a bootstrapped CI of the F-value is feasible for those who do not care for p-values.

[9] But consider that the p-value for life satisfaction is between 0.05 and 0.10. Suppose we use an *a priori* directional hypothesis to claim that life satisfaction is negatively associated with opioid deaths. As mentioned earlier, some researchers claim that directional hypotheses should be assessed with one-tailed p-values. The one-tailed p-value in this situation is $0.088/2 = 0.044$, which might, depending on the decision rule the researcher uses, indicate a "statistically significant" association. This statistical sleight-of-hand provides a good illustration of why we should be cautious about significance tests and avoid an overreliance on them for judging the viability of statistical associations (see the section on best statistical practices in Chapter 1).

are not. This should be guided, preferably, by theory or a conceptual model, rather than by rote estimation of regression models and partial F-tests.[10] Would you judge all the evidence—R^2s, adjusted R^2s, F-tests, and RMSEs—as supporting the notion that the second model does a better job than the first model of predicting opioid deaths? Does your theory dictate that a variable—such as life satisfaction—or a set of variables should be in a model? Ultimately, the researcher must make these decisions based on an *a priori* set of criteria.

Evaluating Model Fit with Information Criterion Measures

An alternative to nested F-tests that some researchers use to judge the relative fit of models involves information criterion (IC) measures.[11] Similar to the adjusted R^2, ICs are designed to balance model fit and model simplicity or parsimony. They give preference to those models that both fit the data well and don't have extraneous "information" in the form of ill-suited explanatory variables.

IC measures are normally used for regression models estimated with ML rather than OLS, but they have been adapted for the latter models. The two most common are Akaike's Information Criterion (AIC) and the Bayesian Information Criterion (BIC),[12] which, as shown in Equations 6.4 and 6.5, may be computed with some results from an LRM estimated with OLS (although alternative equations exist).

$$AIC = n \times \ln(SSE/n) + 2(p + 2) \tag{6.4}$$

$$BIC = n \times \ln(SSE/n) + (p + 2)(\ln(n)) \tag{6.5}$$

In the AIC and BIC equations, SSE is the sum of squared errors, ln is the natural logarithm, n is the sample size, and p is the number of explanatory variables in the particular model. Smaller AICs or BICs indicate better fitting models.

In R, IC measures are available with the AIC and BIC functions (`AIC(model)` and `BIC(model)`). For example, the AICs and BICs from the nested and full models estimated earlier in the chapter are provided in Table 6.1.

---

[10] See Chapter 12 for more information about variable selection procedures.

[11] Erick Suárez et al. (2016), *Applications of Regression Models in Epidemiology*, Hoboken, NJ: Wiley.

[12] AICs and BICs were introduced in Hirotugu Akaike (1973), "Information Theory and an Extension of the Maximum Likelihood Principle," in *The 2nd International Symposium on Information Theory*, B. N. Petrov and F. Csáki (Eds.), and Gideon E. Schwarz (1978), "Estimating the Dimension of a Model," *Annals of Statistics* 6(2): 461–464.

TABLE 6.1

AIC and BIC Measures of Model Fit

Model	AIC	BIC
LRM6.1 (nested)	358.3	365.9
LRM6.2 (full)	350.4	361.9

In both cases the fit statistics are smaller for the full model (LRM6.2) than the nested model (LRM6.1), suggesting that the full model fits the data better than the nested model. LRM6.2 therefore does a better job of predicting state-level opioid deaths per 100,000. But, as emphasized earlier, prediction should not normally be our only concern. We should also consider whether the model is useful for understanding or explaining why some states have more opioid deaths than others.

Some researchers suggest that AICs and BICs may be used to compare non-nested regression models, at least informally.[13] However, the studies underpinning this suggestion have focused chiefly on models estimated with ML rather than OLS. At this point, it seems risky to compare non-nested LRMs estimated with OLS using these IC measures.

Confounding Variables

Let's examine one other issue at this juncture about multiple LRMs. This returns to a point made in Chapter 4 about the need to assess confounding variables. As mentioned there, a confounding variable accounts for the association between an explanatory and an outcome variable (recall the lighter purchases and lung disease example).

To be a bit more precise: a variable is considered a confounder if it meets the following criteria. First, the variable is associated with the explanatory variable. Second, it is associated with the outcome variable independent of the explanatory variable. Third, it is not part of a causal pathway from the explanatory variable to the outcome variable. Returning to the example in Chapter 4, suppose we find that sales of lighters are associated with lung cancer rates. Cigarette smoking serves as a confounder if: (a) it is associated with sales of lighters (highly likely); (b) it is associated with lung cancer rates *independent* of lighter sales (a valid conjecture); and (c) it does not play a role in the causal association between cigarette smoking and cancer rates. The latter claim is also valid since buying more lighters is not likely to directly cause smoking or cancer. It should be clear how a confounding variable casts

---

[13] Peter K. Dunn and Gordon K. Smyth (2018), *Generalized Linear Models with Examples in R*, New York: Springer, chapter 7.

doubt on causal connections between variables: the frequency of lighter purchases does not cause lung cancer rates because cigarette smoking confounds their association.[14]

We'll discuss confounding a bit more in Chapter 12, but for now consider the two LRMs designed to predict the number of opioid deaths per 100,000. Looking back at the coefficients from the two models, in the nested model, average life satisfaction has a partial slope of –2.40, whereas in the full model it has a partial slope of –2.01 The partial slope thus decreased by about 16% when we included the additional variables in the model. We cannot tell at this point whether the decrease is practically or statistically significant (although tools are available for the latter issue), but let's assume it is. We may then tentatively claim that one or both of the new explanatory variables included in the model confounded the association between the average life satisfaction and opioid deaths. They did not confound it completely—the slope would be much closer to zero if this was the case (or it would decrease enough that we'd judge it to be non-essential)—but changed it enough to draw our attention. We usually look for variables that completely account for the association between two variables, but even those that only partially account for it can be interesting. The key question you should always ask yourself, though, is why. Why does the association change when we add a new variable? It could be a random fluctuation in the data, but it might be something important and worthy of further exploration. In this example, perhaps life satisfaction and one of the other explanatory variables, such as the unemployment rate, are associated in an intriguing way. You might ask whether the unemployment rate is associated with average life satisfaction and opioid deaths in a causal fashion or is it simply a confounder?[15]

Chapter Summary

This brief chapter addresses two important topics. First, how can we determine whether one model provides a better fit to the data than another model? In other words, is more of the variance in the outcome variable accounted for by one model than by the other? Of consequence when making such comparisons is whether one model is nested within another. Nested F-tests and IC measures provide evidence regarding whether a full model fits the patterns in the data better than a nested model. Second, we quickly examined

---

[14] It should be clear at this point that we may control for confounding by statistical adjustment using an LRM. But other methods are also available, such as using matching techniques to balance the effects of potential confounding variables (see Chapter 17) or, better yet, by using experimental methods that allow researchers to control the proposed causal factors.

[15] As an exercise, revisit the three criteria described earlier to consider whether the unemployment rate is the confounder.

the issue of confounding variables. We have seen and will revisit this issue in other chapters, yet considering confounding variables is vital for determining if we've estimated an appropriate model and, most importantly, whether the information from the model is useful.

Chapter Exercises

Recall that in the exercises at the end of Chapter 5, we used the *QBPassers2019* *.csv* dataset to investigate factors that predict or account for quarterback ratings. In particular, we examined two LRMs that included the variable rating as the outcome variable and the following explanatory variables:

LRM 1: completion percentage and interceptions per 100 pass attempts.

LRM 2: completion percentage, interceptions per 100 pass attempts, average yards per completion, and touchdowns per 100 pass attempts.

After re-importing the dataset into R, complete the following exercises:

1. Identify which of the two LRMs is the full model and which is the nested model. Explain your answer.

2. Estimate each model in R. Compare the R^2s, the adjusted R^2s, and the RMSEs from the two models. What do these indicate about the relative fit of the models?

3. Estimate a multiple partial (nested) F-test that compares the full and nested models. What do the results indicate? Interpret the p-value associated with this F-test.

4. Estimate the AICs and BICs associated with each model. What do they indicate about the relative fit of the two models?

5. *Challenge*: use R to compute 95% bootstrapped CIs for the R^2 values from each model (hints: R has a native boot function and the car package has a Boot function. An internet search might help you find the appropriate R code). Assuming we wished to make inferences to the "target population," what do these CIs suggest about the relative fit of the two models?

7

Indicator Variables in Linear Regression Models

Chapter 2 includes an example that uses R's t.test function to compare average personal incomes among females and males in the *GSS2018.csv* dataset. The variable female distinguishes between the two groups and, in case you've forgotten, mean income among females is lower than mean income among males. More importantly for our purposes in this chapter, however, is the fact that the female variable is clearly not continuous. Instead, as noted in Chapter 2, this type of variable is discrete or categorical. Such variables abound throughout the sciences.

Categorical variables are often distinguished by whether or not they have a logical order or ranking. Those that cannot be ordered logically are sometimes referred to as *nominal*, whereas those that can be ordered in a sensible manner are called *ordinal*.[1] Examples of nominal or unordered variables include gender, marital status, and political party affiliation. Ordinal or ordered variables include educational milestones such as high school graduate, college graduate, and advanced degree, and Likert scales ("The U.S. should pull out of NATO": strongly agree, agree, disagree, strongly disagree). This chapter discusses the role of categorical explanatory variables in LRMs, with the primary focus on unordered variables.

Imagine a variable with three or more categories, such as marital status that categorizes people's relationship status as single, cohabiting, married, divorced, and widowed. Suppose we wish to determine whether marital status is associated with personal income. How is this type of variable entered into an LRM? We'll return to this issue later, but the straightforward answer is that, in order to facilitate its use in a regression model, this type of variable is transformed into a set of *indicator variables*[2]: variables that take on only a

---

[1] In mathematics, an *ordinal number* identifies the position or order in a list, such as first, second, or third, whereas a *cardinal number* indicates how many of something exists, such as one, two, or three. The classification in social and behavioral sciences of variables into nominal, ordinal, interval, and ratio measures was developed by psychologist S. S. Stevens (1946) (see "On the Theory of Scales of Measurement," *Science*, 103(2684): 677–680).

[2] Indicator variables are also called *dummy variables*, though this term is not meant as an insult. No one is claiming that the variables are dimmer, duller, gormless, or are lower in intelligence than others. Rather, this term is related to the idea that mannequins or ventriloquists' puppets—often known as *dummies*—represent or take the place of real people (see Steven Connor (2000), *Dumbstruck: A Cultural History of Ventriloquism*, New York: Oxford University Press).

DOI: 10.1201/9781003162230-7

limited number of values so that different categories of a discrete variable may be represented in a model. The term indicator variable designates that it "indicates" a group represented in the sample. Two important characteristics are that the groups represented by a set of indicator variables should be (a) mutually exclusive and (b) exhaustive (no category should be left out).

In addition to *t*-tests, if you've ever conducted an ANOVA to determine the mean differences in some outcome for two or more groups, you've seen indicator variables. To extend the example of marital status, a researcher could use an ANOVA model to determine mean income levels among those in the different marital status groups. We'll learn later that using indicator variables in LRMs serves a similar purpose as ANOVA models.

As suggested earlier, research in social, behavioral, and health sciences is brimming with indicator variables. For instance, suppose we wish to determine whether a categorical variable is associated with a continuous outcome variable. Clearly, gender, ethnicity, religious group affiliation, marital status, family structure, and many other variables cannot be represented by continuous, let alone normally distributed, variables. Yet we often wish to include these types of items in regression models. Indicator variables provide an efficient way of including categorical or binary (two-category) variables. The following presents two examples of the coding associated with indicator variables:

Gender: 0 = male; 1 = female

Ethnicity (White, African American, Latinx):

x_1: White = 1; other = 0

x_2: African American = 1; other = 0

x_3: Latinx = 1; other = 0

The first variable, gender, is straightforward since we place people into one of only two groups (though it could be extended).[3] The second set of variables requires an explanation, though. Assume a dataset includes a variable designed to measure ethnicity (to simplify, we'll limit this example to three ethnic groups). It includes three outcomes—White, African American, and Latinx—that are coded 1, 2, and 3. Since this variable is not continuous, nor can it be ordered in a sensible way, we should not include it as is in an LRM (how would we interpret a one-unit increase or decrease in a slope coefficient?). The solution is to create three indicator variables that represent the three groups. The main group represented by each is usually coded as 1, with any sample member not in the group coded as 0. We can extend this

---

[3] Gender is used here in a very general sense to refer to "the biological division into male and female" (commonly referred to as "sex"). However, in sociology and related fields, gender refers to socially and culturally constructed differences attributed to women and men, "cultural ideals and stereotypes" about women and men and about "masculinity and femininity," and the "division of labor in institutions and organizations" that are part of the female/male dichotomy common in most societies (John Scott (2014), *A Dictionary of Sociology*, 4th Ed., New York: Oxford University Press, p.274).

TABLE 7.1

Example of Indicator Variable Coding

Observation	Ethnicity	x_1	x_2	x_3
1	White	1	0	0
2	White	1	0	0
3	African American	0	1	0
4	African American	0	1	0
5	Latinx	0	0	1
6	Latinx	0	0	1

form of indicator variable coding to a variable with any number of unique categories. For example, Table 7.1 illustrates how the three ethnicity indicator variables, x_1, x_2, and x_3, would appear in a dataset (spreadsheet format).

Several other types of coding strategies are available, such as effects coding and contrast coding, but the type shown in Table 7.1 is flexible enough to accommodate many modeling needs. We may also add additional groups (e.g., East Asian, Native American, Aleut, etc.) if the original variable includes codes for them.

A rule to always remember when using indicator variables in a regression model to represent mutually exclusive groups is to include all the indicator variables in the model except one. The omitted indicator variable is called the *reference category* or *comparison group*. Suppose our variable has five categories (represented by k) and we create five indicator variables to represent these categories. We then use four (or $k - 1$) indicator variables in the regression model. If we try to use all five groups, the model will not operate correctly because of perfect collinearity among the indicator variables (see Chapter 10).

How should we choose the reference category? This is up to the researcher, but some guiding principles might be helpful. First, the best choice is to let the conceptual model or hypothesis guide which group to use as the reference category. If, say, we are interested mainly in differences between married and single people, then one of these should be the reference category. Second, many researchers use the most frequent category as the reference category. Third, avoid using a relatively small group as the reference category. Fourth, many statistical software programs, including R, have regression options that automate the creation of indicator variables; in fact, one may use a single categorical variable and programs such as R implicitly includes the $k - 1$ groups. When this is the case, check the software documentation to determine which category is excluded from the model. Some programs exclude the most frequent category; others exclude the highest or lowest numbered category; and still others treat them as characters rather than numbers and omit the first group it comes across in the dataset or alphabetically.

How are indicator variables used in a regression model? Just like any other explanatory variable. Providing the variables with names that are easy to recognize is helpful. So, for example, if we were to create a set of indicator variables representing marital status, we might wish to name them `married`, `cohabits`, `divorced`, `single`, and `widowed`. As we'll learn next, R makes it convenient to use character variables as nominal and ordinal variables.

Let's consider an LRM with only one indicator variable. As we learned from Chapter 2's *t*-test, a variable in the *GSS2018* dataset is called `female`, which is coded into two categories: `male` and `female` (examine its frequency distribution in R using `table(GSS2018$female)`). In the following example, we'll treat it as an explanatory variable, with annual personal income in $10,000s (`pincome`) as the outcome variable.[4] If `female` is the only explanatory variable, the LRM provides the same information as a *t*-test or an ANOVA model and, as we shall see, can be used to estimate mean income levels for males and females.

When R reads the *GSS2018* data file it recognizes the two categories as comprised of character strings or "words": `female` and `male`.[5] In many situations and with some other statistical programs, however, indicator variables are read as numeric variables. For instance, the original dataset from which the *GSS2018* is derived has a variable called `sex` that is coded as 1 = `male` and 2 = `female`. This illustrates a frequent coding situation: many datasets code two-category or binary variables as {1, 2}. We may still use these variables in a regression model, but a better option is to use either a {0, 1} coding scheme or, in R, simply let the program recognize a categorical variable as consisting of characters.[6] In addition, if you browse the *GSS2018* dataset in R you'll find other character variables, such as `legalmarij` ("do you think marijuana should be legal for adults to use?" {no or yes}) and `owngun` ("do you own a gun?" {no or yes}).

Setting up an LRM with indicator variables in R requires no special tools. The setup is the same as with continuous explanatory variables: simply place the indicator variable(s) in the `lm` function after the outcome variable. Let's estimate the model specified earlier and see what R provides (see LRM7.1).

R code
```
LRM7.1 <- lm(pincome ~ female, data=GSS2018)
summary(LRM7.1)
confint(LRM7.1)
```
R output (abbreviated)
```
Residuals:
    Min       1Q    Median      3Q       Max
-10.8818  -8.2214  -0.2214   8.1182   17.7786
```

---

[4] The income variables in the *GSS2018* dataset are not truly continuous, nor are they coded correctly. Rather, they include categories of income that are not in actual $10,000 units, even though we shall label them as such. Nonetheless, in the interests of unfettered learning, we'll treat them as continuous and pretend that the units are accurate.

[5] When inputting data in R, character strings in categorical variables are often called *factor vectors*, although this makes little difference for our analysis work. A handy function for creating distinct indicator variables from a factor vector is available in the `regtools` package: `factorsToDummies`.

[6] We'll define a categorical variable as a binary indicator—as including only two categories—if it is coded as {0, 1} or with two character strings (e.g., `female` and `male`).

```
Coefficients:
             Estimate  Std. Error  t value  Pr(>|t|)
(Intercept)  8.2214    0.2480      33.157   < 2e-16   ***
femalemale   2.6605    0.3698       7.195   8.39e-13  ***
---
Signif. codes: 0 '***' 0.001 '**' 0.01 '*' 0.05 '.' 0.1

Residual standard error: 8.85 on 2313 degrees of freedom
Multiple R-squared: 0.02189, Adjusted R-squared: 0.02147
F-statistic: 51.77 on 1 and 2313 DF, p-value: 8.385e-13

[CIs]          2.5%    97.5%
(Intercept)  7.735    8.708
femalemale   1.935    3.386
```

The output looks similar to others we've requested, with coefficients, standard errors, *t*-values, *p*-values, and 95% CIs. In fact, several of the interpretations are similar, except for the slope and intercept. We learned earlier that the intercept may not be useful in many LRMs because explanatory variables are frequently coded so as to not have an interpretable zero value. Think about how an indicator variable is different, though: it has an interpretable zero value since the omitted reference category is represented, at least implicitly, by zero. But because it no longer makes much sense to refer to a one-unit increase in the explanatory variable—at least not in the same way as with continuous explanatory variables—we need to modify our thinking about the slope coefficient.

Before figuring out how to interpret the model's results, always examine the way the indicator variable is listed in R's summary or confint output. Recall that the variable is labeled female. However, the output lists femalemale. This means that R has chosen one of the categories to place in the model: male. The other category, female, is thus the reference category.[7] The slope coefficient therefore represents something about the personal incomes of males relative to females.

Perhaps writing out the LRM equation will help move us toward a clearer interpretation (see Equation 7.1).

---

[7] To change the reference category, we can do some manipulations in R before estimating the model. Here are the two steps to change the female variable's reference category to male rather than female:

```
GSS2018$female <- factor(GSS2018$female) # change  the  female  variable
                                  into a factor variable
GSS2018$female <- relevel(GSS2018$female, ref="male")

 # use the relevel function to change the reference category to male
```

We may then use the usual lm function to estimate the model and male will be the reference category.

Predicted personal income in $10,000 s = 8.22 + 2.66 (gender = male) (7.1)

Let's use Equation 7.1 to calculate the predicted values for the two groups (see Equations 7.2 and 7.3).

Predicted income for males: 8.22 + 2.66 = 10.88 (7.2)

Predicted income for females: 8.22 + 0.00 = 8.225 (7.3)

Since females comprise the reference category, and therefore we implicitly multiply the slope coefficient by zero, their mean personal income is represented by the intercept, 8.22, which denotes about $82,200. The slope coefficient is the average difference in income between males and females, thus, on average, males report 2.66 units more personal income than females, or about $26,600 (recall, though, that these are not actual income values). The mean of personal income is 9.42, which is close to the average of the two predicted values. It should be for the simple reason that, as foreshadowed earlier, these two predicted values are the mean values for personal income among males and females. You may confirm this using R's t.test function (the same model is estimated in Chapter 2):

R code
```
t.test(pincome ~ female, data=GSS2018)
```

R output (abbreviated)
```
sample estimates:
mean in group female  mean in group male
        8.22135               10.88184
```

Other R functions also provide this type of information. For example, one might use the tapply function in the following manner:

R code
```
tapply(GSS2018$pincome, GSS2018$female, summary)
```

R output (abbreviated; bold font added)
```
$female
   Min.  1st Qu.  Median  Mean   3rd Qu.  Max.
  0.000   0.000    6.000  8.221  16.000   26.000
$male
   Min.  1st Qu.  Median  Mean   3rd Qu.  Max.
  1.00    0.00    13.00   10.88  19.000   26.000
```

Or, with a simple model that consists of the means from only two groups, a bar graph is helpful for visualizing average personal income among males and females (see Figure 7.1).

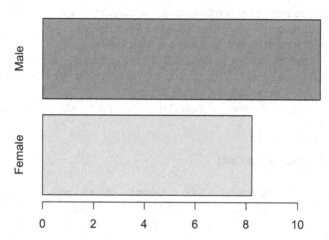

FIGURE 7.1
Bar graph of average personal income among males and females.

Indicator Variables in Multiple LRMs

Since the two-sample *t*-test provides much the same information as the indicator variable approach in a simple LRM, are there any advantages to using indicator variables? A key advantage is that we may include additional explanatory variables in the LRM to estimate predicted means that are adjusted for continuous or other indicator variables. Perhaps you're familiar with analysis of covariance (ANCOVA) models? LRMs offer a direct way to estimate them. But first let's see another nice property of the LRM approach by adding explanatory variables, only this time we'll add indicator variables to explore differences among several groups.

Returning to the personal income model, let's add the race/ethnicity indicator variables contained in the ethnic variable. First, we need to decide which group to omit from the LRM as the reference category. The frequencies for the four groups are 1,496 Whites, 363 African Americans, 344 Latinx people, and 112 from other ethnic groups (based on table(GSS2018$ethnic)). We have not set up a hypothesis that guides the selection of the reference category, so we'll make an arbitrary selection. Let's choose the most frequent category—White—as the reference category. We'll then estimate an LRM that includes female but adds three indicator variables that identify African Americans, Latinx, and members of other racial/ethnic groups to predict personal income (see LRM7.2).

R code
```
GSS2018$female <- relevel(GSS2018$female, ref="female")
 # make females the reference category; see footnote 7
GSS2018$ethnic <- factor(GSS2018$ethnic)
 # changes ethnic to a factor variable
GSS2018$ethnic <- relevel(GSS2018$ethnic, ref="White")
 # uses the relevel function to identify White as the
   reference category for ethnic
LRM7.2 <- lm(pincome ~ female + ethnic, data=GSS2018)
summary(LRM7.2)
confint(LRM7.2)
```

R output (abbreviated)

Coefficients:

	Estimate	Std. Error	t value	Pr(>|t|)	
(Intercept)	8.3368	0.2862	29.131	< 2e-16	***
femalemale	2.6420	0.3700	7.140	1.25e-12	***
ethnicAfrican American	-0.6046	0.5180	-1.167	0.243	
ethnicLatinx	-0.4060	0.5294	-0.767	0.443	
ethnicOther	0.9927	0.8669	1.145	0.252	

Signif. codes: 0 '***' 0.001 '**' 0.01 '*' 0.05 '.' 0.1

Residual standard error: 8.849 on 2310 degrees of freedom
Multiple R-squared: 0.02337, Adjusted R-squared: 0.02168
F-statistic: 13.82 on 4 and 2310 DF, p-value: 3.862e-11

[CIs]	2.5%	97.5%
(Intercept)	7.776	8.898
femalemale	1.916	3.368
ethnicAfricanAmerican	-1.620	0.411
ethnicLatinx	-1.444	0.632
ethnicOther	-0.707	2.693

Considering how we've treated indicator variables so far, what do you think the intercept and slope coefficients measure? As before, let's write out the regression model results so we may predict personal income (see Equation 7.4).

$$\text{Predicted income in } \$10{,}000 = 8.34 + 2.64(\text{gender} = \text{male})$$

$$- 0.60(\text{African American})$$

$$- 0.41(\text{Latinx}) + 0.99(\text{Other race/ethnicity})$$

$$(7.4)$$

TABLE 7.2

Predicted Means for Gender and Ethnic Groups

Group	Computation	Expected Mean
White females	8.34	8.34
White males	8.34 + 2.64	10.98
African American females	8.34 − 0.60	7.73
African American males	8.34 + 2.64 − 0.60	10.37
Latinx females	8.34 − 0.41	7.93
Latinx males	8.34 + 2.64 − 0.41	10.57
Other race/ethnic females	8.34 + 0.99	9.33
Other race/ethnic males	8.34 + 2.64 − 0.99	11.97

Since the intercept in a multiple LRM is the expected value of the outcome variable when all the explanatory variables are set at zero, 8.34 represents the expected mean income for females who are not African American, not Latinx, and are not in the other racial/ethnic category. Who's left? White females. In other words, the intercept represents the mean for the omitted category from both sets of indicator variables: gender and race/ethnicity. The coefficients, once again, represent average differences among the groups and are also called *deviations from the mean*.[8] In particular, we're interested in computing deviations from the mean of the reference category (White females). But we must now combine coefficients to estimate the means for the different groups, as shown in Table 7.2.

We've estimated eight means for each of the mutually exclusive groups defined by the two categorical variables. But suppose we wish to take a significance testing approach to judge presumed differences in the population. Consider the p-values and 95% CIs in the LRM output. As in the earlier model, the `female` coefficient has a relatively small p-value ($p < 0.001$) and a 95% CI that does not include zero {1.9, 3.4}, thus casting doubt on the notion that males and females report, on average, similar personal incomes. The coefficients for African American, Latinx, and other racial/ethnic groups have relatively large p-values ($p = 0.24$, 0.44, and 0.25), however, and their 95% CIs include zero. We would not claim evidence to show, therefore, that the mean income differences between Whites and African Americans differ in the population of adults in the U.S. Furthermore, following this logic, the average income difference between males and females in each racial/ethnic group is presumed identical (2.64).[9]

---

[8] This term was also used in Chapter 2 when defining one of the terms in the formula for the sum of squares. Its use here is different.

[9] If you wish to explore this issue further, the R package `lsmeans` computes multiple comparison tests to determine whether differences across various categories meet certain p-value thresholds.

LRMs with Indicator and Continuous Explanatory Variables

The next situation we'll address involves LRMs that include both indicator and continuous explanatory variables. As mentioned earlier, the well-known ANCOVA model is the analog to this type of regression model. ANCOVA models serve several useful purposes. Their main benefit is that we may compare groups regarding an outcome variable after adjusting for their associations with a set of continuous variables.

In the last chapter we discussed briefly the issue of confounding variables. Here's another example of this phenomenon. Suppose we are studying rates of colon cancer among cities in the U.S. We draw a sample of cities and find that those in Florida, Arizona, and Texas have higher rates of colon cancer than cities in the Northeast, Midwest, and Northwest areas of the country (perhaps we have set up a series of indicator variables that indicate region of the country). Are there unique environmental hazards in warm weather cities that affect the risk of colon cancer? We cannot tell without much more information about the environmental conditions in these cities. Nevertheless, we have not yet considered a key confounding variable that affects analyses of most rates of disease: age. Many warm weather cities—especially in so-called sunshine states such as Florida and Arizona—have an older age structure than cities in colder climates. Age is also associated with colon cancer: it tends to be more common among older adults than among young people.[10] Age is thus associated with both colon cancer rates and region of the country.[11] ANCOVA models are designed to adjust for continuous variables, such as age, that act as confounders.

Multiple LRMs may be used in a similar manner. If we're interested in analyzing differences in some outcome among groups defined by indicator variables, we should consider adjusting for the possible association with conceptually relevant continuous variables. Here's a scenario that's similar to the earlier example but changes the outcome variable. A perception in some circles is that fundamentalist or conservative Christians tend to come from poorer families than do other people. We might surmise—and this is an admittedly crude hypothesis—therefore that family income among fundamentalist Christians is lower on average than in other families in the U.S. Let's test this hypothesis using the *GSS2018* dataset. The next example estimates an LRM with family income (fincome) as the outcome variable and an indicator variable fundamentalist (a recoded version of the GSS fund variable that distinguishes non-fundamentalists and fundamentalist

---

[10] For data from the U.S., see "Cancer Stat Facts: Colorectal Cancer" (https://seer.cancer.gov/statfacts/html/colorect.html).

[11] Epidemiologists and demographers are always attentive to age as a confounder of differences in disease rates across geography and time (see, for example, Randall E. Harris (2019), *Epidemiology of Chronic Disease: Global Perspectives*, 2nd Ed., Burlington, MA: Jones & Bartlett).

Christians). The model also includes `female` as an explanatory variable (see LRM7.3).

R code
```
library(car)
GSS2018$fundamentalist <- recode(GSS2018$fund,
    "c('fundamentalist')='fund'; else='notFund'")
 # use the recode function of the car package to
    collapse the fund variable into two categories of a
    new variable: fundamentalists and non-fundamentalists
GSS2018$fundamentalist <- factor(GSS2018$fundamentalist)
 # make fundamentalist a factor variable
GSS2018$fundamentalist <-
relevel(GSS2018$fundamentalist, ref="notFund")
 # use the relevel function to make notFund the
    reference category
LRM7.3 <- lm(fincome ~ fundamentalist + female,
            data=GSS2018)
summary(LRM7.3)
confint(LRM7.3)
```

R output (abbreviated)
Coefficients:

	Estimate	Std. Error	t value	Pr(>|t|)
(Intercept)	10.97683	0.06846	160.349	<2e-16 ***
fundamentalist				
fund	-0.21201	0.10991	-1.929	0.0539 .
femalemale	0.20261	0.09338	2.170	0.0301 *

```
---
```
Signif. codes: 0 `***` 0.001 `**` 0.01 `*` 0.05 `.` 0.1

Residual standard error: 2.232 on 2312 degrees of freedom
Multiple R-squared: 0.003838, Adjusted R-squared: 0.00298
F-statistic: 4.454 on 2 and 2312 DF, p-value: 0.01173

[CIs]	2.5%	97.5%
(Intercept)	10.843	11.111
fundamentalistfund	-0.428	0.004
femalemale	0.020	0.386

On average, adults in the fundamentalist Christian category report 0.21 units less on the family income scale relative to other adults. This magnitude of the coefficient is close to a conventional threshold for statistical significance ($p = 0.054$, 95% CI: {−0.43, 0.004}). For present purposes, we'll assume it is close enough to conclude that fundamentalist Christians report slightly less family income than others in the U.S.

But think a bit more about this association. Are there other variables that might account for the negative association? Without even knowing much about fundamentalist Christianity in the U.S., I'm sure we can come up with some informed ideas. Let's consider income. What accounts for differences in income in the U.S.? A prime candidate is education. If we were to read a bit about religion in the U.S., we might find research that suggests that conservative or fundamentalist Christians tend to have less formal education than others (although this has changed in recent decades).[12]

A reasonable step, therefore, is to include a variable that assesses formal education to determine whether it affects the results found in LRM7.3. The *GSS2018* dataset includes such a variable, educate, which measures the highest year of formal education conveyed by sample members. Let's estimate another LRM that adds the education variable and determine whether it affects the results (see LRM7.4).

R code
```
LRM7.4 <- lm(fincome ~ fundamentalist + female +
             educate, data=GSS2018)
summary(LRM7.4)
confint(LRM7.4))
```

R output (abbreviated)
Coefficients:

	Estimate	Std. Error	t value	Pr(>|t|)	
(Intercept)	7.95647	0.22022	36.130	<2e-16	***
fundamentalist					
fund	-0.03646	0.10604	-0.344	0.7310	
femalemale	0.22642	0.08951	2.530	0.0115	*
educate	0.21612	0.01504	14.367	<2e-16	***

Signif. codes: 0 '***' 0.001 '**' 0.01 '*' 0.05 '.' 0.1

Residual standard error: 2.139 on 2311 degrees of freedom
Multiple R-squared: 0.08552, Adjusted R-squared: 0.08433
F-statistic: 72.04 on 3 and 2311 DF, p-value: < 2.2e-16

[CIs]	2.5%	97.5%
(Intercept)	7.525	8.388
fundamentalistfund	-0.244	0.171
femalemale	0.051	0.402
educate	0.187	0.246

Once educate is included in the model the coefficient associated with the fundamentalist Christian variable is much closer to zero, the *p*-value

---
[12] For example, see Alfred Darnell and Darren E. Sherkat (1997), "The Impact of Protestant Fundamentalism on Educational Attainment," *American Sociological Review* 62(2): 306–315.

is larger (0.73), and the 95% CI {−0.24, 0.17} includes zero. Given the literature on formal education among fundamentalist Christians in the U.S. and
the results of the two LRMs, we should strongly suspect that education
confounds the association between considering oneself a fundamentalist
Christian and family income.[13]

To illustrate another important issue, how can we estimate means for different groups using LRM7.4? As suggested earlier, a good practice is to begin
by writing out the implied prediction equation (see Equation 7.5).

$$\text{Predicted income} = 7.96 - 0.04(\text{fundamentalist})$$
$$+ 0.23(\text{femalemale}) + 0.22(\text{educate}) \tag{7.5}$$

We've already learned what to do with the indicator variables: simply plug
in a zero or a one and compute the predicted values. But what should we
do with the education variable and its coefficient? Obviously, we could do
something similar with this variable as we did with the indicator variables: plug in some value, compute the products, and add them. Suppose
we think that putting a zero or a one for education is appropriate. What
do these numbers represent? Looking over the characteristics of `educate`,
perhaps by calculating summary statistics in R, we find that it has a minimum value of zero and a maximum value of 20. But notice the few sample
members with less than 11 years of education. Placing a small number,
such as 1, 6, or 7, in the equation is unwise since these values are not represented well in the dataset. A value of 20 might be reasonable, but, for
various reasons, using an "average" value of education in the computations
(e.g., mean, median, or mode) or some other sensible value is better. For
example, in the U.S. a well-recognized educational milestone is graduation
from high school, which is normally denoted as 12 years of formal education. Twelve years is also the most frequent category in the distribution of
`educate`. We thus have justification for using 12 in our computations if we
wish to estimate means for the categories of `fundamental` and `female`
that are adjusted for years of education.

It might seem tempting—though tedious—to simply plug all the appropriate values into the equation and work out the predicted values yourself.
An alternative that is easier, faster, and more accurate is to use R's predict function after estimating the LRM to compute the predicted values
(see Chapter 4 for an example). Unfortunately, the presence of indicator
variables complicates this approach. When using `predict` function, we
need to employ a data frame that indicates what the indicator variables
represent.

---

[13] As an exercise, how would you determine if `educate` is a true confounder of the association between the `fundamentalist` and `pincome` variables? Re-read the definition of a confounder in Chapter 6 and determine how to do this.

R code
```
LRM7.4a <- data.frame(fundamentalist=c("notFund"),
          female=c("male", "female"), educate=12)
 # create a data frame specifying non-fundamentalist
   males and females who report educate = 12 years
predict(LRM7.4, LRM7.4a) # recall that LRM7.4 is the LRM
                          estimated earlier
LRM7.4b <- data.frame(fundamentalist=c("fund"),
          female=c("male", "female"), educate=12)
 # create a data frame specifying fundamentalist males
   and females who report educate = 12 years
predict(LRM7.4, LRM7.4b)
```

R output (abbreviated & annotated)
```
 1     2       # non-fundamentalist Christian
10.77 10.55

 1     2       # fundamentalist Christian
10.74 10.51
```

The predicted family income for males who report they are not fundamentalist Christians and have 12 years of education is approximately 10.77. For males who report they are fundamentalist Christians, predicted personal income is 10.74.[14] The corresponding predicted values for females are 10.55 and 10.51. Once education is considered, the difference between the groups is minimal.

Table 7.3 provides all four of the predicted income values, as well as those from a model that does not statistically adjust for education. To estimate relative differences between adjusted and unadjusted values, calculating the percentage difference is also helpful (this is easy to do since R serves well as

TABLE 7.3

Predicted Means of Family Income

Group	Means[a]	Means adjusted for education[b]
Non-fundamentalist males	11.20	10.77
Non-fundamentalist females	10.96	10.55
Fundamentalist males	10.91	10.74
Fundamentalist females	10.80	10.51

[a] R's aggregate function is used to compute the unadjusted means: aggregate(fincome ~ female + fundamentalist, GSS2018, mean)
[b] Years of education is set at 12 years for each group (see LRM7.4a and LRM7.4b).

---

[14] A straightforward method for computing these various predicted values is with the emmeans package. As an exercise, download this package and try to estimate the predicted values from the model. For Stata users, an R package called margins that is similar to Stata's margins command is available: it estimates marginal effects and predicted values.

a glorified calculator). Choosing, for example, males, Equation 7.6 indicates that adjusting for the association with education reduces the predicted difference in family income by about 90%.

$$\left[\frac{(11.20-10.91)-(10.77-10.74)}{(11.20-10.91)}\right] \times 100 = \left[\frac{0.29-0.03}{0.29}\right] \times 100 \tag{7.6}$$

$$= 90\%$$

In other words, the average difference in family income between fundamentalist and non-fundamentalist males decreases by about 90% once the means are adjusted for education. The main reason that family income differences appear between the two groups is because fundamentalist Christians tend to experience, on average, less formal education than others.[15]

Let's examine one more LRM with indicator variables. This time, we'll consider marital status and family income. Married people are likely to have higher incomes, on average, than single people, whether never married, divorced, or widowed. Families with a married couple have at least the potential to earn two incomes; in fact, about half of married couple families in the U.S. have two wage earners.[16] Therefore, let's estimate an LRM to assess the association between marital status and family income. Marital status in the *GSS2018* dataset is measured by the numeric variable `marital`, which includes four codes—1, 2, 3, and 4—that we'll represent with the following categories: `married`, `widowed`, `div.sep` (divorced or separated), and `never.marr` (never married). After creating a factor variable with `marital`, R uses `married` as the reference category.[17]

The regression coefficients in LRM7.5 support the hypothesis that married people report, on average, higher family income than others ($\bar{x} = 11.67$), although the model fit is poor, with an R^2 of 0.07. The average differences suggest that never married people report the lowest family incomes ($\bar{x} = 11.67 - 1.4 = 10.27$).

R code

```
GSS2018$marital <- factor(GSS2018$marital, levels =
                   c(1,2,3,4), labels = c("married",
                   "widowed", "div.sep", "never.marr"))
  # make marital a factor variable and identify its
    levels
```

---

[15] You may confirm this with a *t*-test: `t.test(educate ~ fundamentalist, data = GSS2018)`.

[16] Bureau of Labor Statistics (2019), *Employment Characteristics of Families – Summary*, Washington, DC: U.S. Department of Labor.

[17] Earlier, we learned that we may choose the specific reference category by using the `relevel` function after creating a factor variable. But `marital` is a numeric variable, so, when we create a factor variable, we give each category a label and R will, by default, use the first level as the reference category.

```
LRM7.5 <- lm(fincome ~ marital, data=GSS2018)
summary(LRM7.5)
confint(LRM7.5)
```

R output (abbreviated)
Coefficients:
```
                 Estimate Std. Error t value   Pr(>|t|)
(Intercept)      11.67482    0.06882  169.647   < 2e-16    ***
maritalwidowed   -0.97130    0.16758   -5.796   7.71e-09   ***
maritaldiv.sep   -0.84643    0.12074   -7.010   3.11e-12   ***
maritalnever.
marr             -1.39730    0.10837  -12.894   < 2e-16    ***
---
Signif. codes: 0 '***' 0.001 '**' 0.01 '*' 0.05 '.' 0.1
```
Residual standard error: 2.155 on 2311 degrees of freedom
Multiple R-squared: 0.07123, Adjusted R-squared: 0.07002
F-statistic: 59.08 on 3 and 2311 DF, p-value: < 2.2e-16
```
[CIs]                     2.5%    97.5%
(Intercept)             11.540   11.810
maritalwidowed          -1.300   -0.643
maritaldiv.sep          -1.083   -0.610
maritalnever.marr       -1.610   -1.185
```

Think some more about variables that are associated with marital status and income. We've seen, for instance, that education is strongly associated with income, but it also may be associated with marital status. Another pertinent variable is age. Is age related to income? The answer is yes since middle aged people earn more on average than younger people, whereas older people, especially in retirement, typically earn less. What about age's association with marital status? It makes sense to argue for an association, especially when the categories include never married and widowed. Never married people tend to be younger, on average, than married people and widowed people tend to be older. Given this information, adding education and age to the model is prudent.

The new model (LRM7.6) is identified as the *full model* to distinguish it from the previous *nested model* (see Chapter 6). It appears to provide a better fit to the data, with an R^2 of 0.14 (adjusted R^2 = 0.14). A nested F-test suggests that the increase in the R^2 from the nested to the full model meets a conventional threshold for statistical significance ($p < 0.05$). The age and education coefficients also have 95% CIs that do not include zero.

R code
```
LRM7.6 <- lm(fincome ~ marital + age + educate,
            data=GSS2018)
summary(LRM7.6)
confint(LRM7.6)
```

R output (abbreviated)
```
Coefficients:
                    Estimate Std. Error t value Pr(>|t|)
(Intercept)         9.188016  0.263532   34.865 < 2e-16  ***
maritalwidowed     -0.612347  0.174716   -3.505 0.000466 ***
maritaldiv.sep     -0.691806  0.117229   -5.901 4.14e-09 ***
maritalnever.marr  -1.405579  0.114050  -12.324 < 2e-16  ***
age                -0.006811  0.002963   -2.298 0.021624 *
educate             0.200951  0.014551   13.810 < 2e-16  ***
---
Signif. codes: 0 '***' 0.001 '**' 0.01 '*' 0.05 '.' 0.1

Residual standard error: 2.07 on 2309 degrees of freedom
Multiple R-squared: 0.1442, Adjusted R-squared: 0.1423
F-statistic: 77.78 on 5 and 2309 DF, p-value: < 2.2e-16
```
```
[CIs]                     2.5%    97.5%
(Intercept)              8.671    9.705
maritalwidowed          -0.955   -0.270
maritaldiv.sep          -0.922   -0.462
maritalnever.marr       -1.629   -1.182
age                     -0.013   -0.001
educate                  0.172    0.230
```

We may have reasonably suspected that the differences by marital status change after adding age and education to the model, but actually the size of the coefficients does not shift much. For example, the never married coefficient stays roughly the same: –1.4. The others change slightly, though. We should conclude, however, that the differences in family income associated with marital status are not accounted for (nor confounded) (at least by much) by age or education.

Thus far, we've seen examples of unordered categorical (nominal) explanatory variables, but LRMs may use ordered (ordinal) explanatory variables in a similar manner. Some researchers tacitly treat ordered variables as continuous by entering them into an LRM as numeric, but this must be done with care since the interpretation of the slope coefficients is often ambiguous. Recall that, unlike continuous variables, the "distances" between ordered categories differ qualitatively and thus may not be consistent, so claiming a "one-unit" difference or increase in its presumed association with an outcome variable is fraught with interpretational limitations. If the variable is limited to only a few categories, the prudent approach is to treat them as a set of indicator variables. For ordered variables with a large number of categories, however, the analyst may simply need to accept the assumption that the "distances" are equivalent and assent to the uncertainty in interpretation.[18]

Let's consider one more issue before concluding this chapter. As discussed in Chapter 4, some researchers use standardized coefficients to compare associations within a model. For instance, since education and age are measured in different units, trying to compare their association with an outcome variable such as family income in a multiple LRM is challenging. By using the standardized coefficients, the argument goes, education and age may be compared in standard deviation units. There may be some merit to this approach, but it fares poorly when considering indicator variables. A one standard deviation unit shift in an indicator variable makes no sense (unless it equals exactly one, which is rarely feasible). Indicator variables shift only from zero to one or one to zero (or, in R, from one character string to another). Interpreting the standardized coefficients associated with indicator variables is thus not appropriate. For example, if we change it to a numeric variable, `female` has a mean of 0.45 and a standard deviation of 0.50. The standardized coefficient from the first LRM in the chapter (LRM7.1) is 0.15.[19] So the mindless interpretation of this coefficient is "a one standard deviation unit increase in gender is associated with a 0.15 standard deviation increase in family income." A one standard deviation unit increase in `female` shifts, for instance, from 0 to 0.50. This is a nonsensical value; a `female` value of 0.50 does not exist, nor can it in the *GSS2018* dataset.[20]

---

[18] Richard Williams (2020) reviews several options when utilizing ordered explanatory variables, including diagnostic tests and promising options using sheaf coefficients that aid interpretation ("Ordinal Independent Variables," available at https://www3.nd.edu/~rwilliam/xsoc73994/OrdinalIndependent.pdf).

[19] As in Chapter 4, we may use the `lm.beta` package to compute standardized regression coefficients:

```
library(lm.beta)
LRM7.7 <- lm(pincome ~ female, data=GSS2018)
LRM7.7.beta <- lm.beta(LRM7.7)
print(LRM7.7.beta)
```

[20] If one is keen to use standardized coefficients, one might make the case that, since the standard deviation of `female` is 0.5, a practical interpretation is "the expected difference in personal income between females and males is about 0.30 (0.15 × 2) standard deviations or z-scores." The following provides R code that estimates such as model by, first, using the `scale` function to create a personal income variable measured in z-scores and then estimating an LRM:

```
GSS2018$z.pincome <- scale(GSS2018$pincome, center = TRUE, scale = TRUE)
summary(lm(z.pincome ~ female, data=GSS2018))
```

This is similar to an approach advocated by Andrew Gelman (2008): place continuous explanatory variables on a common, standardized scale by dividing each by two times its standard deviation. This allows a rough comparison of the effects of continuous *and* indicator variables, as long as the latter type of variable's distribution is not too far from a 50–50 split ("Scaling Regression Inputs by Dividing by Two Standard Deviations," *Statistics in Medicine* 27: 2865–2873). The `arm` package's `standardize` function is designed for this approach.

Chapter Summary

Categorical variables are common in the social, behavioral, and other sciences. We often measure things that are not continuous such as ethnic status, group affiliation, and educational milestones. But they can present some critical issues when used as explanatory variables. This chapter outlines how to use them as indicator variables in LRMs.

In R indicator variables are often measured as character strings, though they can also be numeric variables.[21] Keep in mind an important issue, though: when a categorical variable is represented by a set of indicator variables, like the marital status or race/ethnicity variables used earlier, it's important to choose one of them as a reference or comparison group and leave it out of the model. Preferably the choice is based on theoretical concerns, although many researchers simply use the modal category.

LRMs can easily accommodate indicator variables and continuous variables. This allows the analyst to compute means for groups that are adjusted for the effects of continuous variables such as age or years of education. For instance, LRM7.4 demonstrates that the association between fundamentalist religious affiliation and family income is affected by differences in years of formal education.

Chapter Exercises

The dataset called *Opinion.csv* consists of information from a survey of adults in the U.S., with an oversample of Latinx people. Our objective is to practice estimating and interpreting the results from LRMs that include categorical explanatory variables. The variables in the dataset are

- `id`   Respondent identification number
- `female`   0 = male; 1 = female
- `educate`   Years of formal education (ranges 7–20)
- `income`   Household income in $10,000s (ranges 1–19)
- `immigrant`   "Are you an immigrant to the U.S.?" (0 = no; 1 = yes)

---

[21] The most common coding strategy for mutually exclusive indicator variables is {0, 1}. Using other numbers, such as {1, 2}, is not a good idea unless the analyst is using effects coding or some similar strategy. For a description of alternative coding techniques, see Richard B. Darlington and Andrew F. Hayes (2016), *Regression Analysis and Linear Models: Concepts, Applications, and Implementation*, New York: Guilford Press, chapter 9.

- `SameOpinions`   "I prefer to live where people share my political opinions" (0 = no; 1 = yes)

- `CandidateSame`   "It is important that my congressperson shares my religion and ethnicity" (ranges 3.8–56.1 with higher values indicating stronger agreement with this view)

- Race/ethnicity indicator variables—White is the reference category[22]
    - `AfricanAmerican` (0 = no; 1 = yes)
    - `Latinx` (0 = no; 1 = yes)
    - `OtherEthnic` (0 = no; 1 = yes)

After importing the dataset into R, complete the following exercises.

1. How many of the sample members fall into each of the four racial/ethnic categories represented in the dataset? Check the overall sample size of the dataset to ensure that the four categories identify all the sample members (recall that indicator variables should be exhaustive).

2. Compute the means of the variable `CandidateSame` for males and for females. Compute the means of the variable `CandidateSame` for non-immigrants and for immigrants.

3. Estimate a *t*-test that compares the means of `CandidateSame` for non-immigrants and immigrants. What does the *t*-test suggest about the difference in the means?

4. Estimate an LRM with `CandidateSame` as the outcome variable and the following explanatory variables: `female`, `educate`, `immigrant`, `AfricanAmerican`, `Latinx`, and `OtherEthnic`.

    a. Interpret the slope coefficients associated with `female`, `AfricanAmerican`, `Latinx`, and `educate`.

    b. Interpret the *p*-value and 95% confidence interval associated with the slope coefficient for `immigrant`.

5. Estimate an LRM with `CandidateSame` as the outcome variable and the following explanatory variables: `female`, `educate`, `immigrant`, `AfricanAmerican`, `Latinx`, `OtherEthnic`, `income`, and `SameOpinions`. Interpret the R^2 and the *p*-value for the *F*-statistic from the model.

6. Estimate a multiple partial (nested) *F*-test that compares the LRMs in exercises 4 and 5.

---

[22] This variable does not appear in the dataset because once we know that an observation has a zero for the other three race/ethnicity indicator variables, we know for certain it represents a White person.

a. What does the *F*-test suggest about the two models?

b. Compute the AICs and BICs for the two models. What do they suggest about the two models? Is the evidence they provide consistent or inconsistent with the evidence provided by the *F*-test? Why or why not?

7. *Challenge*: estimate LRMs to compute predicted means of CandidateSame for African American, Latinx, Other Ethnic, and White members of the sample:

a. Without adjusting for years of formal education.

b. When years of formal education are set at 12 years.

c. When years of formal education are set at 16 years.

Compute the percentage differences for each group based on whether they report 12 or 16 years of formal education. Which group's predicted mean differs by the greatest amount as formal education shifts from 12 to 16 years?

8

Independence

We've now learned how to estimate and interpret the results of LRMs, how to assess and compare the fit of models, and how to include categorical variables as explanatory variables. As discussed in Chapters 3 and 4, a set of assumptions should also be satisfied if we are to gain clear and beneficial information from LRMs. This chapter and several hereafter provide details about these assumptions, including how to judge whether they are satisfied and what to do if they are not.

To begin our examination of these issues, recall that the independence assumption states that the errors in predicting Y with the Xs are independent of one another, which may be represented as $\text{cov}(\varepsilon_i, \varepsilon_j) = 0$. Rather than considering the errors, though, we'll start with a discussion of the observations in a dataset.

A common reason the errors of prediction are not independent is that the observations are not independent. Think about how the samples that yield the data with which to estimate LRMs are collected. A simple random sample (SRS), the main workhorse of statistics, is designed so the observations—whether people, chipmunks, heads of lettuce, or bacteriophages—are sampled independently of one another. This normally leads to observations that do not influence others. But researchers, rather than collecting an SRS, often sample students from the same schools, companies from the same city, chipmunks from the same forest habitat, or bacteriophages from the same sewage treatment plant.

Many samples are also selected in stages, such as by choosing a sample of states first, then school districts, then schools, and then students (see Figure 8.1). This creates *nesting* or *clustering* among the observations.[1] In Chapter 15, we address models that are based on nested data, but take the nesting structure explicitly into account. In this context, the data are also called *multilevel* or *hierarchical* (see Chapter 15 for more information).

When samples are collected across spatial units or time, observations may also not be independent, even in populations, since those nearer to one another in space or time tend to be more closely related in some way than

---

[1] One of the earliest examples of clustered or multistage survey sampling was by the statisticians Corrado Gini and Luigi Galvani when they administered the 1921 Italian General Census, although they may have adopted their methods from Russian and Polish scientists (Jerzy Neyman (1934), "On the Two Different Aspects of the Representative Method: The Method of Stratified Sampling and the Method of Purposive Selection," *Journal of the Royal Statistical Society* 97(4): 558–625).

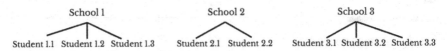

FIGURE 8.1
Multilevel data example.

those that are farther apart. For example, counties or provinces that share a border tend to be more alike than counties or provinces that are far away from one another, which suggests a lack of independence.

A key problem is that when we attempt to predict an outcome with one or more explanatory variables, yet the observations are not independent, the errors of prediction are often dependent in some manner. The independence assumption is, therefore, not met. When this occurs, the standard error formula does not lead to the correct results and the estimated standard errors are (typically) *biased downward*: smaller than they would have been under independence. We may thus be misled into thinking that regression coefficients are more precise than they actually are in the absence of dependence.

A helpful way of thinking about why a lack of independence affects standard errors is to realize, under an SRS strategy, that each observation is assumed to provide an equal amount of information to the computation of a statistic. For instance, recall that the formula for the standard error of the mean ($\bar{x} / \sqrt{n}$) or slope coefficient includes the sample size such that larger samples yield smaller standard errors. When relying on these formulas, we assume that each n contributes the same amount of "information" about the variables of interest. But suppose we conduct a research project that asks college students in a sample to judge the quality of their statistics teachers and we wish to infer something about the mean quality of teachers in the target population. If some students have the same statistics teacher, they will likely provide similar, and thus partial, information about their teacher based on qualities that the project does not measure. But we assume that they are providing the same amount of information as others who provide data on unique teachers. We think that each student contributes one "unit" of information—that the n increases by one—when actually those who share a teacher might provide only partial information. Therefore, the actual n— in terms of how much information is available—used in the standard error formula is inflated. When the errors of prediction are not independent, not as much information is available and the standard errors tend to be smaller relative to those under independence. Fortunately, as shown later in the chapter, adjustments to standard errors that take into account dependence are available.

The independence assumption also has a couple of consequences that are elaborated in Chapters 12 and 13. This involves whether an association exists between the errors of prediction and the explanatory variables (Xs). In classical regression, the Xs are fixed or under the control of the researcher (the technically proficient say the Xs are *nonstochastic*). Most social science applications rely on observational data, however, so Xs are not fixed, but rather are random (*stochastic*) and cannot be manipulated. This implies, for instance, that they can take on different values from sample to sample and this variation is not under the control of the researcher. We can therefore only assume that the Xs and the errors of prediction are also independent: $\text{cov}(\mathbf{X}, \varepsilon_i) = 0$. When they are associated, though, we may be left with inefficient, or even biased, estimates of the LRM slope coefficients. This can occur when we have omitted important explanatory variables from the model (see Chapter 12) or when we don't have accurate measures of the variables (see Chapter 13).

After discussing one way to determine and adjust for the dependence of the errors of prediction, this chapter reviews a couple of issues involving (a) data collected over time, such as on the same units over several months or years, and (b) data collected across spatial units, such as schools or cities. The first type of data is called *longitudinal*, *panel*, or *time-series data*, whereas the second is known as *spatial data*.

Determining Dependence

Determining whether the observations are dependent in some way is often a simple task because, based on the nature of the data collection, observations are clearly either dependent or independent. As implied earlier, observations in a sample are independent if collected using an SRS.[2] For example, suppose we wish to sample adults in the U.S. regarding their political preferences. If we had some way of listing all adults in the U.S., we could then randomly select a sample of them and, in all likelihood, they would comprise a set of independent observations.

However, many large surveys collect data in stages, thus leading to some dependence. A common approach is to use clustering when samples are drawn. This might involve choosing a random sample of neighborhoods first, followed by a random selection of adults from each neighborhood. In this case, the individuals are no longer independent, but may share characteristics—many of which are unobservable or not measured and thus part of the error term—because they live in the same area.

But how much of this dependence should concern us? Though no strict guidelines exist, methods are available to determine the effect of clustering on the estimated standard errors. The most common approach is to

---

[2] As with any random process, dependence of observations is possible, but less likely with an SRS than with other sampling schemes.

examine the ratio of standard errors—those that assume independence and those that assume dependence, which is the general basis for the *design effect*. Designated d^2, it is defined as ratio of the variance of a statistic computed under the sample design (e.g., clustered design) divided by the variance of a statistic assuming an SRS. Thus, each statistic, such as an arithmetic mean or a regression coefficient, has its own design effect. A design effect of two or more usually indicates that the clustering is substantial enough to warrant some type of statistical adjustment. Important exceptions to this general rule exist, however, such as when the number of clusters is relatively small.[3]

Let's consider an example to understand how to compute and evaluate d^2. The dataset *MultiLevel.csv* includes information from samples of individuals who live in several U.S. towns. The individuals are represented by an individual-level identification variable (id), with each town identified by the idcomm variable. The simplest way to determine the design effects for various statistics is to use the R package survey. This package allows the analysis of various forms of survey data, some of which can be complex. The data used here are relatively simple, though, collected using a two-stage cluster design.

Browse the *MultiLevel* dataset in R or another program to get a sense of how the data are organized. Notice, for example, the multiple ids for each idcomm and that some of the variables are constant within idcomm, such as pop2000 (the population of each town).

Next, after activating the survey package, we need to create an object with the svydesign function—clus _ multilevel in the following example— that includes information about the organization of the data. The documentation for the survey package provides more information about options for creating an object. Since we have a cluster design, use the following R code:

R code
```
library(survey) # load the package into memory
library(psych) # we'll use this package to get relevant
                statistics from the original dataset
clus_multilevel <- svydesign(ids = ~idcomm, strata =
                NULL, fpc = NULL, data = MultiLevel)
                # idcomm is the cluster variable;
                don't worry about R's warning
                concerning the weights
summary(clus_multilevel) # provides information about
                the sample
```

The survey package has a host of functions, including those for descriptive statistics. The summary function informs us that the dataset includes 99

---

[3] Specific details about different types of sampling strategies and analysis techniques for each are provided in Robert Groves et al. (2009), *Survey Methodology*, 2nd Ed., New York: Wiley.

clusters or towns. Before estimating an LRM, let's examine some information about one of the variables in the dataset, commlength, which measures the length of time respondents have lived in their towns. We'll examine the mean and standard error of this variable with and without correction for the clustering (and potential nonindependence of errors) of observations.

R code

```
svymean(~commlength, design = clus_multilevel)
  # compute the mean and standard error adjusted for
    clustering
describe(MultiLevel$commlength) # information from the
                                  original dataset not
                                  adjusted for
                                  clustering
```

R output (abbreviated)

```
           mean   SE
commlength 33.26  0.53 # adjusted for clustering

  n        mean   se
9859       33.26  0.23 # not adjusted for clustering
```

The estimated mean is identical using the two methods. But examine the standard errors. What is the design effect based on their distinct values? Keep in mind that the design effect is the ratio of the variances, yet R provides standard errors. But, by taking the ratio of the two standard errors that R provides, we can compute the *root design effect* (often abbreviated mysteriously as DEFT) and then square it to get the design effect. Thus, the DEFT is $0.53/0.23 = 2.3$, and the design effect (d^2) is $2.3^2 = 5.29$.[4] Since d^2 is greater than two, we may infer that the clustering is sizeable enough to warrant some type of statistical adjustment.

Example of Adjustment for Clustering

Lack of independence is often just a nuisance based on the way the sample is collected. But we would still like to have correct standard errors from an LRM that uses the sample. In this situation, the easiest correction is to (a) tell R that the data are based on a clustered sampling design (or some other complex design), as shown in the last section or (b) use a R package's function

---

[4] A simpler way to compute the design effect is with the deff function in the Hmisc package:

```
deff(MultiLevel$commlength, MultiLevel$idcomm)
```

R returns a design effect of 5.62, slightly different from ours due to rounding error.

formulated to adjust standard errors for clustering. For instance, as we'll learn in Chapter 15, the R package lme4 is designed to estimate regression models with clustered data, though it utilizes a different estimation technique than OLS.

As a simple example of adjusting for clustering, we'll use, once again, the *MultiLevel* dataset. Suppose we're interested in predicting the variable income with the variable male, but also wish to consider whether respondents are married. First, what does an LRM estimated with OLS that ignores the clustering by towns provide (see LRM8.1)?

LRM with No Adjustment for Clustering

R code[5]
```
LRM8.1 <- lm(income ~ male + married, data =
             MultiLevel)
summary(LRM8.1)
confint(LRM8.1)
```

R output (abbreviated)
```
Coefficients:
            Estimate Std. Error t value Pr(>|t|)
(Intercept) 2.93097  0.03635    80.63   <2e-16 ***
malemale    0.51873  0.03942    13.16   <2e-16 ***
marriedyes  1.74758  0.04222    41.39   <2e-16 ***
---
Signif. codes:  0 '***' 0.001 '**' 0.01 '*' 0.05 '.' 0.1

[CIs]          2.5%   97.5%
(Intercept)   2.860   3.002
malemale      0.441   0.596
marriedyes    1.665   1.830
```

LRM8.1 suggests that males and married people tend to report higher incomes than females or people who are not married. But the standard errors are probably not accurate—likely too small or biased downward—because of the nonindependence of observations (and the consequent, in all likelihood, nonindependence of the errors). Let's use the survey package to adjust the standard errors for the possible dependence due to sample members being clustered by town. The package has a built-in

---

[5] If the psych package is still active, you may need to enter the R function detach("package:psych", unload=TRUE) before executing these functions. The psych package might "mask" the income variable in the *MultiLevel* dataset. This means that because the psych package includes a function or object called income, any similarly named variable in an active dataset will be suppressed.

function, svyglm, for estimating LRMs, as shown in the next example (see LRM8.2).

LRM That Adjusts for Clustering

R code
```
LRM8.2 <- svyglm(income ~ male + married, design =
        clus_multilevel)
            # glm is short for generalized linear model,
                of which linear regression is one type and
                the default for the svyglm function
summary(LRM8.2)
confint(LRM8.2)
```

R output (abbreviated)
```
Survey design:
svydesign(ids = ~idcomm, strata = NULL, fpc = NULL, data
        = MultiLevel)

Coefficients:
             Estimate Std. Error t value   Pr(>|t|)
(Intercept) 2.93097    0.04062     72.16    <2e-16 ***
malemale     0.51873    0.04207     12.33    <2e-16 ***
marriedyes   1.74758    0.04416     39.57    <2e-16 ***
---
Signif. codes: 0 '***' 0.001 '**' 0.01 '*' 0.05 '.' 0.1

(Dispersion parameter for gaussian family taken to be
3.643145)

[CIs]          2.5%   97.5%
(Intercept)    2.851  3.011
malemale       0.436  0.601
marriedyes     1.661  1.834
```

Compare the standard errors from LRM8.1 and LRM8.2: those from the latter model are slightly larger than those from the former model. The confidence intervals of the male and married coefficients are also wider. This is a consequence of the lack of independence in these data. When adjusted for dependence, the standard errors are suitably larger. They do not have to be, however. In rare situations, models have negative dependence of errors and the standard errors are too large in the standard LRM. We won't discuss this issue because it happens so infrequently. Chapter 15 furnishes additional examples of how to examine dependence in the *MultiLevel* dataset.

Serial Correlation

In addition to data based on clustered or multistage sample designs, two other common situations arise in which the errors of prediction do not tend to be independent: (a) when data are collected over time and (b) when data are collected across space. Data collected over time come in several types. First, *repeated cross-sectional data* are those from different samples, but that are repeated. An example is the General Social Survey (GSS; we've used its data in previous chapters). About every two years, a representative sample of adults in the U.S. is drawn by the research organization, NORC, and sample members are asked questions about various issues. Even though each sample is made up of different people, the data are useful for tracking aggregate-level changes in attitudes and behaviors in the U.S. One issue that can be explored with GSS data, for example, is whether public support for the death penalty has shifted in the past several decades.

Second, data collected from the same individuals (these can be people, animals, or other individual units) over time are known as *longitudinal* or *panel data*. Third, data collected on the same aggregated unit (e.g., a city or a stock market) over time are called *time-series data*. Time-series data are often limited to one unit, such as the city of Detroit or the NASDAQ, whereas data collected on several aggregate units over time are termed *cross-sectional time-series data* (e.g., data from 50 U.S. cities, 2000–2020).

One way to think about a lack of independence is to consider the nature of the errors of prediction in data collected over time: errors from the time periods that are closer together typically have stronger relationships than those farther apart. For example, if we collect information on crime rates in New York City over a 25-year period and try to predict them based on unemployment rates over the same period, the errors of prediction are probably more alike in 1990 and 1991 than in 1990 and 2010. The errors are, therefore, correlated differentially depending on time. Another name for this form of dependence is *serial correlation*.

As with other forms of dependence, the main consequence of serial correlation is incorrect standard errors. The slopes, on average, are still unbiased, but the standard errors tend to be too small and, consequently, the *p*-values are smaller and the CIs narrower. When judging the results of a model using a classic inferential approach, researchers may thus be misled into thinking that a slope coefficient is "statistically significant" or otherwise notable.

The scatter plot displayed in Figure 8.2 illustrates a classic example of serial correlation. Notice the snaking pattern of the residuals around the linear fit line. This is the typical appearance of serial correlation and is the consequence of the stronger association among errors closer together than among those farther apart in time. Of course, such a pattern is rarely so apparent; a more common situation, especially with large samples, is to find an unrecognizable pattern to the residuals. We thus need additional tools to determine if serial correlation is an issue for an LRM.

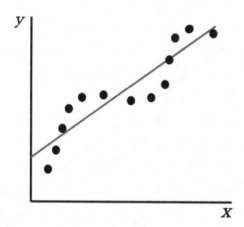

FIGURE 8.2
Scatter plot representing serial correlation.

TABLE 8.1

First Few Rows of the Colombia Dataset

Index	Year	HomicideRate	Unemploy	Poverty
1	2000	66.5	20.5	53.7
2	2001	68.6	40.3	60.5
3	2002	68.9	40.9	53.2
4	2003	53.8	41.5	50.9
5	2004	44.8	42.1	50.8
6	2005	39.6	42.6	46.7

Let's observe an example of serial correlation. The dataset, *Colombia2000 _16.csv*, includes information from the nation of Colombia on homicide rates, unemployment, poverty, and other issues. Table 8.1 shows a few rows and columns from the dataset, which illustrate that it represents a classic time series. Examine homicides over time and you'll notice a decreasing trend. But let's also look at a scatter plot of the homicide rate (y) by year (x) to see if we can uncover evidence of serial correlation. What do you see in Figure 8.3? Does it look similar to Figure 8.2?

R code for Figure 8.3
```
head(Colombia2000_16)
plot(homicideRate ~ year, data=Colombia2000_16,
     xlab="Year", ylab="Homicides per 100,000")
lines(homicideRate ~ year, data=Colombia2000_16,
      type="l")
abline(lm(homicideRate ~ year, data=Colombia2000_16),
      col="red")))
```

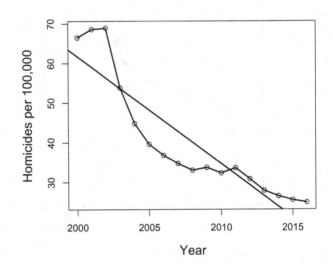

FIGURE 8.3
Scatter plot of homicides deaths in Colombia by year with linear fit line to assess serial correlation.

Suppose we wish to use one of the variables in the dataset to predict the number of homicides. A common question among criminologists is whether poverty is part of a social environment where homicides are more likely.[6] We'll thus consider whether the percent of people who live in poverty predicts the number of homicides per 100,000 population. What does LRM8.3 suggest?

Linear Regression Model

R code
```
LRM8.3 <- lm(homicideRate ~ poverty, data =
              Colombia2000_16)
summary(LRM8.3)
confint(LRM8.3)
```

R output (abbreviated)
```
Coefficients:
              Estimate  Std. Error  t value  Pr(>|t|)
(Intercept) -18.9556     7.1415     -2.654   0.018     *
poverty       1.4079     0.1661      8.479   4.18e-07 ***
---
Signif. codes: 0 '***' 0.001 '**' 0.01 '*' 0.05 '.' 0.1
```

---

[6] See, for example, William A. Pridemore (2011), "Poverty Matters: A Reassessment of the Inequality-Homicide Relationship in Cross-National Studies," *British Journal of Criminology* 51(5): 739–772.

```
Residual standard error: 6.479 on 15 degrees of freedom
Multiple R-squared: 0.8274, Adjusted R-squared: 0.8159
F-statistic: 71.89 on 1 and 15 DF, p-value: 4.178e-07

[CIs]                2.5%     97.5%
(Intercept)     -34.177    -3.734
poverty           1.054     1.762
```

The results of LRM8.3 indicate that poverty is positively associated with the number of homicides that Colombia experienced from 2000 to 2016 (β = 1.41; 95% CI: {1.05, 1.76}). But we also suspect that serial correlation is a problem. How should we investigate this issue? A plot of the residuals by the predicted values is a useful diagnostic tool. Recall from Chapter 4 that the residuals are computed as $(y_i - \hat{y}_i)$, with the predicted values represented by $\hat{y}_i$. The predicted values are based on the LRM: $\hat{y}_i = \alpha + \beta_1 x_1 + \beta_2 x_2 + \ldots$. Plug in particular values of the xs and slope coefficients to get predicted values for each observation (recall that R saves the predicted (fitted) values and the residuals as part of the lm object created when we save an LRM; look under the Global Environment window in RStudio).

After estimating LRM8.3, we'll use a function from the R package olsrr to create a graph that plots the residuals—in this case, the *studentized deleted residuals*[7]—by the predicted values. The horizontal line at which the residuals equal zero provides a reference that allows patterns to be detected more easily.

R code for Figure 8.4
```
library(olsrr)
ols_plot_resid_stud_fit(LRM8.3)
```

Does Figure 8.4 furnish evidence of serial correlation? It may be difficult to tell based on the graph, but begin on the left side and track to the right. Consider where the points fall relative to the horizontal line at residuals = 0. Notice how the residuals decrease and then, after the predicted values reach about 50, begin to increase, with the last four observations representing positive values. A pattern such as this signifies serial correlation.

Looking for patterns among the residuals and predicted values can be challenging, especially with large datasets. Fortunately, some numeric tests are available for determining serial correlation. The most common is called

---

[7] A problem that can occur in computing residuals is when the data include extreme values that have an undue influence on the regression line or surface. To compensate for this potential issue, deleted residuals are calculated based on an LRM that removes observations one at a time, each time re-estimating the regression model on the other $n - 1$ observations. The deleted residuals are then standardized based on Student's *t* distribution, hence the term studentized deleted residuals. Many experts recommend their use, especially because LRMs can be influenced quite a bit by extreme values (see Fox (2016), *op. cit.*, chapter 11). Chapter 14 provides additional details about these residuals and extreme values.

Studentized Residuals vs. Predicted Values

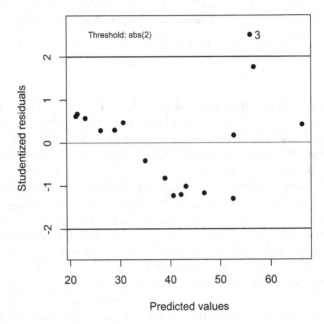

FIGURE 8.4
Scatter plot of studentized residuals by predicted values to assess serial correlation from LRM8.3.

the *Durbin–Watson test*,[8] which is calculated by considering the squared difference of the residuals adjacent in time relative to the sum of the squared residuals (see Equation 8.1).

$$d = \frac{\sum_{2}^{n} \left(\hat{\varepsilon}_{ti} - \hat{\varepsilon}_{ti-1} \right)^{2}}{\sum_{1}^{n} \hat{\varepsilon}_{ti}^{2}} \tag{8.1}$$

The t subscript indicates time. One commonly used criterion is that the closer the d value is to two, the less likely the model is affected by serial correlation.[9] The theoretical limits of d are zero and four, with values closer to zero indicating positive serial correlation (the most common type) and values closer to four indicating negative serial correlation (a rare situation). To use the Durbin–Watson test, first compute the d value and then compare it to the values from special tables of d values available in many regression textbooks or online. Some statistical programs provide values from these tables as part of their output. The values in the table include an upper limit and a lower

---

[8] The test was introduced in James Durbin and Geoffrey S. Watson (1971), "Testing for Serial Correlation in Least Squares Regression. III," *Biometrika* 58(1): 1–19.
[9] Don't confuse this d value with the d^2 used earlier in the chapter to designate the design effect.

limit based on the number of coefficients (explanatory variables + 1 or $\{k + 1\}$) and the sample size (n). R also provides a significance test, but one concerned with whether the autocorrelation statistic (not d) is greater than zero. In any event, the following criteria apply to the use of the d values from an LRM:

If $d_{model} < d_{lower\ limit}$, then positive serial correlation is present

If $d_{model} > d_{upper\ limit}$, then no positive serial correlation is present

If $d_{upper\ limit} < d_{model} < d_{lower\ limit}$, then the test is inconclusive

Returning to the LRM example, we may ask R for the d statistic by first activating the car package and then using the durbinWatsonTest function, in which the lm function may be embedded.

R code
```
library(car)
durbinWatsonTest(lm(homicideRate ~ poverty, data =
                Colombia2000_16))
```

R output (abbreviated)
```
lag Autocorrelation D-W Statistic   p-value
1         0.564            0.696     0 [< 0.001]
Alternative hypothesis: rho != 0
```

The d (D-W Statistic) value is 0.70. The Autocorrelation term is also listed in the output, with a value of 0.56.[10] The significance test is based on the null hypothesis that the implied (auto)correlation in the errors from one year to the next (e.g., 2001 → 2002) is zero. The test suggests, however, that the model is probably affected by positively correlated errors in the presumed population. We should still verify this by checking a table of Durbin–Watson values to determine if d falls outside the boundaries for $\{k + 1\} = 2$ and $n = 16$. A table found online furnishes the following boundaries for a 5% test: $\{0.74, 1.25\}$. The results of the poverty–homicide LRM thus provide evidence of serial correlation among the residuals ($d_{model}(0.70) < d_{lower\ limit}(0.74)$).

Solutions for Serial Correlation

In addition to representing that the errors of prediction are not independent, serially correlated errors may also be a symptom of incorrect model specification (see Chapter 12): an important variable may have been left out

---

[10] This statistic is computed as

$$\rho = \frac{cov\left(\hat{\varepsilon}_t, \hat{\varepsilon}_{t-1}\right)}{\sqrt{var\left(\hat{\varepsilon}_t\right) \times var\left(\hat{\varepsilon}_{t-1}\right)}}$$

of the model. The most obvious possibility is a time variable. What happens in the Colombia homicide example when we add year to the model? The *d* value decreases (*d* = 0.54), which provides little evidence that we've attenuated serial correlation. A scatter plot of the studentized deleted residuals by the predicted values also suggests that a problem remains.

The best approach to minimizing the effects of serial correlation is to use a statistical technique designed for data collected over time. The methods of time-series analysis are suitable but are beyond the scope of this presentation.[11] Among the several useful R packages that estimate time-series models is forecast. But the background to understand and utilize these models is sizeable and is not presented here.

Two popular regression-based approaches are aimed at diminishing the effects of serial correlation: Prais–Winsten regression and Cochran–Orcutt regression.[12] Both models use a lagged period of the data (*t* − 1) to come up with better estimates of the coefficients. Let's see an example of a Prais–Winsten regression model using the prais package (LRM8.5) and compare it to the original LRM estimated with OLS (LRM8.4).

Linear Regression Model (OLS)

R code
```
LRM8.4 <- lm(homicideRate ~ poverty, data =
             Colombia2000_16)
summary(LRM8.4)
confint(LRM8.4)
```

R output (abbreviated)
```
Coefficients:
            Estimate  Std. Error  t value  Pr(>|t|)
(Intercept) -18.9556     7.1415    -2.654    0.018    *
poverty       1.4079     0.1661     8.479   4.18e-07 ***
---
Signif. codes: 0 '***' 0.001 '**' 0.01 '*' 0.05 '.' 0.1

Residual standard error: 6.479 on 15 degrees of freedom
Multiple R-squared: 0.8274, Adjusted R-squared: 0.8159
F-statistic: 71.89 on 1 and 15 DF, p-value: 4.178e-07
```

[11] A helpful guide with many examples is Robert H. Shumway and David S. Stoffer (2010), *Time Series Analysis and Its Applications: With R Examples*, New York: Springer.

[12] The Prais–Winsten method was introduced in Sigbert J. Prais and Christopher B. Winsten (1954), "Trend Estimators and Serial Correlation," Chicago: Cowles Commission Discussion Paper. The Cochran–Orcutt approach was presented in Donald Cochrane and Guy H. Orcutt (1949), "Application of Least Squares Regression to Relationships Containing Auto-Correlated Error Terms," *Journal of the American Statistical Association* 44(245): 32–61.

```
[CIs]              2.5%    97.5%
(Intercept)     -34.177   -3.734
poverty           1.054    1.762
```

Prais–Winsten Regression Model

R code
```
library(prais)
LRM8.5 <- prais_winsten(homicideRate ~ poverty, data =
        Colombia2000_16)
summary(LRM8.5)
```

R output (abbreviated)
```
AR(1) coefficient rho after 13 Iterations: 0.6351
Coefficients:
              Estimate Std. Error t value  Pr(>|t|)
(Intercept) -10.1427    11.1090    -0.913   0.375674
poverty       1.2247     0.2558     4.789   0.000239 ***
---
Signif. codes: 0 '***' 0.001 '**' 0.01 '*' 0.05 '.' 0.1

Residual standard error: 4.976 on 15 degrees of freedom
Multiple R-squared: 0.8019, Adjusted R-squared: 0.7887
F-statistic: 60.73 on 1 and 15 DF, p-value: 1.187e-06

Durbin-Watson statistic (original): 0.6957
Durbin-Watson statistic (transformed): 1.8739
```

The Prais–Winsten model seems to have diminished the effects of serial correlation. The d statistic (transformed) is 1.87, above the upper bound of 1.25, thus indicating that positive serial correlation is no longer an issue. We also continue to find a statistically significant association between poverty and homicides per 100,000 in Colombia. Yet the slope is slightly smaller and the standard error is larger in the new model.[13] The R^2 is also smaller in the Prais–Winsten model (0.80 vs. 0.83). A question remains, however: why is the association between poverty and homicides positive?

---

[13] Although, as noted earlier, serial correlation primarily affects the standard errors of slope coefficients, the Prais–Winsten model results in different values because (a) it discards the first observation in the time series and (b) it is not estimated with OLS, but rather with a different estimation routine.

Generalized Estimating Equations for Longitudinal Data

Some useful techniques for other types of data collected over time, in particular, longitudinal or panel data, are also available. A set of techniques widely used by the research community, for example, is known as *generalized estimating equations* (GEEs). A limitation of the Cochran–Orcutt and Prais–Winsten regression approaches is that, without adjustment, they are limited to models with a *first-order autoregressive process*, or AR(1). This assumes that the immediately previous residuals have a much stronger direct association with the current residuals (e.g., 2010 → 2011) than other preceding residuals. The matrix in Equation 8.2 depicts this assumption, with the symbol ρ *(rho)* used for correlation coefficient for each pair of years, $t - t + 2$. Recall that correlations fall between zero and one.

$$\begin{bmatrix} 1 & \rho & \rho^2 \\ \rho & 1 & \rho \\ \rho^2 & \rho & 1 \end{bmatrix} \tag{8.2}$$

But other previous residuals (e.g., 2008 and 2009) may also have a strong effect on the current residuals. When the correlation structure of the errors is not specified correctly, the standard errors might still be biased. GEEs offer substantial flexibility to model various types of correlation structures, not just AR(1). An R package called gee estimates GEE models.

Let's examine a GEE. The dataset *Esteem.csv* includes information collected over a seven-year period from a sample of adolescents. It includes measures of family relationships, self-esteem, stressful life events, and several other issues. Before considering a regression model, look at how the data are set up using R's View function or by examining Table 8.2: each

TABLE 8.2

First Few Rows of the Esteem Dataset

Newid	Year	Age	Stress	Cohes
13	1	11	−6.84	6.88
13	2	12	−6.14	6.15
13	3	13	−3.42	−9.28
13	4	14	−5.64	−7.41
		⋮		
14	1	...		
14	2	...		
14	3	...		
14	4	...		

adolescent provides up to seven years of observations (if the identification number—in this dataset called newid—is the same then the rows of data are from the same adolescent). In a sense, the data consist of two sample sizes: the number of youth represented ($n = 750$) and the number of total observations, or $n \times t$ ($750 \times 7 = 5{,}250$). Has the key assumption addressed in this chapter been met in this type of dataset? The observations are not independent because the same people contribute more than one observation, so the errors of prediction are likely correlated, which is a common occurrence with longitudinal data.

R code (abbreviated)
```
head(Esteem)
```

Let's first set up an LRM. The model is designed to predict self-esteem (esteem) among these adolescents using family cohesion (cohesion; a measure of family closeness) and the number of stressful life events (stress) experienced by the adolescents in the previous year.

The results of LRM8.6 suggest a positive association between family cohesion and self-esteem, but a negative association between stressful life events and self-esteem. Now let's see what a GEE indicates about these associations. The GEE model set up is familiar except for two modifications. First, tell the function that each person (or unit) in the dataset is identified by a variable: newid. Second, inform it what correlation structure to use for the residuals (recall that the Prais–Winsten model assumed an AR(1) structure). In LRM8.7, we'll use the AR(1) structure. This is what the "AR-M", Mv=1 portion of the gee function implies. Mv can be modified to estimate an AR(2), AR(3), or other patterns.

Linear Regression Model (OLS)

R code
```
LRM8.6 <- lm(esteem ~ cohesion + stress, data=Esteem)
summary(LRM8.6)
confint(LRM8.6)
```

R output (abbreviated)
Coefficients:

	Estimate	Std. Error	t value	Pr(>|t|)	
(Intercept)	2.894e-06	0.087	0.000	1	
cohesion	0.254	0.009	28.995	<2e-16	***
stress	-0.149	0.016	-9.147	<2e-16	***

Signif. codes: 0 '***' 0.001 '**' 0.01 '*' 0.05 '.' 0.1

Residual standard error: 6.307 on 5247 degrees of freedom

```
Multiple R-squared: 0.1704, Adjusted R-squared: 0.17
F-statistic: 538.7 on 2 and 5247 DF, p-value: < 2.2e-16

[CIs]              2.5%   97.5%
(Intercept)     -0.171   0.171
cohesion         0.237   0.271
stress          -0.181  -0.117
```

The slope coefficients are closer to zero in the GEE model than in the OLS model. However, they are also interpreted slightly differently because they are based on a *population-average* interpretation: the slope coefficients represent the difference in self-esteem experienced by an "average person" with, say, a family cohesion score one unit higher than the family cohesion score of another "average person." Another way to understand this is to imagine we select at random an adolescent with a family cohesion score of, say, 1.5. We'd expect this adolescent to have a self-esteem score about 0.20 units higher than another adolescent selected at random with a family cohesion score of 0.5. In general, we presume that the comparisons inherent in the GEE model occur at a population level.[14]

General Estimating Equation (GEE) Model with AR(1) Pattern

R code[15]
```
library(gee)
LRM8.7 <- gee(esteem ~ cohesion + stress, id = newid,
              data = Esteem, corstr = "AR-M", Mv = 1)
summary(LRM8.7)
```

R output (abbreviated)
```
Coefficients:
            Estimate Naive S.E. Naive z  Rob. S.E. Rob. z
(Intercept)  0.000    0.144     0.001    0.163     0.001
cohesion     0.202    0.009    21.777    0.012    16.663
stress      -0.071    0.015    -4.734    0.017    -4.203
Estimated Scale Parameter: 40.339
Number of Iterations: 4
```

---

[14] For information about how to interpret population-average compared to similar models, see Josephy C. Gardiner, Zhehui Luo, and Lee Anne Roman (2009), "Fixed Effects, Random Effects and GEE: What Are the Differences?" *Statistics in Medicine* 28(2): 221–239.

[15] R's gee function does not provide *p*-values by default. But they are simple to compute with R's pnorm function. For instance, what is the *p*-value associated with the coefficient that gauges the stress-esteem association in LRM8.7? In R: 2*pnorm(-abs(-4.20)), where −4.20 is the robust z-value from the output. R returns 2.67e-05, which equates to $p \cong 0.00003$. We could take a similar approach to compute confidence intervals, though some complications might be present that fall beyond the scope of this presentation.

Note also that the standard errors from the GEE model are larger than those from the OLS model, which reflects likely serial correlation of the errors. For example, the robust standard error (Rob S.E.) for the cohesion coefficient is 0.012 in the GEE model, whereas the corresponding standard error in the OLS model is 0.009.[16] But none of the standard errors are demonstrably different across the two models.

Both family cohesion and stressful life events are associated with self-esteem, but the AR(1) model might not be the best choice. Is it reasonable to assume that residuals in previous years affect residuals in directly subsequent years, but that other years don't have as much of an effect? Perhaps not; other years may also be highly influential. A GEE model can be modified to assume an *unstructured* pattern among the residuals. This allows the model to estimate the correlation structure rather than presuming that the residuals are uncorrelated (OLS) or follow an AR(1) pattern. A depiction of this presumed correlation structure is provided in Equation 8.3.

$$\begin{bmatrix} 1 & \rho_{12} & \rho_{13} \\ \rho_{21} & 1 & \rho_{23} \\ \rho_{31} & \rho_{32} & 1 \end{bmatrix} \tag{8.3}$$

The correlations across the time periods are all potentially different. The GEE model with an unstructured pattern is executed with the R code that produces LRM8.8.

General Estimating Equation (GEE) Model with Unstructured Pattern

R code
```
LRM8.8 <- gee(formula = esteem ~ cohesion + stress, id
              = newid, data=Esteem, corstr =
              "unstructured")
summary(LRM8.8)
```

R output (abbreviated)
```
Coefficients:
              Estimate   NaiveS.E.  Naive z    Rob.SE   Rob z
(Intercept)   5.117e-06  0.163      3.131e-05  0.161    3.183e-09
cohes         0.197      0.009      21.770     0.012    16.184
stress        -0.077     0.015      -4.596     0.017    -4.021

Estimated Scale Parameter: 40.424
Number of Iterations: 5
```

---

[16] R also outputs a Working Correlation that provides the presumed correlation matrix of the residuals. Can you work out the AR(1) structure from these correlations?

The first thing to notice is the minimal differences between the AR(1) model and the unstructured model. The assumption of an AR(1) pattern among the residuals might therefore be reasonable. Examine the Working Correlation output provided by R to determine the correlation patterns from the model. But also keep in mind that if we do not take into consideration the longitudinal nature of the data, we overestimate the association between family cohesion and self-esteem and between stressful life events and self-esteem (compare the slope coefficients from each model).

Like other regression models, analysts should evaluate GEEs models for fit and assess assumptions such as collinearity, homoscedastic errors, and linearity. For model fit, measures similar to R^2 are available, but some experts recommend QICs (*quasi-information criterions*), which are analogous to AICs.[17] An R package called MuMIn includes a QIC function. For example, the QICs for LRM8.7 and LRM8.8 are 5,188 and 5,176, which suggest the model with the unstructured correlation pattern fits the data slightly better than the model with the AR(1) correlation pattern.

Assessing some of the assumptions of GEE models is challenging because residuals occur within individuals and across individuals, so examining each type is necessary. Methods for achieving this are available, but are beyond the scope of this book.[18] The assessment methods used in Chapter 15 for examining residuals, however, may be adapted for use with GEE models.

An important advantage of longitudinal relative to cross-sectional data is that we may observe changes within individuals over time and, thus, individuals can serve as their own controls. This allows us to get closer to identifying causal associations among variables. Recall Mill's criteria for causation (see Chapter 2), one of which was that other factors that might account for the causal relationship must be eliminated. By individuals serving as their own controls, differences across individuals that might account for an association in the model are adjusted out.[19] However, we should also recall another of Mill's criteria: the "cause" must precede the "effect." In the model predicting self-esteem (LRM8.8), the explanatory variables were measured at the same time as the outcome variable, so their time-ordering is ambiguous. One way

---

[17] See Wei Pan (2001), "Akaike's Information Criterion in Generalized Estimating Equations," *Biometrics* 57(1): 120–125.

[18] For more information, see Sohee Oh, K. C. Carriere, and Taesung Park (2008), "Model Diagnostic Plots for Repeated Measures Data Using the Generalized Estimating Equations Approach," *Computational Statistics & Data Analysis* 53(1): 222–232.

[19] Some ambiguity exists here when using GEE models, however. Recall that GEE coefficients are population-average estimates, so they do not explicitly compare differences within individuals over time. An alternative model for longitudinal data, the random-effects model, allows a within-person interpretation and may therefore be more suitable for causal interpretations. For instance, the interpretation of a random-effects coefficient for the family cohesion–self-esteem association involves the increase in self-esteem for each one-unit increase in family cohesion for the same individual. R's plm package is useful for estimating random-effects models. The multilevel models in Chapter 15 are also suitable for estimating these models, though their estimation algorithm is different.

to overcome this limitation is to *lag* the variables so that, for instance, family cohesion and stressful life events are measured in the year preceding self-esteem.[20] R's plyr package has a function, ddply, that is useful for creating lagged variables. The following provides code that extends the previous GEE model by using lagged versions of family cohesion and stressful life events.

R code
```
library(plyr)
esteem <- ddply(Esteem, .(newid), transform, stress.lag
               = c(NA, stress[-length(stress)])) #
               stress lagged
esteem <- ddply(Esteem, .(newid), transform, cohes.lag
               = c(NA, cohesion[-length(cohesion)])) #
               cohesion lagged
 # these two lines of code create lagged values of the
   variables;
 # note that the first year for each observation is
   missing for the lagged variables

# estimate a GEE model predicting current year's esteem
  with the previous (lagged) year's cohesion and stress
LRM8.9 <- gee(formula = esteem ~ cohes.lag + stress.lag,
              id=newid, data=esteem, corstr =
              "unstructured")
summary(LRM8.9)
```

R output (abbreviated)
```
Coefficients:
```

	Estimate	Naive S.E.	Naive z	Robust S.E.	Robust z
(Intercept)	1.48e-04	0.188	7.89e-04	0.187	7.93e-04
cohes.lag	0.074	0.010	7.075	0.012	6.349
stress.lag	-0.032	0.017	-1.914	0.017	-1.843

LRM8.9 demonstrates that using measures of the previous year's explanatory variables to predict self-esteem results in more modest associations. However, we should also question whether a one-year time lag is appropriate.

---

[20] Some researchers also create a lagged version of the outcome variable (e.g., esteem$_{t-1}$) and use it as an explanatory variable in longitudinal models, but this leads to problems satisfying the independence assumption. One alternative is to employ *change scores* ($y_t^* = y_t - y_{t-1}$) as the outcome variable. But both approaches may be affected by regression to the mean (see Chapter 3, fn. 14) (Adrian G. Barnett et al. (2004), "Regression to the Mean: What It Is and How to Deal with It," *International Journal of Epidemiology* 34(1): 215–220). Books on longitudinal data analysis provide more information about these and related issues (e.g., Jeffrey D. Long (2011), *Longitudinal Data Analysis for the Behavioral Sciences Using R*, Los Angeles, CA: Sage; Judith D. Singer and John B. Willett (2003), *Applied Longitudinal Data Analysis*, New York: Oxford University Press).

Is this too far in advance? Would it be better to measure family cohesion and stressful events so that they are closer in time to self-esteem? One of the challenges with longitudinal data is determining the correct timing of measurement.

GEE models provide a valuable alternative to LRMs estimated with OLS. When analyzing longitudinal data—which offer some important advantages over cross-sectional data—GEE models are a preferred approach. Other models are also available, including, as mentioned earlier random-effects models and a related method known as fixed-effects models.[21]

Spatial Autocorrelation

A second independence challenge is known as *spatial autocorrelation*. This problem occurs when the units in the data are spatial areas, such as neighborhoods, counties, states, or nations.[22] It should be evident that the errors of prediction are likely to be more similar in adjacent or nearby units than in more distant units. When analyzing spatial data, assessing the likelihood of spatial autocorrelation is thus a good idea because, as with serial correlation, the standard errors are affected.[23]

The Durbin–Watson statistic (d) is not useful for assessing spatial autocorrelation. However, a standard test known as *Moran's I* offers an alternative.[24] Assuming we transform the variable into z-scores (this simplifies the formula), Equation 8.4 depicts how Moran's I is calculated.

$$I = \frac{n \sum_{i}^{n} \sum_{j}^{n} w_{ij} z_i z_j}{(n-1) \sum_{i}^{n} \sum_{j}^{n} w_{ij}} \tag{8.4}$$

The equation includes the n spatial units; w_{ij} is a measure of the "distance" between the specific units, indexed by i and j. Two spatial units that are close together and exhibit similar scores on the variable (z) make a positive

---

[21] A review of these various models is beyond the scope of this chapter, but an excellent start is James Hardin and Joseph Hilbe (2012), *Generalized Estimating Equations*, 2nd Ed., Boca Raton, FL: CRC Press. Books on econometrics provide thorough overviews of fixed-effects and random-effects models (e.g., Hanck et al. (2019), *op. cit.*).

[22] The *MultiLevel* dataset includes towns as a unit of observation. Although we considered only the nesting of individuals by town in the example earlier in this chapter, if some of the towns share a border, then we could also consider spatial autocorrelation as a potential source of inefficiency in the estimates.

[23] The information in this section is based on Peter A. Rogerson (2006), *Statistical Methods for Geography*, 2nd Ed., Thousand Oaks, CA: Sage.

[24] Introduced in P. A. P. Moran (1950), "Notes on Continuous Stochastic Phenomena," *Biometrika* 37(1/2): 17–23.

contribution to Moran's *I*. If the nearest units tend to have scores that correlate more than units that are far apart, then Moran's *I* is larger. In fact, its theoretical values are –1 and 1, much like Pearson's correlation coefficients, with higher values indicating positive spatial autocorrelation. Negative spatial autocorrelation may also occur, indicated when Moran's *I* is close to –1, though this is rare.

An important issue is to figure out the w_{ij}s because the Moran's *I* value is highly reliant on them. The simplest approach is to create a binary variable with a value of one if the spatial units are adjacent (or share a border) and a zero if they are not, which is called *binary connectivity*. Another way is to compute actual distance measures (e.g., two kilometers), although the specific points from which to compute the distance must be specified. A commonly used distance measure is from the center of one region to the center of another (e.g., the city center of Los Angeles to the city center of San Diego). When using distance, the w_{ij} in Moran's *I* is the inverse of the distance so that units closer together receive a larger weight.

As an example of how to compute Moran's *I*, consider the map in Figure 8.5. It consists of data from five counties in the state of Hypothetical. The numbers represent a measure of the number of crimes committed in each county in the last year adjusted for their population sizes. We wish to determine the degree of spatial autocorrelation before going any further in the analysis.

To compute Moran's *I*, we first need to decide on a system of weights for the distances. To ease the computations, let's use the binary connectivity

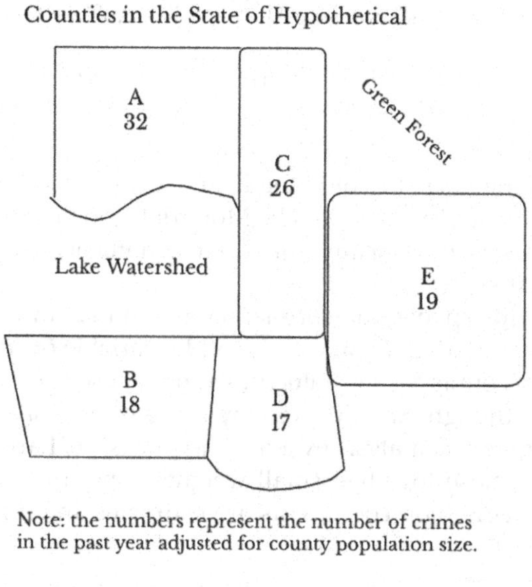

Counties in the State of Hypothetical

Note: the numbers represent the number of crimes
in the past year adjusted for county population size.

FIGURE 8.5
Fabricated counties in the state of Hypothetical.

approach where a one is assigned to adjacent counties (e.g., A and C) and a zero is assigned to nonadjacent counties (e.g., A and B). A simple way to see these weights is with a matrix where the entries are the w_{ij}s (see Equation 8.5).

$$
W = \begin{matrix} A \\ B \\ C \\ D \\ E \end{matrix} \begin{bmatrix} 0 & 0 & 1 & 0 & 0 \\ 0 & 0 & 0 & 1 & 0 \\ 1 & 0 & 0 & 1 & 1 \\ 0 & 1 & 1 & 0 & 1 \\ 0 & 0 & 1 & 1 & 0 \end{bmatrix} \tag{8.5}
$$

This matrix is symmetric, with the same pattern above the diagonal as below the diagonal. For example, a one is listed in row 1, column 3 and a one is listed in row 3, column 1, which indicate that counties A and C share a border. The overall mean of the crime variable is 22.4, with a standard deviation of 6.43. To compute Moran's I, we may save some steps by converting the specific crime values into z-scores.[25] We then add the products of each pair of z-scores and multiply this sum by the sample size (5) to obtain the numerator (the nonadjacent pairs, since they have zero weights, are omitted from the computation):

$$
5 \times [AC + BD + CA + CD + CE + DB + DC + DE + EC + ED]
$$

$$
= 5 \times \big[(1.494 \times 0.56) + (-0.685 \times -0.84) + (0.56 \times 1.494)
$$

$$
+ (0.56 \times -0.84) + (0.56 \times -0.53) + (-0.84 \times -0.685) + (-0.84 \times 0.56)
$$

$$
+ (-0.84 \times -0.53) + (-0.53 \times 0.56) + (-0.53 \times -0.84) \big]
$$

$$
= \mathbf{10.9}
$$

The sum of the weights is 10 (count the 1s in the W matrix), so the denominator in the equation is {4 × 10} = 40. The Moran's I value is therefore 10.9/40 = 0.273. This suggests a modest amount of spatial autocorrelation among these county crime statistics.

The solutions for spatial autocorrelation are similar in logic to the solutions for serial correlation. First, we may add a variable or set of variables to the model that accounts for the autocorrelation. These types of variables are difficult to find, though. Second, we may use *geographically weighted regression*, which weights the analysis by a distance measure. Large weights apply to units that are close together; small weights apply to those that are farther apart. The regression coefficients are estimated in an iterative fashion

---

[25] The z-scores (calculations omitted), listed in order from counties A to E, are {1.494, −0.685, 0.56, −0.84, and −0.53}.

after finding the optimal weights. Third, spatial regression models that are designed specifically to address spatial data and autocorrelation are available. The research area of *Geographical Information Systems* (GIS) includes a host of spatial regression approaches.[26] R has several user-written packages available for spatial analysis (e.g., `spatialreg`, `lagsarlm`). For a thorough overview of spatial models in R, see the RSpatial website (https://www.rspatial.org).

Chapter Summary

We now know about some of the risks when the errors of prediction (and indirectly, the observations) are not independent. The main consequence is incorrect standard errors, which are usually too small, and, thus, inefficient slope coefficients. Since a common goal of regression modeling is to estimate results that allow inferences to a population, obtaining correct standard errors is important.[27] An essential task that we should always think about, therefore, is the nature of the data and variables to evaluate the likelihood of a lack of independent observations/errors, serial correlation, and spatial autocorrelation.

Several diagnostic tools and solutions are available when faced with dependence issues. The following summarizes several that are discussed in this chapter:

1. Always check the design effects for complex survey data. If you are interested only in individual associations but have data that are part of a nested design, adjusting for clustering is easy in R. Using data based on a multistage sampling design is also feasible in R.

2. A helpful graph plots the studentized deleted residuals by the predicted values from an LRM. For data collected over time, in particular, a snaking pattern is often indicative of serial correlation.

3. The Durbin–Watson d statistic is a useful numeric test for serial correlation.

4. If you don't know the source of the serial correlation, use a model specifically designed to correct this problem. Prais–Winsten regression provides better estimates than OLS regression in the presence of serially correlated residuals, but is limited, without modification, to an AR(1) autocorrelation pattern.

---

[26] Rogerson (2006), *op. cit.*, provides a relatively painless overview of spatial statistics and regression models.

[27] As discussed in earlier chapters, though, other aspects of the model are also important as evidence for determining the implications of the results.

5. For longitudinal data, a frequently used approach is to employ one of the GEE models. They provide a flexible way to estimate longitudinal associations among variables.

6. Data collected over spatial units present the same conceptual problems as data collected over time, though the techniques for diagnosing spatial autocorrelation and for adjusting the models are different. Moran's *I* is a widely used diagnostic test for spatial autocorrelation. And many regression routines intended specifically for spatial data are available to the interested researcher.

Chapter Exercises

The dataset *USCounties2010.csv* includes data from 2,257 counties in the U.S. It also includes state and county identifiers. The following exercises are designed to show the implications of a lack of independence for LRMs. The variables in the dataset include

- State — State name
- StateCode — A numeric code that identifies each state
- County — County name
- CountyCode — A numeric code that identifies each county
- BingeDrink — The percentage of county residents who engage in heavy alcohol use
- TeenBirths — The percentage of babies born to teenage mothers
- PrimaryCare — The number of primary care physicians and physician's assistants per 10,000 county residents
- PerHSGrads — The percentage of county residents who are high school graduates
- PerSingleParentHH — The percentage of households that are headed by a single parent
- ViolentCrime — The number of violent crimes per 100,000 county residents
- UnhealthyFoods — The percentage of county residents who have limited access to healthy foods
- PerFastFood — The percentage of restaurants in the county that are fast food establishments

After importing the dataset into R, complete the following exercises.

1. Estimate the mean and standard error of `ViolentCrime` before and after adjusting for the clustering of observations by state. What do the results suggest about the degree of clustering and how it affects the summary information about `ViolentCrime`?

2. Estimate the design effects of the means of the `ViolentCrime` and `BingeDrink` variables. What do these suggest about the degree of clustering in the data?

3. Estimate an LRM with `ViolentCrime` as the outcome variable and `PerSingleParentHH` as the explanatory variable. Estimate the same model but adjust for the clustering of observations by state. Request R's `summary` and `confint` results for both models. How do the results of the models differ?

4. Estimate a multiple LRM with `ViolentCrime` as the outcome variable and the following explanatory variables: `PerSingleParentHH`, `BingeDrink`, `TeenBirths`, and `PerHSGrads`. Estimate the same multiple LRM but adjust for the clustering of observations by state. What are the similarities and differences between the two models? Suppose you wish to infer the model's results to a hypothetical target population. Do you reach any different conclusions once you adjust for clustering by states?

5. *Challenge*: estimate the design effects for the slope coefficients from the multiple LRM in exercise 4 that adjusts for clustering by states.

9

Homoscedasticity

A critical assumption of LRMs is that the variance of the errors of prediction is constant for all combinations of the explanatory variables. Substantial effort has been directed toward studying this assumption and what it means for linear models. The example discussed in Chapter 3 (see Figure 3.3) illustrates the presumed—and imaginary—association between a nation's percent labor union and public expenditures. To visualize this association in a statistical model, we assume that the nations in the sample have labor union values that represent labor union values for the population. Similarly, when we collect information from a sample of adults in the U.S., we assume that each adult represents some number of adults in the population of U.S. residents. Each adult's value on some variable of interest represents, inferentially speaking, the mean value of the variable in the subpopulation she or he represents. We'll take for granted that these sample members' values are a good representation of the means for the subpopulations and ask instead about their variability? Is it similar across sample members who have different values for some set of explanatory variables?

To think about the issue of variability clearly, consider the model in Chapter 5 designed to predict first-year college GPAs (LRM5.3). The model includes the following variables believed to account for GPA: SAT math scores, SAT reading/writing scores, and high school math grades. Several assumptions underlie the suitability of the LRM. Let's focus on just one explanatory variable—high school math grades—to understand the assumption addressed in this chapter. We assume that the variability in the errors in predicting college GPA is the same for different values of high school math grades. This means, for instance, that the errors of prediction have roughly the same degree of dispersion whether math grades are mostly in the C range or mostly in the A range.[1] Do you find this to be a reasonable conjecture?

The name of this assumption is *homoscedasticity*. The term *homo* implies *same* and the term *scedastic* means *scatter* (from the Greek *skédasi*, which means *scattering*). The errors of prediction are thus assumed to be homoscedastic or have the same scatter. The alternative is that the errors do not have the same or similar variability: they are *heteroscedastic* (*hetero* implies *different*). Figure 9.1 provides an example of heteroscedastic errors: the presumed variability of the errors in predicting y increases with hypothetical values of

---

[1] In a model with only two variables, a threshold test is to consider the ratio of the largest to the smallest variance: if it falls above 1.5, evidence of non-constant or unequal variance exists. For example, imagine that the error variance of those who received As is 2.6 and of those who received Cs is 1.2. Since 2.6/1.2 = 2.2 we have evidence of non-constant variance.

DOI: 10.1201/9781003162230-9

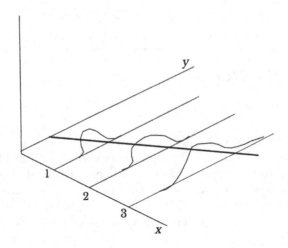

FIGURE 9.1
Illustration of heteroscedastic errors.

x. Imagine that $x = 1$ indicates math grades in the C range and $x = 3$ indicates math grades in the A range: the variability in the errors predicting college GPA is larger among those who received As in high school math, a clear example of heteroscedasticity.

The main consequence of heteroscedastic errors is that, although the slope coefficients are, according to statistical theory, unbiased, they are *inefficient*. Slope coefficients that are inefficient are, on average, imprecise and have larger standard errors than efficient slope coefficients. The degree of imprecision is rarely known, although it seems to be especially severe in small samples. Yet, because of its effect on standard errors, it can affect significance testing even in moderately sized samples.

Heteroscedastic errors are a frequent phenomenon in LRMs and seem to be particularly common in cross-sectional data, or data that are collected at one point in time. The following example provides an illustration of heteroscedasticity. The dataset, *Income2018.csv*, has information on sample members' annual incomes in $1,000s and how many years of formal education they completed. A sensible prediction is that people who attend school longer tend to earn more income. The results of a simple LRM support this proposition: each one-year increase education is associated with $5,350 more in annual income (95% CI: {$5,035, $5,666}) (see LRM9.1).

R code
```
LRM9.1 <- lm(income ~ education, data=Income2018)
summary(LRM9.1)
confint(LRM9.1)
```

R output (abbreviated)
```
Coefficients:
```

```
            Estimate   Std. Error   t value    Pr(>|t|)
(Intercept) -12.254      2.216       -5.528     3e-05 ***
education     5.350      0.150       35.658    <2e-16 ***
---
Signif. codes: 0 '***' 0.001 '**' 0.01 '*' 0.05 '.' 0.1

Residual standard error: 3.527 on 18 degrees of freedom
Multiple R-squared: 0.986, Adjusted R-squared: 0.9853
F-statistic: 1272 on 1 and 18 DF, p-value: < 2.2e-11

[CIs]              2.5%     97.5%
(Intercept)     -16.910    -7.597
education         5.035     5.666
```

A graphical test to consider after estimating LRM9.1 is a *partial residual plot*, which is useful for assessing homoscedastic errors. It plots some form of the residuals (e.g., raw, studentized) by the explanatory variable. For instance, let's plot the raw or untransformed residuals $(y_i - \hat{y}_i)$ by education. The results of requesting this from R are provided in Figure 9.2 (the reference line at $y = 0$ helps highlight the pattern).

R code for Figure 9.2
```
resid.LRM9.1 <- resid(LRM9.1) # save the residuals from
                    LRM9.1
plot(Income2018$education, resid.LRM9.1,
    ylab="Residuals", xlab="Years of formal education")
abline(0,0, col="red") # add a horizontal reference line
```

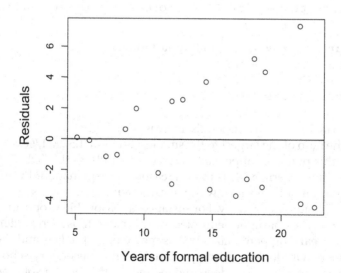

FIGURE 9.2
Partial residual plot of years of formal education to assess homoscedastic errors from LRM9.1.

The residuals spread out at higher values of education, which represents increasing variability and offers a prototypical example of how heteroscedasticity appears in a partial residual plot. An opposite pattern also indicates heteroscedasticity: less variability of the residuals at higher values of the explanatory variable. Homoscedastic residuals appear as a random scatter, with no discernible association in a partial residual plot.

Let's consider a larger dataset—*GSS2018.csv*—that includes information on income among a sample of adults in the U.S. to determine whether the variability of the LRM residuals differs for those with differing levels of education (see LRM9.2). Once again, we ask R to save the residuals from an LRM that uses years of education (educate) to predict personal income (pincome). The R code then requests the variances of the residuals (resid) for those who reported 12 years of education, 13–15 years of education, and 16 or more years of education.

R code
```
LRM9.2 <- lm(pincome ~ educate, data=GSS2018)
GSS2018$resid <- resid(LRM9.2)
GSSEduc1 <- GSS2018[which(GSS2018$educate==12), ]
   # select those with 12 years of education
var(GSSEduc1$resid)
GSSEduc2 <- GSS2018[which(GSS2018$educate > 12 &
                      GSS2018$educate < 16), ]
   # select those between 13 and 15 years of education
var(GSSEduc2$resid)
GSSEduc3 <- GSS2018[which(GSS2018$educate > 15), ]
   # select those with 16 or more years of education
var(GSSEduc3$resid)
```

R output (abbreviated and annotated)
```
12 years:          66.8
13-15 years:       71.6
16 or more years:  90.1
```

The variability of the residuals increases at higher years of education, though the ratio of the largest to the smallest is below 1.5 (90.1/66.8 = 1.35; see fn. 1). At this point, an important inquiry is to address the "why" question: why does this pattern exist? Is there a reasonable explanation? By answering this question in a thoughtful way, we can often figure out a solution to the problem of heteroscedasticity. For instance, a reasonable proposition is that at least two types of higher-educated people exist: those who enter careers with higher earning potentials, such as medicine or the law, and those who enter careers with lower earning potential, such as secondary school teachers or social workers. How does understanding this offer a solution to heteroscedasticity, though? If we have a variable in the dataset that measures occupation, we can include it in the model and determine if it accounts for the heteroscedasticity among the residuals. This illustrates that, as with many problems that affect LRMs, a good theory—or simply a reasonable idea—goes a long way.

Another common reason for heteroscedasticity is that one or more of the variables requires a transformation. Transformations are often useful when continuous variables are skewed (see Chapter 11). Yet this might not be apparent when examining a scatter plot between the outcome and one of the explanatory variables. Previous research may provide a clue, however. For instance, measures of income are positively skewed in many samples since relatively few people make very high incomes. Researchers often normalize income measures by taking their square roots or natural logarithms, which may be all that is needed to eliminate heteroscedastic errors, though it doesn't work in some situations.

As an exercise, take the natural logarithm of income (income) from the *Income2018* dataset,[2] re-estimate the LRM using this transformed variable, and construct a partial residual plot. What does the plot show? The results look unusual,[3] though they do not represent a classic heteroscedastic pattern. Before taking this too far, however, we should assess if income is skewed in these data (short answer: no.) How can we determine this?

Assessing Homoscedasticity in Multiple LRMs

Evaluating whether the errors are homoscedastic or heteroscedastic in a multiple LRM is only a bit more cumbersome. If you suspect that a particular explanatory variable is involved in a heteroscedasticity issue (draw upon your conceptual model and knowledge of previous research!), then construct a partial residual plot to diagnose it (see the R code for Figure 9.2). When the model includes multiple explanatory variables, though, it may not be apparent if one or more of them is inducing heteroscedastic errors. An alternative to a partial residual plot is to graph some form of the residuals against the predicted values from the model ($\hat{y}_i$). Executing the R function plot after estimating an LRM with the lm function provides several post-regression graphs, including a scatter plot of the raw or unstandardized residuals by the predicted values. Some experts recommend plotting the studentized deleted residuals (*y*-axis) against the standardized predicted values (*x*-axis) because

---

[2] If we wish the logged values of income to be part of the existing dataset, one approach in R is

```
Income2018$log.income <- log(Income2018$income)
```

As we'll learn later, we may also embed the transformation in the lm function.

[3] The residuals follow an inverted U-shaped pattern, which may indicate other issues we don't have the space to investigate (but see Chapter 11).

these residuals tend to have more constant variance than the raw or standardized residuals. As shown in Chapter 8, a plot using these residuals is available as part of R's olsrr package (see Figure 8.4). Regardless of the type of residual used, we still look for a megaphone- or cone-shaped pattern in the scatter plot (hourglass shapes may also occur). Remember, though, that in order to accept the assumption that the errors are homoscedastic, we wish to see a random pattern.

Let's consider a multiple LRM using the *StateData2018* dataset. Our objective is to predict gun ownership per 100,000 residents (GunsPerCapita), so it serves as the outcome variable. The four explanatory variables are the number of robberies per 100,000 population, the number of state prisoners per 100,000 adult residents, the percent of residents living below the poverty level, and the unemployment rate (see LRM9.3). After estimating the model in R, we'll rely on two graphs that the plot function provides. The first plots the residuals on the *y*-axis against the fitted/predicted values on the *x*-axis. The second (the third among the four furnished by R), called a *Scale-Location plot*, provides the square root of the absolute value of the residuals on the *y*-axis and the fitted/predicted values on the *x*-axis. Heteroscedastic errors are suggested by larger residuals at higher points along the *x*-axis.

The results suggest a modest fit of the model to the data, with an R^2 of 0.25. Robberies per 100,000, the poverty rate, and the unemployment rate are negatively associated, whereas the number of state prisoners per 100,000 adult residents is positively associated, with gun ownership.

R code for Figures 9.3 and 9.4

```
LRM9.3 <- lm(GunsPerCapita ~ RobberyRate + PrisonRate +
             PerPoverty + UnemployRate,
             data=StateData2018)
summary(LRM9.3)
confint(LRM9.3)
plot(LRM9.3) # provides four diagnostic graphs in R's
               plots window; we're interested in the
               first and third graphs
```

R output (abbreviated)

```
Coefficients:
               Estimate   Std. Error   t value   Pr(>|t|)
(Intercept)    17.20644   22.06651      0.780    0.43962
RobberyRate    -0.33796    0.11075     -3.052    0.00381  **
PrisonRate      0.06960    0.02474      2.813    0.00724  **
PerPoverty     -5.17894    2.24488     -2.307    0.02571  *
UnemployRate   12.67320    6.28390      2.017    0.04971  *
---
Signif. codes:  0 '***' 0.001 '**' 0.01 '*' 0.05 '.' 0.1
```

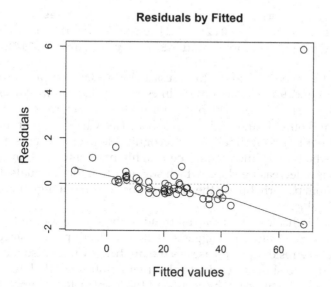

FIGURE 9.3
Residuals by fitted plot to assess homoscedastic errors from LRM9.3.

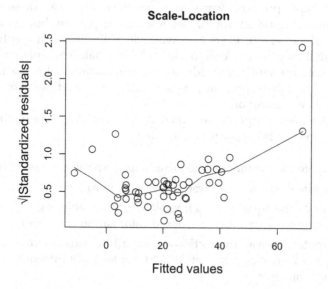

FIGURE 9.4
Square root of the absolute value of the residuals by fitted values to assess homoscedastic errors from LRM9.3.

```
Residual standard error: 28.38 on 45 degrees of freedom
Multiple R-squared: 0.25, Adjusted R-squared: 0.1833
F-statistic: 3.75 on 4 and 45 DF, p-value: 0.0102
```

The scatter plots designed to assess homoscedasticity are provided in Figures 9.3 and 9.4. The nonlinear lines in the figures are lowess (*locally weighted regression*) lines, which can be useful for determining heteroscedasticity and other issues with the residuals (see Chapter 11 for more information about lowess fit lines). If no recognizable pattern occurs among the residuals, the lowess lines should be roughly horizontal. But, in Figure 9.3, the line has a decreasing slope, which suggests that the residuals do not follow a random pattern. But do they increase or decrease at higher levels of the predicted values?

Figure 9.4 indicates an increasing trend in the square root of the absolute values of the residuals. This suggests a heteroscedastic pattern since the relative size of the residuals increases as the predicted values also increase. But both graphs also show a residual value that is substantially larger than the others (look at the upper right portion of the graphs). Reviewing the residuals in the dataset, this observation is in row 50 and represents Wyoming. It deserves further scrutiny.

The two graphs provide a way to visualize the residuals and assess whether they follow a homoscedastic or heteroscedastic pattern, but, as described next, several numeric tests are also available. A limitation of graphical techniques is that they are difficult to use with large datasets (imagine trying to visualize patterns with hundreds of points in a scatter plot), so alternative techniques for diagnosing heteroscedasticity are beneficial. We'll discuss three of the most common.

The first numeric approach is called *White's test*.[4] After estimating a multiple LRM, the test takes the following steps:

1. Compute the unstandardized residuals from the regression model.

2. Compute the square of these values (e.g., `resid²`).

3. Compute the square of each explanatory variable (e.g., x_1^2, x_2^2). Do not compute squared values of indicator variables, though.

4. Compute two-way interactions using all the explanatory variables (e.g., $x_1 \times x_2$; $x_1 \times x_3$; $x_2 \times x_3$; etc.). (See Chapter 11 for information about interaction terms.)

5. Estimate a new regression model with the squared values of the residuals as the outcome variable and the variables computed in (3) and (4) as the explanatory variables (make sure to include their constituent terms: x_1, x_2, etc.).

---

[4] See Halbert White (1980), "A Heteroskedasticity-Consistent Covariance Matrix Estimator and a Direct Test for Heteroskedasticity," *Econometrica* 48(4): 817–838.

6. Two methods are then available to identify whether or not the errors are homoscedastic:

   a. Examine the R^2 from the model in (5): if its p-value—based on the F-statistic—is above a predetermined threshold (e.g., $p > 0.05$), then the errors are presumed homoscedastic.

   b. Use the test statistic shown in Equation 9.1.

$$nR^2 \sim \chi_k^2 \qquad (9.1)$$

where n = sample size, R^2 is from the model, and k = the number of explanatory variables.

This test statistic is distributed as a χ^2 variable, so compare its value to a χ^2 value with k degrees of freedom. Once again, a p-value above a predetermined threshold indicates homoscedasticity.

An advantage of White's test is that it tests not only for homoscedastic errors, but also provides evidence about potential nonlinearities in the associations (see Chapter 11). An R package called het.test furnishes White's test but does so only with a vector autoregressive (VAR) model.[5] We won't consider this model, though, since alternatives to White's test are available.

The second diagnostic approach is known as *Glejser's test*.[6] The simplest way to understand how this test works is to ponder the appearance of heteroscedastic errors in a scatter plot (see Figures 9.2–9.4). Recall that residuals have a mean of zero, with negative and positive values. Suppose we folded over the residuals along the line marked by $y = 0$ so that the negative values were pulled up to be positive. This, in effect, is what happens when we take the absolute value of the residuals, or $|\hat{\varepsilon}_i|$, as R did to create Figure 9.4. Assuming a cone-shaped pattern, what would the association between the predicted values and the absolute values of the residuals look like? It would be positive if the residuals fan out to the right and negative if they fan out to the left. The scatter plot in Figure 9.5, based roughly on Figure 9.2, demonstrates this. The presumed linear fit line indicates a positive association.[7] This, in brief, is how Glejser's test is performed: compute the absolute values of the residuals and estimate an LRM with these new residuals as the outcome variable and the original explanatory variables. An advantage of this

---

[5] As the name implies, the VAR model is designed for time-series data and uses previous values of a variable to estimate trends in subsequent values. The model is appropriate for multiple time-series (see Hanck et al. (2019), *op. cit.*, chapter 16), but is not well suited to the LRM examples.

[6] See Herbert Glejser (1969), "A New Test for Heteroskedasticity," *Journal of the American Statistical Association* 64(235): 315–323.

[7] Figure 9.4 also depicts this type of pattern: a positive slope for the lowess fit line once the predicted values exceed 20.

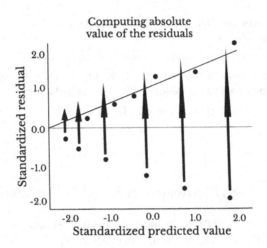

FIGURE 9.5
Effect of standardizing the residuals in a residual by predicted values plot.

test is that it allows researchers to isolate particular explanatory variables that are inducing heteroscedasticity.

Let's continue the example of gun ownership. After estimating the LRM and creating a new variable based on the unstandardized residuals, we'll examine the homoscedasticity issue using Glejser's test. First, compute the absolute values of the residuals and call them `absresid`. The term `absresid` is an arbitrary name assigned to the variable. Follow this with a new LRM that regresses the absolute value of the residuals on the explanatory variables (see LRM9.3r).

R code
```
absresid <- abs(resid(LRM9.3))
  # compute the absolute value of the residuals and
    regress them on the explanatory variables
LRM9.3r <- lm(absresid ~ RobberyRate + PrisonRate +
              PerPoverty + UnemployRate, data =
              StateData2018)
summary(LRM9.3r))
```

R output (abbreviated): Glejser's Test
```
Coefficients:
              Estimate Std. Error  t value   Pr(>|t|)
(Intercept)    1.23051  16.11879    0.076    0.93949
RobberyRate   -0.24669   0.08090   -3.049    0.00383 **
PrisonRate     0.04601   0.01807    2.546    0.01438 *
PerPoverty    -4.14117   1.63980   -2.525    0.01515 *
UnemployRate  13.93088   4.59016    3.035    0.00399 **
---
Signif. codes:  0 '***' 0.001 '**' 0.01 '*' 0.05 '.' 0.1
```

Assuming we wish to use $p < 0.05$ as the predetermined threshold, Glejser's test provides evidence of heteroscedasticity. For instance, the negative association between robberies (RobberyRate) and the absolute value of the residuals suggests a decrease in the variability of the residuals as robberies increase. Yet states with higher unemployment rates tend to have more variability among the residuals from LRM9.3. A good way to assess the patterns further is to create partial residual plots for each explanatory variable.[8]

The third test for homoscedastic errors is a variation of Glejser's test known as the *Breusch–Pagan test*.[9] The difference is that the squared residuals are used rather than the absolute values of the residuals. This test is implemented in the lmtest package's bptest function. The null and alternative hypotheses are the same as in White's test. As shown next, using a threshold of $p < 0.05$, the test provides evidence that casts doubt on the null hypothesis, so heteroscedasticity is likely.

R code
```
library(lmtest)
bptest(LRM9.3)
```

R output
```
studentized Breusch-Pagan test
data: LRM9.3
BP = 10.809, df = 4, p-value = 0.028794
```

We now have several pieces of evidence supporting heteroscedastic errors, though the graphs depict patterns that might indicate other issues. Two diagnostic tests—Glejser's and Breusch–Pagan—suggest some degree of heteroscedasticity. A worthwhile step is to explore whether the explanatory variables that Glejser's test indicates are inducing heteroscedasticity. But we should also consider the extreme residual in Figures 9.3 and 9.4: observation 50, which has a residual value greater than 150 and represents Wyoming. This state also has the highest number of guns per 100,000 residents (229), whereas the next states—New Mexico (47) and New Hampshire (47)—have substantially lower residuals. Omitting Wyoming from the model, which is not generally a good practice (see Chapter 14), changes the LRM results

---

[8] Consider, for example, the R code:

```
plot(StateData2018$UnemployRate, resid(LRM9.3), ylab="Residuals", xlab =
"Unemployment Rate");abline(0,0, col="red")
```
[9] See T. S. Breusch and A. R. Pagan (1979), "A Simple Test for Heteroskedasticity and Random Coefficient Variation," *Econometrica* 47(5): 1287–1294.

markedly, and, according to the Breusch–Pagan test, eliminates heteroscedasticity.[10] This reinforces the notion that one observation is affecting the results. It might be tempting to drop the observation to attenuate heteroscedasticity in the LRM, but some better solutions are provided in the next section.

What to Do About Heteroscedasticity

We've already mentioned one possible solution to heteroscedasticity: figure out why it occurs. What variable or variables might influence it? In the gun ownership example, for instance, Wyoming constitutes an extreme value that appears to produce the evidence for heteroscedastic errors. And recall what an extreme value might indicate: a non-normal distribution. If we examine, say, a kernel density plot of GunsPerCapita, we would find substantial right skew. As mentioned earlier in the chapter, using the logarithm of a skewed variable in an LRM might attenuate heteroscedasticity. Therefore, let's estimate an LRM that uses as the outcome variable the natural logarithm of GunsPerCapita and reevaluate the diagnostic tests (see LRM9.3b).

R code
```
LRM9.3b <- lm(log(GunsPerCapita) ~ RobberyRate +
              PrisonRate + PerPoverty + UnemployRate,
              data = StateData2018)
summary(LRM9.3b)
bptest(LRM9.3b)
```

R output (abbreviated)
```
Coefficients:
              Estimate  Std. Error  t value  Pr(>|t|)
(Intercept)   2.3164188  0.4432101    5.226  4.31e-06 ***
RobberyRate  -0.0074932  0.0022244   -3.369  0.00156  **
PrisonRate    0.0022319  0.0004969    4.492  4.89e-05 ***
PerPoverty   -0.1133531  0.0450889   -2.514  0.01558  *
UnemployRate  0.2327910  0.1262133    1.844  0.07171  .
---
Signif. codes:  0 '***' 0.001 '**' 0.01 '*' 0.05 '.' 0.1
```

---

[10] For example:
```
LRM9.3a <- lm(GunsPerCapita ~ RobberyRate + PrisonRate + PerPoverty +
              UnemployRate, data=StateData2018[-50, ])
# remove Wyoming (row 50) from the model
summary(LRM9.3a)
bptest(LRM9.3a)
```

```
Residual standard error: 0.5699 on 45 degrees of freedom
Multiple R-squared: 0.3832, Adjusted R-squared: 0.3284
F-statistic: 6.99 on 4 and 45 DF, p-value: 0.0001824

        studentized Breusch-Pagan test
data: LRM9.3b
BP = 1.9382, df = 4, p-value = 0.7471
```

Not only does the model fit improve (R^2: 0.25 → 0.38), but the Breusch–Pagan test suggests that heteroscedasticity is no longer an issue in the model.

Evidence for heteroscedastic errors can also stem from nonlinear associations among variables. For instance, suppose a nonlinear association exists between the unemployment rate or the poverty rate and gun ownership. There may also appear to be heteroscedastic errors, but they could be accounted for by the nonlinear association.[11] Note that these solutions involve the proper specification of the model: are the correct variables and the correct forms of the variables in the model? Heteroscedasticity is often merely a symptom of improper model specification (see Chapters 11 and 12).

Some situations arise, however, when other characteristics of the model that don't usually interest the researcher induce heteroscedasticity. This problem emerges frequently when we analyze *repeated cross-sectional data*. Recall from Chapter 8 that the GSS is an example of this type of dataset. Although we've employed only one year of data, the GSS has been collected for about 50 years. A different sample of adults in the U.S. is surveyed every couple of years; hence, it involves cross-sectional data that are repeated over time.

One of the problems with such data is that the sample size may vary from one survey to the next. Since sample size affects variability (e.g., larger samples generally yield smaller standard errors), differing sample sizes across years—assuming one wishes to analyze data from several years—induce heteroscedasticity. Some analysts merely include a variable that gauges year in their models, hoping it will minimize heteroscedasticity. A preferred approach, however, is to use *weighted least squares* (WLS) to adjust for differing sample sizes. WLS uses a weight function to adjust the standard errors for heteroscedasticity. The general formulas for WLS are provided in Equations 9.2 and 9.3.

$$\text{SSE} = \Sigma \left(\frac{1}{s_i^2} \right) \left(\left(y_i - \bar{y}' \right)^2 \right) \tag{9.2}$$

$$\beta_1 = \frac{\Sigma \frac{1}{s_i^2} \left(\left(x_i - \bar{x}' \right) \left(y_i - \bar{y}' \right) \right)}{\Sigma \frac{1}{s_i^2} \left(x_i - \bar{x}' \right)^2} \quad \text{where } \bar{y}' = \frac{\Sigma \frac{y_i}{s_i^2}}{\Sigma \frac{1}{s_i^2}} \text{ and } \bar{x}' = \frac{\Sigma \frac{x_i}{s_i^2}}{\Sigma \frac{1}{s_i^2}} \tag{9.3}$$

---

[11] Chapter 11 provides more information on nonlinear associations in LRMs.

The estimator for the slope (β_1) in Equation 9.3 is for a simple WLS regression model but extending it to multiple WLS models is not difficult. The key for WLS is the weight function, or $1/s_i^2$.[12] When using repeated cross-sectional data, the inverse of the sample size often works well to correct standard errors for heteroscedasticity.[13]

In many situations—especially those that do not involve repeated cross-sectional data—analysts estimate the weight function using the inverse of the squared residuals (or absolute value of the residuals) from the OLS estimated LRM (weights cannot be negative, so only positive values are feasible). Yet, we may be able to do better since we know a bit about which explanatory variables are implicated in heteroscedasticity (we'll ignore the skewed nature of the gun ownership variable). Recall from the last section that one of the variables associated with more variation in the residuals is the unemployment rate. Let's use this information to create a weight from the following equation: $|residuals(\text{LRM9.3})| = \alpha + \beta(unemployment\ rate)$. The predicted values from this equation are estimates of the residual standard error, which is an estimate of the standard deviation of the residuals based on this explanatory variable (also called the RMSE; see Chapter 5). The recommended weight function is $1/s^2$, so we simply need to take the predicted values, square them, and then compute the inverse to yield the weight function.

To organize this example, let's first re-estimate the original LRM (see LRM9.3).

R code
```
LRM9.3 <- lm(GunsPerCapita ~ RobberyRate + PrisonRate +
             PerPoverty + UnemployRate,
             data=StateData2018)
summary(LRM9.3)
confint(LRM9.3)
```

R output (abbreviated)
Coefficients:

	Estimate	Std. Error	t value	Pr(>|t|)	
(Intercept)	17.20644	22.06651	0.780	0.43962	
RobberyRate	-0.33796	0.11075	-3.052	0.00381	**
PrisonRate	0.06960	0.02474	2.813	0.00724	**
PerPoverty	-5.17894	2.24488	-2.307	0.02571	*
UnemployRate	12.67320	6.28390	2.017	0.04971	*

---

[12] The inverse of the variance shown here is the standard formula used in WLS. However, which variance to use is based on what the analyst thinks is causing the heteroscedasticity issue. In some situations, the variance of a particular variable or the residuals from an LRM might be suitable. Deciding what type of weight to use is challenging yet vital when estimating a WLS model.

[13] See, for example, John P. Hoffmann (2013), "Declining Religious Authority? Confidence in the Leaders of Religious Organizations, 1973–2010," *Review of Religious Research* 55(1): 1–25.

Signif. codes: 0 `***' 0.001 `**' 0.01 `*' 0.05 `.' 0.1

Residual standard error: 28.38 on 45 degrees of freedom
Multiple R-squared: 0.25, Adjusted R-squared: 0.1833
F-statistic: 3.75 on 4 and 45 DF, p-value: 0.01024

Second, we'll estimate the model using the weights. The computation of the weights is intentionally drawn out in the following R code to clarify each step. As you may recall, though, R functions may be embedded, so the following code could be written in more compact form.[14]

R code

```
StateData2018$abs.resid <- abs(residuals(LRM9.3))
 # compute the absolute value of the residuals from the
   original LRM model
firstwt <- lm(abs.resid ~ UnemployRate, data =
              StateData2018)
 # regress the absolute value of residuals on the
   unemployment rate
StateData2018$secondwt <- predict(firstwt)
 # save the predicted values from the model; this
   provides the estimated standard deviations of the
   residuals
StateData2018$secondwt <- predict(firstwt)^2 # square
                                              them
StateData2018$wt <- 1/StateData2018$secondwt
 # take their inverse
LRM9.4 <- lm(GunsPerCapita ~ RobberyRate + PrisonRate +
             PerPoverty + UnemployRate, weight=wt, data
             = StateData2018)
 # estimate the WLS model
summary(LRM9.4)
confint(LRM9.4)
```

R output (abbreviated)
Coefficients:

	Estimate	Std. Error	t value	Pr(>|t|)	
(Intercept)	13.16483	15.70522	0.838	0.40632	
RobberyRate	-0.30826	0.09185	-3.356	0.00161	**
PrisonRate	0.06669	0.02056	3.244	0.00222	**
PerPoverty	-6.61363	2.21105	-2.991	0.00450	**
UnemployRate	18.75392	5.96003	3.147	0.00293	**

Signif. codes: 0 `***' 0.001 `**' 0.01 `*' 0.05 `.' 0.1

---

[14] This is only one way of computing weights when the variable causing the issue is known. Other ways are feasible, but most result in a similar WLS model.

```
Residual standard error: 1.808 on 45 degrees of freedom
Multiple R-squared: 0.3206, Adjusted R-squared: 0.2602
F-statistic: 5.309 on 4 and 45 DF, p-value: 0.001372
```

What are some differences between the original OLS estimated model (LRM9.3) and the WLS estimated model (LRM9.4)? The model fit appears to be better in the latter model (R^2: $0.25 \rightarrow 0.32$). The WLS model also has smaller standard errors and thus larger t-values (in absolute value terms). For example, the t-value for the unemployment rate increases from 2.0 in the OLS model to 3.1 in the WLS model. A good idea, though, is to check the WLS model for homoscedastic errors using the Breusch–Pagan test or the plots used earlier. You'll find evidence of heteroscedasticity. This points to the challenge of identifying the correct weights for a particular model. Furthermore, since the heteroscedastic errors can be traced mainly to one observation, focusing on it remains a reasonable diagnostic strategy.

The other solutions to heteroscedastic errors do not require us to find the specific source of the problem, but are general solutions designed to adjust the standard errors, which, you will recall, tend to be affected by heteroscedasticity. One solution is called *White's correction, White's estimator,* or the *Huber–White sandwich estimator*[15], and the other is called the *Newey–West estimator.*[16] Both involve substantial matrix manipulation that is beyond the scope of this chapter. Fortunately, R has functions available that provide these approaches. For example, the `lmtest` package includes a function—`coeftest`—that, coupled with the sandwich package, furnishes the Huber–White sandwich estimator. This estimator is also called a *robust* estimator because it is relatively unaffected by extreme values in the data (see Chapter 14), such as those that might induce heteroscedasticity.

R code
```
library(sandwich)
library(lmtest)
```

---

[15] The founding documents for this estimator include Friedhelm Eicker (1967), "Limit Theorems for Regression with Unequal and Dependent Errors," *Proceedings of the Fifth Berkeley Symposium on Mathematical Statistics and Probability,* pp.59–82; Peter J. Huber (1967), "The Behavior of Maximum Likelihood Estimates under Nonstandard Conditions," *Proceedings of the Fifth Berkeley Symposium on Mathematical Statistics and Probability,* pp.221–233; and White (1980), *op. cit.* Unfortunately, Professor Eicker is not given credit in many of the designations for the estimator.

[16] Introduced in Whitney K. Newey and Kenneth D. West (1987), "A Simple, Positive Semi-Definite, Heteroskedasticity and Autocorrelation Consistent Covariance Matrix," *Econometrica* 55(3): 703–709.

```
coeftest(LRM9.3, vcovHC(LRM9.3, type="HC3"))
 # HC3 is one method among several[17]
```

R output (abbreviated)

	Estimate	Std. Error	t value	Pr(>|t|)
(Intercept)	17.206445	21.948681	0.7839	0.4372
RobberyRate	-0.337956	0.298433	-1.1324	0.2634
PrisonRate	0.069600	0.054409	1.2792	0.2074
PerPoverty	-5.178940	5.764268	-0.8985	0.3737
UnemployRate	12.673199	18.068778	0.7014	0.4867

```
---
```

Signif. codes: 0 '***' 0.001 '**' 0.01 '*' 0.05 '.' 0.1

The differences in the standard errors from the original model and the model that adjusts the standard errors are dramatic, with substantially larger standard errors in the latter model. Consider the unemployment rate: the standard error increases from 6.2 to 18.1, whereas for percent poverty it increases from 1.6 to 5.8. Why this occurs is not clear, though it likely involves the down-weighting of the relevant data from Wyoming. One lesson to take away from this analysis, therefore, is that a single extreme observation can have enormous consequences for an LRM. We explore some of additional implications of extreme values in Chapter 14.

The final method we'll discuss to adjust the standard errors for heteroscedasticity is called *bootstrapping*. Briefly, bootstrapping—which is mentioned in Chapter 5—is a resampling method that draws numerous random subsamples with replacement from the dataset, estimates a model (e.g., LRM) or computes a statistic (e.g., median) with each subsample, and then furnishes the average of the estimates.[18] If the model is affected by heteroscedasticity, bootstrapping tends to diminish the influence of areas of the model's data with larger variability, which should produce more robust results.[19]

In R, we may utilize the car package's Boot function to obtain bootstrapped standard errors based on LRM9.3 (see LRM9.3.boot).

---

[17] HC is short for heteroscedasticity consistent. Other algorithms, including HC0, HC1, HC2, and HC4 are also available. HC3 and HC4 work well with small samples. Differences among these are described in Achim Zeileis (2006), "Econometric Computing with HC and HAC Covariance Matrix Estimators," available at https://cran.r-project.org/web/packages/sandwich/vignettes/sandwich.pdf.

[18] Bootstrapping methods were introduced in the late 1970s by statistician Bradley Efron, who was inspired by earlier work on *jackknife resampling*—drawing repeated samples and dropping one observation in each to determine the behavior of estimates—by Maurice Quenouille, John Tukey, and Rupert Miller (see Bradley Efron (1979), "Bootstrap Methods: Another Look at the Jackknife," *Annals of Statistics* 7(1): 1–26).

[19] More information about bootstrap methods in R is available in Michael R. Chernick and Robert A. LaBudde (2014), *An Introduction to Bootstrap Methods with Applications to R*, Hoboken, NJ: Wiley. For an overview of resampling methods in general, including the bootstrap, see Laura M. Chihara and Tim C. Hesterberg (2011), *Mathematical Statistics with Resampling and R*, Hoboken, NJ: Wiley.

R code
```
library(car)
set.seed(39445) # for replication
LRM9.3.boot <- Boot(LRM9.3, R=500)
  # requests 500 samples with replacement
summary(LRM9.3.boot)
S(LRM9.3, vcov.=vcov(LRM9.3.boot))
Confint(LRM9.3.boot, level=.95)
```

R output (abbreviated)
```
Number of bootstrap replications R = 500
              original     bootBias      bootSE     bootMed
(Intercept)    17.20644   -4.2519202   19.232680   13.098724
RobberyRate    -0.33796   -0.0112576    0.231306   -0.368656
PrisonRate      0.06960    0.0021549    0.043114    0.075968
PerPoverty     -5.17894   -0.4804675    4.651256   -5.486310
UnemployRate   12.67320    2.6439019   14.256789   12.477520
```

```
Standard errors computed by vcov(LRM9.3.boot)
Coefficients:
             Estimate Std. Error t value Pr(>|t|)
(Intercept)  17.20644   19.23268    0.895    0.376
RobberyRate  -0.33796    0.23131   -1.461    0.151
PrisonRate    0.06960    0.04311    1.614    0.113
PerPoverty   -5.17894    4.65126   -1.113    0.271
UnemployRate 12.67320   14.25679    0.889    0.379
```

The first section of the R printout provides the slope coefficients from the original model; the bootstrap bias (bootBias), which is the difference between the slope coefficients for the full sample and the mean of the slope coefficients from the bootstrapped samples; the bootstrapped standard errors (bootSE); and the median slope coefficient from the bootstrapped samples (bootMed). The second section furnishes the results of the model with the bootstrapped standard errors and the *t*-values and *p*-values computed from them. Similar to the earlier model that employs the Huber–White estimator, those from the bootstrapping approach result in larger standard errors for each slope coefficient.

Some experts suggest that, because heteroscedasticity is such a common issue in LRMs, we should always use a correction method such as Huber–White or bootstrapping. If no heteroscedasticity is present, the results of the corrected and uncorrected models are similar. But if heteroscedasticity does occur, then the results of a standard LRM can be misleading.[20] In the case of the model of gun ownership estimated in this chapter, however, a simple solution to the heteroscedasticity issue is to use the natural logarithm of the outcome variable (see LRM9.3b). Thus, it is rarely a good idea to go to one of the standard error adjustment procedures without first examining the characteristics of the variables used in the LRM.

Chapter Summary

We now know about some of the risks of heteroscedasticity and its consequences: inefficient slope coefficients, which are manifest in incorrect standard errors. Obtaining homoscedastic errors is also key to whether the LRM's estimates are BLUE (see Chapter 4). As suggested earlier, heteroscedasticity may also be produced by skewed variables, nonlinear associations, or important variables that are left out of the model. An important exercise, therefore, is to continually think about the data and the variables in the dataset to evaluate the likelihood of heteroscedasticity.

Several diagnostic tools and solutions exist for heteroscedasticity. Here's a summary of several that are available in R.

1. Always plot the residuals by the predicted values after estimating an LRM (some experts prefer the studentized deleted residuals by the standardized predicted values). Partial residual plots are also helpful. A cone-shaped pattern to the residuals, one that either spreads out or narrows down at higher values of the predicted values, is indicative of heteroscedastic errors. A random pattern suggests homoscedastic errors.

2. Examine the distributions of the variables. As with the gun ownership model, a highly skewed outcome variable—or non-normality among the explanatory variables—might induce heteroscedasticity.

3. Several statistical tests for heteroscedasticity are available. White's test, Glejser's test, and the Breusch–Pagan test offer simple diagnostic methods after estimating an LRM.

4. The best solution for heteroscedasticity is to figure out why it occurs. Have you left a variable out of the model? Is there a pattern among the variables you haven't considered? Are there skewed variables? If

---

[20] Additional information about this suggestion is found in J. Scott Long and Laurie H. Ervin (2000), "Using Heteroscedasticity Consistent Standard Errors in the Linear Regression Model," *American Statistician* 54(3): 217–224.

you know a variable is inducing heteroscedasticity, consider a transformation or a WLS regression model.

5. If you don't know the source, or if you are simply worried about heteroscedasticity, then use a model specifically designed to correct this problem. The Huber–White sandwich estimator for standard errors works well. An alternative is to bootstrap the standard errors. The R sandwich package or the car package's Boot function provides a simple way to minimize the effects of heteroscedasticity on the standard errors.

Chapter Exercises

The dataset *GPATests.csv* includes data from 31 college sophomores who recently completed their freshman year, along with information about their performance on high school tests and family background. The following exercises are designed to demonstrate heteroscedastic errors and how they influence an LRM. The variables in the dataset include

- GPA   Freshman year grade point average
- MathTests   Average scores on high school math tests
- EnglishTests   Average scores on high school English tests
- FamilySES   Socieconomic status of the students' parents

After importing the dataset into R, complete the following exercises.

1. Estimate an LRM with GPA as the outcome variable and MathTests, EnglishTests, and FamilySES as the explanatory variables.
   a. Interpret the slope coefficients for MathTests and EnglishTests.
   b. Interpret the *p*-value and 95% confidence interval associated with the FamilySES slope coefficient.
2. Check the residuals by fitted values plot and the scale-location plot for evidence regarding whether the errors are homoscedastic. What do these plots suggest?
3. Create partial residual plots for each explanatory variable. What do these plots suggest?

4. Estimate the Breusch–Pagan test and Glejser's test based on the LRM in exercise 1.

   a. What do these tests suggest about whether the errors are homoscedastic?

   b. If you find evidence for heteroscedastic errors, which explanatory variables are implicated?

5. Re-estimate the model but with Huber–White sandwich estimators of the standard errors. What differences do you find between this model and the model estimated in exercise 1?

6. Re-estimate the model but bootstrap the standard errors. What differences do you find between this model and the model estimated in exercise 1?

7. *Challenge*: a quick fix is available that diminishes the evidence of heteroscedastic errors (hint: it involves a transformation of a variable). Identify and utilize this quick fix and re-estimate the model. Demonstrate that it worked by providing evidence that favors homoscedastic errors.

10

Collinearity and Multicollinearity

The third assumption of LRMs introduced in Chapter 4 states that no perfect collinearity occurs among the explanatory variables. The two components of the word *collinearity* are *co*, or together, and *linear*, or the implication of a straight line or a flat plane. The latter component is tied to the geometric relationship between two variables in space. Since we have not discussed the geometric bases of statistical models, we'll discuss collinearity as an issue of covariance.[1]

We're not, at this point, interested in the covariability of any particular explanatory variable with the outcome variable. Most researchers hope to find high covariability between at least one of the explanatory variables and the outcome variable. Collinearity concerns covariability among the explanatory variables. Recall that in Chapter 4 covariability was depicted as overlapping circles, with statistical adjustment in LRMs represented by the unique overlap between an explanatory and outcome variable (see Figure 4.4). But overlapping circles also come in handy for representing collinearity.

Figure 10.1 illustrates what happens when two explanatory variables are highly correlated. Suppose we wish to determine the unique association between x_1 and y, but x_1 and x_2 have this degree of overlap—indicative of very high covariation. Where is the association of x_1 and y controlling for the association with x_2? Notice the thin sliver where x_1 overlaps with y but does not include any portion of x_2. Assume we wish to estimate this overlap with an LRM. Do you see any potential problems?

$$se\left(\hat{\beta}_i\right) = \sqrt{\frac{\sum\left(y_i - \hat{y}_i\right)^2}{\sum\left(x_i - \bar{x}\right)^2\left(1 - R_i^2\right)\left(n - k - 1\right)}} \tag{10.1}$$

Recall the formula for the standard error of the slope coefficient (see Equation 10.1). As mentioned in Chapter 4, the tolerance $\left(1 - R_i^2\right)$ in the denominator is based on an auxiliary regression model that includes x_i as the outcome variable and all the other explanatory variables in the LRM as predictors. What will the tolerance be if x_1 and x_2 have a large overlap? We cannot be certain with a figure, but rest assured it will be small. What happens to the standard error of the slope as the tolerance decreases? Suppose, for example, that the correlation between x_1

---

[1] For those interested in the geometry of linear models, see Thomas D. Wickens (1995), *The Geometry of Multivariate Statistics*, New York: Lawrence Erlbaum Associates. Chapter 5 discusses collinearity.

DOI: 10.1201/9781003162230-10

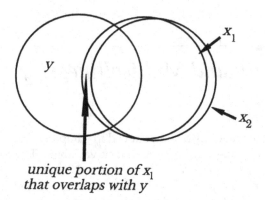

unique portion of x_1
that overlaps with y

FIGURE 10.1
Illustration of collinearity with variables represented with overlapping circles.

and x_2 is 0.95. Then, assuming an LRM with only two explanatory variables, the tolerance is $(1 - 0.95^2) = 0.098$. But if the correlation between another sample's x_1 and x_2 is much lower, say 0.25, the tolerance is also lower: $(1 - 0.25^2) = 0.438$. As the tolerance decreases, the standard error increases if other conditions, such as the sums of squares of x_1 and y, are held constant.

The assumption of linear regression, though, states that no *perfect* collinearity is present in the model. Perfect collinearity is represented by completely overlapping circles among explanatory variables, such as a correlation between x_1 and x_2 of 1.0 or −1.0. What happens to the standard error when perfect collinearity occurs? The tolerance in this situation is {1 − 1}, which means that the denominator in the standard error equation is zero. We cannot compute the standard error: it is not defined or is infinite. But rarely do we find a perfectly collinear association, unless we create a variable ourselves or accidentally rename a variable and enter it in an LRM with its original twin sibling. Statistical software such as R also recognizes when two explanatory variables are perfectly correlated and excludes one from the model. What is not uncommon, however, is to find high collinearity, such as Figure 10.1 represents.

Books on statistical theory emphasize that the main problem incurred in the presence of high or perfect collinearity is biased or unstable standard errors. Slope coefficients are considered unbiased, at least when the model is estimable. Practical experience indicates, however, that slopes coefficients can be volatile when two (or more) variables are highly collinear. Let's assess a model that manifests this problem.

The *StateData2018.csv* dataset includes three variables that we suspect are associated with violent crimes per 100,000 residents (`ViolentCrimeRate`): percent of children living below the poverty level (`PerChildPoverty`), the number of suicides per 100,000 residents (`SuicideRate`), and the age-adjusted number of suicides per 100,000 residents (`SuicRateAgeAdj`). Yet something is peculiar about the latter two variables: both address the state's number of suicides. Why is one called *age-adjusted*? As noted in Chapter 7, demographers and epidemiologists are well aware that several health outcomes, such as disease rates or the prevalence of some health problems, are influenced by the age distribution. Some states, for instance, may have a higher prevalence of particular types of cancer because their residents are, on average, older. Since older people are more likely to suffer from certain types of cancer, states with older populations have more cases of these cancers. Therefore, epidemiologists usually compute the *age-adjusted* prevalence or rate (see Chapter 7, fns. 10–11). In essence, age adjustment controls for the effects of age before one estimates a regression model. If age has a large association with a health problem, then the original prevalence of the variable and the age-adjusted prevalence might differ considerably. What about the two suicide variables in *StateData2018*? A correlation matrix provides useful evidence to judge the association between the age structure of states and the number of suicides.

R code
```
sub.StateData <- StateData2018[c("SuicideRate",
                "SuicRateAgeAdj", "PerChildPoverty")]
  # subset the StateData2018 data frame to enter into the
    cor function
cor(sub.StateData)
```

R output (abbreviated)

	SuicideRate	SuicRateAgeAdj	PerChildPoverty
SuicideRate	1.000	0.987	-0.107
SuicRateAgeAdj	0.987	1.000	-0.053
PerChildPoverty	-0.107	-0.053	1.000

The correlation between the number of suicides and the age-adjusted number of suicides is 0.99, which is close to the maximum value of one. We thus have two variables with much of their variability overlapping. Age structure is thus only modestly associated with the state-level prevalence of suicide since the distributions of the two variables are so similar (request a scatter plot of the two variables to confirm this). But suppose we include `SuicideRate` and `SuicRateAgeAdj` in an LRM designed to predict violent crimes per 100,000. LRM10.1 is clearly an artificial example since the two variables are measuring virtually the same thing. But let's use them nonetheless to explore the effects of collinearity in an LRM.

R code[2]
```
LRM10.1 <- lm(ViolentCrimeRate ~ SuicideRate +
              SuicRateAgeAdj + PerChildPoverty, data =
              StateData2018)
```

---

[2] If we've created a subset of only the variables used in the model, a time-saving step in R is LRM10.1 <- lm(ViolentCrimeRate ~ ., data=subset.usdata). R recognizes that we wish to use the variables other than the outcome (ViolentCrimeRate) as explanatory.

```
summary (LRM10.1)
confint (LRM10.1)
```

R output (abbreviated)
Coefficients:
```
                 Estimate   Std. Error   t value   Pr(>|t|)
(Intercept)       140.485      89.426      1.571    0.12305
SuicideRate       -31.701      22.043     -1.438    0.15717
SuicRateAgeAdj     33.864      22.835      1.483    0.14489
PerChildPoverty    11.597       3.628      3.196    0.00252 **
---
Signif. codes: 0 '***' 0.001 '**' 0.01 '*' 0.05 '.' 0.1

Residual standard error: 113 on 46 degrees of freedom
Multiple R-squared: 0.2772, Adjusted R-squared: 0.23
F-statistic: 5.879 on 3 and 46 DF, p-value: 0.001744

[CIs]                 2.5%       97.5%
(Intercept)        -39.520     320.491
SuicideRate        -76.072      12.675
SuicRateAgeAdj     -12.100      79.829
PerChildPoverty      4.293      18.900
```

Scrutinize the slope coefficients for the two suicide variables: one is negative (−31.7) and the other is positive (33.9). Statistical theory suggests that these slopes are based on an unbiased estimator in the presence of non-perfect collinearity,[3] but you can see they are untenable. The variables measure the same phenomenon, yet suggest that as one increases violent crimes tend to decrease, whereas as the other increases violent crimes tend to increase (though the 95% CIs of their slope coefficients are wide and include zero). Both interpretations of the association between suicides and violent crimes cannot be correct.

An LRM that omits one state, Mississippi, is displayed in LRM10.1.NoMiss.

R code
```
LRM10.1.NoMiss <- lm(ViolentCrimeRate ~ SuicideRate +
                     SuicRateAgeAdj + PerChildPoverty,
                     data = StateData2018[-24,])
 # omit row 24 from the data - Mississippi
summary(LRM10.1.NoMiss)
confint(LRM10.1.NoMiss))
```

---

[3] Remember that the property of unbiasedness refers to whether the mean of the sampling distribution of a statistic equals the parameter it is intended to estimate in the population. In other words, does it hit the target, on average, when one takes repeated samples from the population (see Figure 2.3)? Keep in mind, though, that any particular sample will rarely yield a statistic (e.g., $\hat{\beta}$) that precisely matches the parameter.

R output (abbreviated)
```
Coefficients:
                  Estimate  Std. Error  t value  Pr(>|t|)
(Intercept)       116.686      84.987     1.373  0.176563
SuicideRate       -43.506      21.328    -2.040  0.047263 *
SuicRateAgeAdj     45.654      22.058     2.070  0.044245 *
PerChildPoverty    13.938       3.548     3.929  0.000291 ***
---
Signif. codes: 0 '***' 0.001 '**' 0.01 '*' 0.05 '.' 0.1

Residual standard error: 106.8 on 45 degrees of freedom
Multiple R-squared: 0.3653, Adjusted R-squared: 0.3229
F-statistic: 8.632 on 3 and 45 DF, p-value: 0.0001232

[CIs]                  2.5%     97.5%
(Intercept)         -54.488   287.859
SuicideRate         -86.464    -0.549
SuicRateAgeAdj        1.228    90.081
PerChildPoverty       6.792    21.083
```

The slope coefficients and standard errors of the two suicide variables shift quite a bit from one model to the next, whereas those for percent child poverty move by only a skosh. Although this instability may be because the state of Mississippi has unusual values on the suicide variables (see Chapter 14), it likely results from collinearity between the two variables. Collinearity often creates unstable coefficients that shift substantially with minor changes to the model.

We probably would not estimate a model with two variables that are so similar. Even though we used these models to illustrate a point, we should consider whether two (or, as we shall see, more) explanatory variables are highly correlated *before* estimating the model. Let's now examine an LRM that includes only one suicide variable (see LRM10.2).

R code
```
LRM10.2 <- lm(ViolentCrimeRate ~ SuicideRate
                + PerChildPoverty, data=StateData2018)
summary(LRM10.2)
confint(LRM10.2)
```

R output (abbreviated)
```
Coefficients:
                  Estimate  Std. Error  t value  Pr(>|t|)
(Intercept)       110.1471    88.1585     1.249  0.217698
SuicideRate         0.6246     3.3270     0.188  0.851895
PerChildPoverty    13.4045     3.4608     3.873  0.000331 ***
---
Signif. codes: 0 '***' 0.001 '**' 0.01 '*' 0.05 '.' 0.1
```

```
Residual standard error: 114.5 on 47 degrees of freedom
Multiple R-squared: 0.2426, Adjusted R-squared: 0.2104
F-statistic: 7.527 on 2 and 47 DF, p-value: 0.001459

[CIs]                    2.5%      97.5%
(Intercept)           -67.205    287.499
SuicideRate            -6.069      7.318
PerChildPoverty         6.442     20.367
```

The results suggest a positive association between the number of suicides per 100,000 residents and the number of violent crimes per 100,000 residents (β = 0.62) after statistically adjusting for the association with percent child poverty. Although the 95% CI includes zero (−6.1, 7.3), notice the differences from the earlier models. More importantly, consider the standard error in this model, which is much smaller than the standard errors from the other models. This demonstrates how collinearity inflates standard errors. The slopes and standard errors related to percent child poverty are roughly the same across the models because this variable's correlation with the suicide variables is low.

Multicollinearity

The tolerance from the auxiliary regression equation is based on the proportion of variance among one explanatory variable explained by the other explanatory variables. In other words, the tolerance is derived from the R^2 of an LRM. I hope it is clear that we should consider overlapping variability between not only two variables but among all the explanatory variables. Suppose, for instance, that an LRM includes four explanatory variables, x_1, x_2, x_3, and x_4. The largest bivariate correlation between any two variables is 0.4. If we were to use only the two variables with the largest correlation in an LRM, the tolerance in the standard error formula would not be any smaller than $(1 - 0.4^2 =)$ 0.84. A tolerance of this magnitude is not likely to create instability in the model's coefficients. But some set of explanatory variables might be a *linear combination* of the others.[4] If, for instance, x_4 is predicted perfectly or nearly perfectly by x_1, x_2, and x_3, then the tolerance will approach zero and the same problems ensue: overly large standard errors and unstable slopes. In terms of the overlapping circles (see Figure 10.2), multicollinearity appears as two or more variables encompass another explanatory variable.

Examining only the bivariate correlations among a set of explanatory variables is, therefore, rarely sufficient for determining whether multicollinearity

---

[4] The term *linear dependence* is an alternative name for collinearity since, as implied here, explanatory variables that share a substantial linear association are collinear.

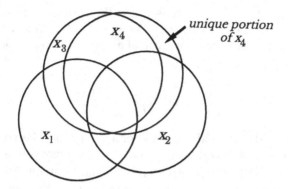

FIGURE 10.2
Illustration of multicollinearity with variables represented with overlapping circles.

is present. Any of the bivariate correlations may be below thresholds that warn of collinearity, but they won't reveal a potential problem if some combination of explanatory variables predicts another explanatory variable.

How to Detect Collinearity and Multicollinearity

Since bivariate correlation matrices are of limited use for detecting multicollinearity, we should find alternative diagnostic tools. First, though, consider a criterion that researchers have used to gauge collinearity: estimate a correlation matrix and then look for bivariate correlations that match or exceed $|0.7|$. If any are identified, collinearity might be affecting the LRM results. We should not accept this threshold without considering the sample size, however. All else being equal, collinearity is less likely as the sample size gets larger because greater variability is available to share. Although $|0.7|$ provides a general point of reference, other tools should be considered as well.

But what about multicollinearity? Are diagnostic tests available that assess its presence? How would one go about devising such tests? A hint is in the earlier discussion of the tolerance formula in the standard error equation (10.1). Can we use this information to assess the possibility of multicollinearity? A simple step is to regress each explanatory variable on all the others (the auxiliary regressions described earlier) and compute an R^2 for each model. We need a threshold criterion, though: how high must an R^2 be before we suspect multicollinearity? Perhaps we should use the same criterion mentioned earlier: 0.7. Since the auxiliary LRM provides an R^2, though, perhaps it should be the square of the threshold value: $0.7^2 = 0.49$. Some researchers use criteria that specify the threshold for the auxiliary R^2 and the tolerance, but a measure used more frequently is called the *variance inflation factor* (VIF), which has become the test *de jour* for assessing multicollinearity.[5] The VIF is defined as the inverse of the tolerance of an auxiliary regression model (see Equation 10.2).

$$\text{VIF} = \frac{1}{\left(1 - R_i^2\right)} \tag{10.2}$$

A commonly used threshold is to look for a VIF greater than or equal to 9 or 10 as indicative of multicollinearity.[6] Others recommend that if the square root of the VIF ($\sqrt{\text{VIF}}$) is greater than or equal to about three, we should suspect sufficient multicollinearity to raise concerns. One way to understand the square root of the VIF is to note that it represents the increased width of the 95% CI for the slope. For instance, a VIF of nine suggests that the 95% CI for the slope is three times as wide as it would be if the explanatory variables had a correlation of zero.[7] Alternatively, the $\sqrt{\text{VIF}}$ denotes the increase in the standard error of a slope coefficient due to associations with other explanatory variables. As mentioned earlier, though, don't forget to consider the sample size: LRMs are affected less by collinearity or multicollinearity as the sample size increases. All else being equal, a model that uses a sample size of 10,000 is less sensitive to multicollinearity—at least using conventional criteria—than a model with a sample size of 100. Unfortunately, the decision rules that have emerged for VIFs do not normally consider the sample size.

Another standard diagnostic test for multicollinearity involves *condition indices* and *variance proportions* (VPs). Before demonstrating how to use this test, let's take a quick journey into the intersection of linear algebra and principal components analysis (PCA) for some terminology.[8] The *principal components* (PCs) of a set of variables are a reduced set of components that are linear combinations of the variables. PCs are one type of *latent variable* (see Chapter 13), yet those from a set of variables are uncorrelated with one another. If we have a set of, say, ten explanatory variables, we might reduce

---

[5] This does not mean there aren't critics. See, for example, José Dias Curto and José Castro Pinto (2011), "The Corrected VIF (CVIF)," *Journal of Applied Statistics* 38(7): 1499–1507.

[6] The VIF may bring to mind the design effect described in Chapter 8 since both gauge how much the variance measures in an LRM are affected by dependence. Some researchers equate the two by using the terms VIF and design effect as synonyms and liken their effects on the standard errors from LRMs (see F. Y. Hsieh et al. (2003), "An Overview of Variance Inflation Factors for Sample-Size Calculation," *Evaluation & the Health Professions* 26(3): 239–257).

[7] See Fox and Weisberg (2018), *op. cit.*, chapter 8.

[8] PCA is a data reduction technique that takes a set of variables and attempts to reduce them to a "latent" variable that measures how closely related they are to one another. This method is useful for determining if a set of variables are indicators of some concept. As discussed later, PCA is also helpful when a set of explanatory variables is highly collinear. Karl Pearson, of Pearson's *r* fame, introduced PCA in 1901 ("On Lines and Planes of Closest Fit to Systems of Points in Space," *London, Edinburgh, and Dublin Philosophical Magazine and Journal of Science* 2(11): 559–572).

them to a set of, perhaps, two or three uncorrelated PCs. An *eigenvalue* is the variance of a PC. Briefly, then, eigenvalues are a measure of joint variability among a set of explanatory variables. As the eigenvalue gets smaller, more shared variability exists among the variables, with a value of zero indicating perfect collinearity. An interesting property of eigenvalues is that they sum to equal the number of variables plus the intercept, or $k + 1$, when used to assess an LRM.

But how is this useful for judging multicollinearity? We may use eigenvalues to compute *condition indices*, which are another measure of joint variability among explanatory variables. They are used in conjunction with VPs to assess multicollinearity in LRMs (they may also be used with other regression models). VPs assess the amount of variability in each explanatory variable (as well as the intercept) that is accounted for by the *dimension*, another term for the PC. Let's now see an example of VIFs, condition indices, and VPs. First, re-estimate the LRM that includes both suicide variables and percent child poverty (see LRM10.3). Then use the vif function from the car package to calculate the VIFs. Finally, execute the ols_eigen_cindex function available in the olsrr package.[9]

R code
```
library(car)
library(olsrr)
LRM10.3 <- lm(ViolentCrimeRate ~ SuicideRate +
              SuicRateAgeAdj + PerChildPoverty,
              data=StateData2018)
 # there's no need to output the results
vif(LRM10.3)
ols_eigen_cindex(LRM10.3)
```

R output (abbreviated)
```
SuicideRate SuicRateAgeAdj PerChildPoverty
   45.539         45.144           1.140
```

	Eigen.	CI	intercept	Suicide Rate	SuicRate AgeAdj	PerChild Poverty
1	3.866	1.000	0.002	0.000	0.000	0.004
2	0.109	5.970	0.012	0.003	0.003	0.314
3	0.025	12.441	0.900	0.001	0.004	0.559
4	0.001	69.216	0.085	0.995	0.993	0.123

Examine the VIFs first. Two of them—those associated with suicides and age-adjusted suicides—are well above the threshold of 10 (or $\sqrt{\text{VIF}} > 3$), which provides strong evidence of collinearity involving these two variables.

---

[9] An alternative is to use the perturb package's colldiag or perturb function to examine LRMs for multicollinearity issues.

Second, examine the eigenvalues and condition indices. The condition index column is arranged from the smallest to the largest value. Its entries are calculated based on Equation 10.3.

$$\text{Condition index} = \sqrt{\frac{\text{largest eigenvalue}}{\text{eigenvalue}_i}} \rightarrow \sqrt{\frac{3.866}{0.109}} = 5.96 \qquad (10.3)$$

As shown, the condition index in row 2 is derived by dividing the largest eigenvalue, 3.866, by its eigenvalue, 0.109, and taking the square root of the quotient. Larger condition indices specify that more of the joint variability of the variables is accounted for by the dimension.

The VPs are furnished in columns 3–6. Larger VPs suggest a greater proportion of a variable's variance is accounted for by the dimension gauged by the row's condition index. To determine whether sufficient multicollinearity is present to warrant concern, look for any condition index greater than or equal to 30. If one is identified, look across its row and find any VP that exceeds 0.50. A condition index greater than or equal to 30 coupled with a VP greater than or equal to 0.50 indicates a collinearity or multicollinearity issue that should be analyzed further. An advantage of condition indices and VPs relative to VIFs is that they identify the specific variables that are highly (multi)collinear. They are also useful for determining if the intercept is involved in a collinearity issue.[10]

According to the collinearity diagnostics for LRM10.3, one condition index, associated with row 4, exceeds 30: 69.22. Looking across this row, note that two VPs exceed 0.50: those associated with suicides (0.995) and age-adjusted suicides (0.993). We therefore have compelling evidence that these two variables are highly collinear. Child poverty, on the other hand, has a VP of 0.12 in the fourth row. Its VP is 0.56 in row three, but the row's condition index is only 12.44, so we do not conclude that the child poverty variable is implicated in a collinearity issue.

What to Do About Collinearity and Multicollinearity

Now that we've learned how to detect collinearity issues in an LRM, let's examine some possible solutions. The first solution is mentioned early in the chapter. Recall that collinearity and multicollinearity have less influence on LRMs as the sample size gets larger because larger samples tend to have more variability with which to estimate models. If a model is beset with

---

[10] Researchers are rarely interested in collinearity issues involving the intercept since they are seldom concerned with intercepts from LRMs. A common situation when a collinearity issue implicating the intercept occurs, though, is when the model explains a large proportion of the variance in the outcome (e.g., $R^2 > 0.95$).

collinearity problems, then, if possible, one should add more observations to the sample. This is rarely feasible outside of a laboratory, however. Survey data are typically of a fixed sample size because of cost constraints, so adding more observations is seldom practical. But in this era of online sampling, it may be feasible to recruit more respondents to a study.

A second solution is to consider combining collinear variables. Suppose that some subset of explanatory variables is involved in a multicollinearity problem. Inspect these variables: are they measuring a similar phenomenon? Would it be practical to combine them in some way? Highly collinear variables are often indirectly measuring some underlying concept, so using them to create a latent variable—such as with PCA—is a common solution to multicollinearity. Rather than including the collinear variables in the regression model, we may place the latent variable in the model. We may then assess its association with the outcome variable.[11]

A third solution is to use a technique known as *ridge regression*. Rarely recommended as a general solution, this model may offer some benefit when nothing else seems to work or in specific situations, which we won't go into here. The general idea underlying ridge regression is to create (artificially) more variability among the explanatory variables. Recall that the diagonals of a variance–covariance matrix include the variances of the explanatory variables (see Equation 4.5). Since collinearity or multicollinearity involve, in one sense, too little variability in a set of explanatory variables, ridge regression involves adding a constant to the variances, thus increasing variability and decreasing the effects of multicollinearity. The choice of a constant may be arbitrary or based on a specific formula. This might sound promising, but ridge regression also results in biased slope coefficients. But the bias may be less than in the OLS estimates made unstable by multicollinearity. A key problem with ridge regression, though, is that it seems to change slope coefficients with larger standard errors more than those with smaller standard errors. Assuming a classic inferential approach, the analyst should therefore have a well-conceived conceptual model to determine which explanatory variable's slope coefficients are and are not meaningful.[12]

A fourth solution, which is not normally recommended, is to use an automated variable selection procedure to choose the "best" set of predictive variables. These methods typically drop one or more of the variables involved in collinearity or multicollinearity problems. They also retain the

---

[11] PCs are one type of latent variable, but others are also available. The R function `princomp` and R packages `psych`, `lavaan`, `sem`, and `FactoMineR` have capabilities for estimating various types of latent variables. For more information about latent variables in R, see W. Holmes Finch and Brian F. French (2015), *Latent Variable Modeling with R*, New York: Routledge; and A. Alexander Beaujean (2014), *Latent Variable Modeling Using R: A Step-By-Step Guide*, New York: Routledge. Chapter 13 presents additional information about latent variables and LRMs.

[12] See Chatterjee and Hadi (2012), *op cit.*, chapter 11, for a discussion of ridge regression. Ridge regression is available in R's `ridge` package, in the `regtools` package (`ridgelm`), and in the `lmridge` package (see Muhammad Imdad Ullah et al. (2018), "lmridge: A Comprehensive R Package for Ridge Regression," *R Journal* 10(2): 326–348).

explanatory variables that are the strongest predictors of the outcome variable (Chapter 12 furnishes an example). However, these methods also take important conceptual work out of the hands of the analyst and should generally be avoided.[13,14]

The last solution is recommended least often but is probably employed most frequently by those who use regression models: omit one or more of the variables involved in the collinearity or multicollinearity issue. Taking such a step should be seen as a last resort and the analyst should be able to provide a clear reason why a particular variable is removed. This approach is most useful when one of the collinear variables measures virtually the same phenomenon as other collinear variables (such as was the case with the two suicide variables examined in this chapter). Two variables rarely measure the same thing, however. Collinearity is more often caused by recoding problems, such as when one variable is created from another, or due to a small sample size.

Collinearity and multicollinearity present some potentially serious problems for LRMs. Biased standard errors are a nuisance, but, when coupled with unstable slope coefficients, the results of models are not trustworthy. What should we do about this problem? The best approach is to consider carefully the source of collinearity or multicollinearity and then use the most reasonable solution available. Alas, there are no easy answers.

Chapter Summary

LRMs are difficult to estimate when explanatory variables are strongly related to one another, when their shared variability is particularly high. If the model is based on a small sample size, collinearity and multicollinearity can also present nettlesome obstacles to attaining valid estimates. A good idea is to carefully consider what the explanatory variables are measuring and their association with one another before estimating an LRM. Researchers should also test for multicollinearity even if they have little

---

[13] See Fox (2016), *op. cit.*, chapter 13, for more information on automated variable selection procedures and why researchers should be wary of them.

[14] With the explosion of data science, data analytics, and machine learning there has been more emphasis on automated variable selection tools because they can be efficient for establishing a highly predictive model with large datasets. Many data scientists are concerned mainly with predicting an outcome variable accurately, so creating a successful prediction model is their goal. For example, companies regularly compile very large datasets with extensive customer information. One of their main objectives is to predict how customers behave when confronted with various options (e.g., advertising, shelf location, time of day, or webpage design), but are less concerned with *why* customers behave in certain ways (explanation). Rather, in order to set or meet, say, certain sales thresholds, analysts are simply interested in predicting if customers buy certain products under particular conditions.

reason to suspect its presence. Diagnosing a model is simple with R's `vif` or `ols_eigen_cindex` functions.

If multicollinearity is present in the model, don't immediately drop one or more of the variables. Instead, consider the other solutions described in this chapter. One of them might be feasible. In addition, think about issues such as collinearity during the study design phase, including throughout the sampling and instrument creation stages. Though outside the scope of this book, considering whether collinearity issues might be induced by the study design should always be on one's mind.

Chapter Exercises

The dataset *StateHealth2015.csv* includes data from 50 U.S. states, the District of Columbia, Guam, and the U.S. Virgin Islands. It provides information on the percent of adult residents in different types of living arrangements (e.g., married, widowed, divorced, separated, and never married), who graduated from college, who did not have a paid job for the previous year, and several health and nutrition indicators. We shall explore these data using an LRM and consider some diagnostics regarding collinearity and multicollinearity.

Once you've imported the dataset into R, complete the following exercises.

1. Estimate a linear regression model that uses `dailysmoker` (percent of adult residents who are daily cigarette smokers) as the outcome variable and the following explanatory variables: `married`, `divorced`, `separated`, `nevermarried`, `college`, `nojob1year`, `bingedrink`, and `fruit _ vegetables`.

   a. Interpret the slope coefficient associated with the variable `nojob1year` (percent of adult residents who were not employed for the past year).

   b. Interpret the CI associated with the variable `nojob1year`.

   c. Interpret the multiple R^2 from the model.

2. Estimate a correlation matrix with the explanatory variables from the LRM estimated in exercise 1. Using Pearson's $r \geq |0.7|$ as a threshold, identify any of the pairs of variables that we should consider as potentially producing a collinearity situation.

3. Compute the VIFs from the model estimated in exercise 1. Describe whether the VIFs indicate a collinearity or multicollinearity issue

with the model (make sure you provide specific information from the VIFs and which, if any, variables are implicated).

4. If you find any variables that are collinear, construct a scatter plot or scatter plots of each pair of these variables. Include linear fit lines in the scatter plot(s). What does (do) the scatter plot(s) indicate about the association(s) among these explanatory variables and whether we should be concerned about the model because of their covariability?

5. Compute the condition indices and VPs from the model estimated in exercise 1.

   a. How many, if any, collinearity and multicollinearity issues are shown by these measures?

   b. If you find any issues, what are they? Be specific about what the condition indices and VPs indicate and which variables are implicated.

6. If the evidence from the condition indices and VPs does show one or more collinearity or multicollinearity issues, discuss one approach you could take to diminish the problem. (Note: assume that adding more data is not feasible in this situation because each U.S. state and territory, with a few exceptions, is already represented.)

7. *Challenge*: assuming you found evidence of a multicollinearity issue in exercise 5, estimate a PCA of the implicated variables (and any others that it is reasonable to include). Use the results of the PCA to create a PC (a latent variable) that underlies the set of variables and re-estimate the model in exercise 1 but use the PC rather than its set of constituent variables. Interpret the slope coefficient associated with the PC. (Hint: you need to identify what the PC measures.)

11

Normality, Linearity, and Interaction Effects

In Chapter 4 we discussed two assumptions of LRMs: normality and linearity. The first is concerned with whether the errors of prediction follow a normal distribution. Even if not as vital as other assumptions, satisfying normality is important for ensuring that the estimators are efficient. This means that, when the normality assumption is met, the OLS estimators yield estimates with the smallest variance among all unbiased estimators.[1] Second, the linearity assumption concerns whether the mean value of Y for each specific combination of the Xs is a linear function of the Xs. Remember that an LRM attempts, using least squares techniques, to fit a straight line or flat surface to a set of data. But, as shown in Figure 11.1, when the association between X and Y is not linear—rather, when it exhibits curvature or is *curvilinear*—the OLS estimates will not yield satisfactory results.

Curvilinear and other associations among variables that are not represented well by a straight line are called *nonlinear* associations. Although making comparisons across the social and behavioral sciences is difficult, nonlinear associations are likely the norm, with linear associations the exception.[2] Yet linear associations are much simpler to conceptualize and model using statistical procedures such as correlations and LRMs. A growing literature addresses nonlinear regression techniques, and modern-day statistical software provides convenient tools for estimating nonlinear associations, but the way we think about associations among variables still tends to be linear. Perhaps this involves the limitations of human cognition, but it might also mean we just need more experience with nonlinear thinking and modeling.

---

[1] Chapter 4 mentions an additional assumption frequently lumped in with the normality assumption: the expected mean of the errors is zero (this is also called *zero conditional mean*). We won't address this assumption in detail since (a) it is difficult to test and (b) its main consequence involves the correct estimation of the intercept, which is rarely a principal concern among users of LRMs.

[2] In a wise book, Gerald van Belle (2008, p.10) cautions, "Beware of linear models ... a statistical model such as a linear model is a good first start only" (*Statistical Rules of Thumb*, New York: Wiley). This is similar to a point made by the eminent British economist John Maynard Keynes more than 80 years ago. In a review of a book on investment and business cycles, Keynes (1939, p.564) observes that it is an "improbable postulate to suppose that all economic forces are [linearly related]. ... indeed, it is ridiculous" ("Official Papers," *The Economic Journal* 49(195): 558–577). Moreover, recall that the claim that the OLS estimators are BLUE (see Chapter 4) addresses *linear estimators* or those that gauge straight-line/flat surface associations.

DOI: 10.1201/9781003162230-11

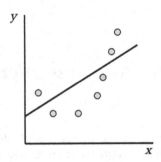

FIGURE 11.1
Illustration of a curvilinear association.

After discussing the normality assumption in more detail, we'll address the interesting topic of linear and nonlinear associations. We'll also assess another way of considering associations among explanatory variables by introducing *interaction effects*.

Are the Errors of Prediction Normally Distributed?

As mentioned earlier, we assume the errors of prediction are normally distributed. Some call this a "soft" assumption because it is not required for the OLS estimators to be BLUE. Still, normality is worth discussing since, when satisfied, the OLS estimators are the *most efficient* unbiased estimators. Recall that the errors of prediction (or their representatives in a sample—the residuals) are presumed to measure everything about the variability of the outcome variable once we've considered the part that is accounted for by the explanatory variables. We hope that all that is left is random noise, but this is not always the case because important explanatory variables may have been omitted. We still assume, though, that the variability that remains is normally distributed; in other words, it follows a normal (Gaussian) distribution. This assumption is examined after estimating the LRM, even though many analysts assess whether the outcome variable itself is normally distributed.[3] Since the latter step is taken so frequently, let's see how we might examine the "normality" of a variable. As noted in Chapter 2, the most convenient methods for inspecting its distribution are with a *histogram*, a *box*

---

[3] Whether a variable is normally distributed is usually a question for continuous variables. When a variable is non-continuous or discrete (such as a categorical variable), researchers are still interested in *how* it is distributed. Chapter 16 addresses one type of outcome variable that follows a binomial distribution. For additional information about regression models for non-continuous outcome variables, see John P. Hoffmann (2016), *Regression Models for Categorical, Count, and Related Variables*, Oakland, CA: University of California Press.

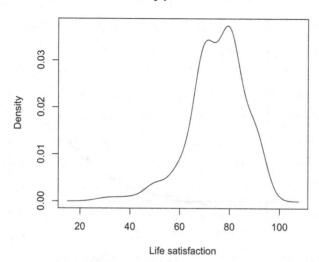

FIGURE 11.2
Kernel density plot of life satisfaction.

plot, a *kernel density plot*, or a *normal q-q plot*. Figure 11.2 provides an example of a kernel density plot in R that furnishes the distribution of the life satisfaction variable (`lifesatis`) from the *GSS2018.csv* dataset. The distribution has a slight left skew to it. However, much of its density is close to a normal (Gaussian) distribution.

Another useful graph for assessing the distribution of a variable is a normal *q-q* plot, which statisticians also call a *normal probability plot* or a *Gaussian rankit plot*. Here's how it works. First, the statistical software determines the mean and standard deviation of the variable. Second, it generates a simulated variable that follows a normal distribution but has the same mean and standard deviation as the observed variable. Third, it orders the simulated variable and the observed variable from their lowest to highest values. Fourth, it creates a scatter plot with the simulated variable on the *x*-axis and the observed variable on the *y*-axis. If the observed variable is normally distributed, the data points line up along a diagonal line drawn by the software. If not normally distributed, the points deviate from the diagonal line. The direction of deviation provides some guidance as to how the observed variable is distributed. Let's inspect a couple of normal *q-q* plots in R (see Figures 11.3 and 11.4) and then examine the residuals from an LRM.

R code for Figure 11.2
```
plot(density(na.omit(GSS2018$lifesatis), adjust = 2),
    main = "Kernel density plot of life satisfaction",
    xlab="Life satisfaction")
```

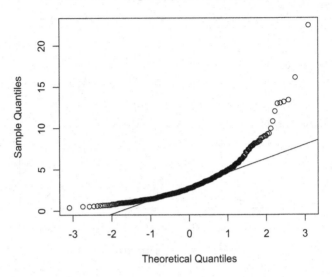

FIGURE 11.3
A normal *q-q* plot of a right-skewed variable.

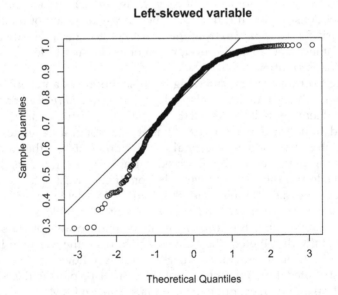

FIGURE 11.4
A normal *q-q* plot of a left-skewed variable.

```
# na.omit removes the missing values from the variable
  before creating the graph; adjust smooths the line so
  it doesn't appear too bumpy
```

Figure 11.3 provides a normal q-q plot of a right-skewed variable. The plot shows that the observed variable's observations (sample quantiles) manifest a much longer right tail than a normal distribution. One way to *normalize* such a distribution (this is simply shorthand for saying we are trying to transform it into a normally distributed variable) is to take its square root or its natural logarithm ($\log_e$) to "pull in" the long right tail.[4] Regression models with logged outcome variables are so widespread in applied statistics that they are called *log-linear* models (see, for example, LRM9.3b in Chapter 9). Also common in economics, the interpretation of a regression coefficient in a log-linear regression model is known as *semi-elasticity* and is easily transformable so its interpretation is based on a percent difference/change in y for each one-unit difference/change in x.

Figure 11.4 illustrates a variable that has an overly short right tail and a long left tail: it manifests left skew. In order to normalize it, we might square or exponentiate its values to stretch them out. Exponentiating a variable involves raising the irrational number e to the power of the observed value, e^{xi}.

As an exercise, take the gross state product variable (GrossStateProduct), which is right skewed, from the *StateData2018.csv* dataset and transform it using the square root function (sqrt) and the natural logarithm function (log) in R. Then, evaluate these two transformed variables using a normal q-q plot (qqnorm and qqline functions in R). What do the distributions look like?

We have not exhausted the ways we might transform a variable to induce a normal distribution. In addition, transformed variables may be placed in an LRM just like any other outcome variable, but, as suggested by the semi-elasticity discussion, the interpretations of the coefficients change to account for the transformation. For example, if we estimate an LRM with the *GSS2018.csv* dataset and use the squared values of lifesatis as the outcome variable and pincome as the explanatory variable, the slope indicates differences in *squared units* of life satisfaction that are associated with each one-unit difference in personal income.[5]

---

[4] Remember that when we take the square root or the natural logarithm of a variable, we need to be careful of negative and zero values. Recall that the natural logarithm of zero is undefined. The square root and natural logarithm of negative numbers are also undefined (absent imaginary numbers). If you attempt to take the natural logarithm or square root of a variable's negative value, R returns a missing value along with a warning. One solution is to add a constant to a variable with zero or negative values so that it has only positive values.

[5] Numeric tests are also available for determining whether a variable or the residuals from a regression model follow a normal distribution. Three of the most common are the *Shapiro–Wilk* test, the *Kolmogorov–Smirnov* test, and the *Anderson–Darling* test. These tests are not, however, particularly helpful with large samples and tend to be sensitive to extreme values. Visual methods typically work just as well, if not better. For more information about these tests and how well they perform, see Bee Wah Yah and Chiaw Hock Sim (2011), "Comparisons of Various Types of Normality Tests," *Journal of Statistical Computation and Simulation* 81(12): 2141–2155.

We've discussed thus far what we should do to test whether the outcome variable is normally distributed before estimating an LRM. But the normality assumption addresses the errors of prediction rather than the y variable. We may use a normal q-q plot to determine their distribution (although some analysts prefer a kernel density plot), which is available after estimating the LRM using the plot function (recall this function also provides the graphical tests for heteroscedasticity introduced in Chapter 9). Let's estimate an LRM with the *StataDate2018* dataset and examine the residuals. Opioid deaths per 100,000 residents serve as the outcome variable, with the following explanatory variables: average life satisfaction, percentage of residents of ages 0–18, percent of residents who are White, and the unemployment rate (see LRM11.1). We'll then request a post-model plot.

R code
```
LRM11.1 <- lm(OpioidODDeathRate ~ LifeSatis +
            PerAge0_18 + PerWhite + UnemployRate,
            data=StateData2018)
plot(LRM11.1, 2) # request the second plot only, which
                    is shown in Figure 11.5
```

The graph provided in the plots window is the normal q-q plot of the standardized residuals (see Figure 11.5).[6] The residuals from the model fall away from the diagonal line near the top of the distribution. Recall from Figure 11.3 that this suggests a right-skewed variable. Depending on the sample size, the distribution might be sufficient for some researchers, but can we make the residuals fall closer to a normal distribution? Let's examine the residuals from a model that uses the natural logarithm of opioid deaths as the outcome variable and the same explanatory variables (see LRM11.1a).

R code
```
LRM11.1a <- lm(log(OpioidODDeathRate) ~ LifeSatis +
            PerAge0_18 + PerWhite + UnemployRate,
            data=StateData2018)
 # the natural logarithm function (log) is embedded in
    the lm function so that the outcome variable is the
    log of opioid deaths per 100,000
plot(LRM11.1a, 2) # produces Figure 11.6
```

Figure 11.6 furnishes the normal q-q plot from the updated model. The residuals on the upper end of the distribution, with the exception of the observation in the upper right of the graph (it is observation 44), fall close to

---

[6] A similar plot is available with the car package's qqPlot function:

```
library(car)
qqPlot(LRM11.1, ylab="Studentized residuals", envelope=FALSE)
```

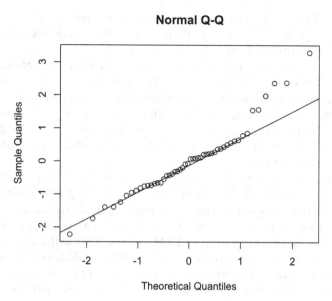

FIGURE 11.5
A normal *q-q* plot of the standardized residuals from LRM11.1.

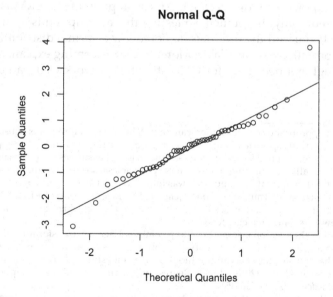

FIGURE 11.6
A normal *q-q* plot of the standardized residuals from LRM11.1a.

the diagonal line, but an observation on the lower end falls off the line. This model appears to do a better job than the previous model of satisfying the normality assumption, however.[7]

A useful exercise is to explore other variables that demonstrate substantial skew to practice normalization approaches. For instance, as mentioned earlier, the gross state product variable from the *StateData2018* dataset is skewed to the right. Try estimating a model with this variable and its various transformations as outcome variables and population density (PopDensity) as the explanatory variable. Can you find a transformation that works well to normalize the residuals?

You will rarely find an outcome variable in the social and behavioral sciences that has observations that fall directly on the diagonal line of a normal *q-q* plot or that look as symmetric as a true normally distributed variable. But remember that the goal is to find a transformation, assuming one is needed, which allows the residuals from an LRM to come close to the diagonal line in a normal *q-q* plot or appear somewhat like a normal distribution. Moreover, as mentioned earlier, the normality assumption is not as critical in larger samples; the OLS estimates still tend to be accurate even when the errors are not normal.[8] In any event, scan the functions available in R; many promising transformations are available.[9]

In general, then, one should examine the distribution of the residuals with a normal *q-q* or kernel density plot. An assumption of the LRM is that the errors are normally distributed, so testing this assumption is an important exercise.[10] If you find the residuals do not follow a normal distribution, myriad transformations are available. Moreover, transforming explanatory variables, though not necessary for OLS estimators to work well, can be useful

---

[7] Finding an appropriate transformation can be difficult and often requires substantial trial and error. A useful approach is to consider the boxcox function in R's MASS package. It allows the estimation of a series of regression models and examines transformations of the variables—usually the outcome, but explanatory variables also—that result in residuals distributed closest to normal. The approach was introduced in George P. Box and David R. Cox (1964), "An Analysis of Transformations," *Journal of the Royal Statistical Society: Series B* 26(2): 211–243. For more information, see Simon Sheather (2009), *A Modern Approach to Regression with R*, New York: Springer, chapter 6.

[8] For this and other reasons, some experts recommend against transforming the outcome variable as a general strategy to satisfy the normality assumption. For example, research suggests that transformations can produce biased point estimates in LRMs (see Amand F. Schmidt and Chris Finan (2018), "Linear Regression and the Normality Assumption," *Journal of Clinical Epidemiology* 98: 146–151).

[9] For advice about how to use these various transformations and numerous other ways to think visually about the distribution of variables, an excellent resource is William S. Cleveland (1993), *Visualizing Data*, Summit, NJ: Hobart Press.

[10] As already mentioned, having a large sample obviates normality problems in LRMs. How large is difficult to pin down but seems to be about 100 or more observations. Yet this depends on the degree of non-normality. At least 500 observations or more is optimal (see Lumley et al. (2002), *op. cit.*).

in some situations. With additional experience examining normal q-q plots, you'll find that selecting an appropriate transformation becomes easier. [11]

Nonlinearities

As suggested earlier, variables can be associated in a nonlinear manner in many more ways than in a linear fashion. As its name implies, though, the LRM assumes linear associations between the explanatory variables and the outcome variable. Fortunately, straightforward ways of adapting the linear model are available so it accommodates some types of nonlinear associations.

Two general approaches to nonlinear analysis are used in regression modeling. We'll begin with the more complex approach, although we'll end up learning about the easier approach. The first type of nonlinear regression analysis proposes that the estimated parameters are nonlinear; in other words, they do not imply a straight-line pattern. For instance, consider Equation 11.1.

$$y_i = \alpha + \beta_1 x_1 + \ln\left(\beta_2 x_2\right) + \beta_3^2 x_3 + \hat{\varepsilon}_i \tag{11.1}$$

Observe that the betas are not simple, linear terms; rather, they include *higher-order terms* (e.g., β_3^2), along with a logarithmic transformation (recall that *ln* implies the natural or Naperian logarithm). Nonlinear regression routines that are designed specifically for situations when the estimated parameters are hypothesized to be nonlinear are available in R and other software. Guiding hypotheses can be difficult to formulate, however, so we shall not cover these routines. [12] Another promising avenue, but one we will also not discuss, is called *nonparametric regression*. A useful class of these regression models is known as *generalized additive models* (GAMs), which allow the analyst to fit various types of associations—linear and nonlinear—among the outcome and explanatory variables. [13]

In this section, we'll focus on a simpler approach that adapts the LRM so it allows nonlinear associations among the explanatory and outcome

---

[11] In some situations, however, it may be difficult to find an appropriate transformation and, since we have not met the normality assumption, the LRM estimates are not efficient. The bootstrap is a useful method for estimating standard errors in the presence of non-normal residuals. Bootstrapping may be implemented in R with its boot package or with the car package's Boot function (see Fox and Weisberg (2018), *op. cit.*, chapter 5). Examples of bootstrapping an LRM are provided in Chapters 9 and 14.

[12] Information about these models and how to estimate them in R is available in Christian Ritz and Jens Carl Streibig (2008), *Nonlinear Regression with R*, New York: Springer.

[13] For an overview of GAMs in R, see Simon Wood (2006), *Generalized Additive Models: An Introduction with R*, Boca Raton, FL: Chapman & Hall/CRC Press. The R package gam is designed to estimate some of these models.

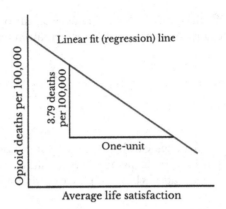

FIGURE 11.7
Linear fit line illustration of regression slope interpretation.

variables. To begin to understand how we may do this, consider that when we estimate an LRM and interpret the slope, we assume that the magnitude of change in *y* associated with each unit shift in *x* is constant. In other words, the slope is the same regardless of where in the *y* distribution the shift is presumed to take place. Suppose we find, like in Chapter 3 (Figure 3.5), that the slope of the regression line representing the average life satisfaction–opioid deaths association is −3.79. By now, we should be able to interpret this number effortlessly: each one-unit increase or difference in average life satisfaction is associated with a 3.79 decrease in the number of opioid deaths per 100,000. Keeping this in mind, consider Figure 11.7.

The fit line assumes that the difference in opioid deaths per 100,000 residents is the same at all sections of the hypothetical continuum of life satisfaction. Suppose, though, that in states with low average life satisfaction each one-unit difference is associated with a decrease of two opioid deaths per 100,000 residents, whereas in high life satisfaction states each one-unit difference is associated with a decrease of eight opioid deaths per 100,000 residents. Perhaps states with more people satisfied with the lives also have fewer people who turn to opioids for pain management or to alleviate life stresses. Whatever the hypothesized mechanism, these sorts of patterns are legion and recommend against simple conceptual models and hypotheses that suggest only linear associations.

Let's now consider a common example of a nonlinear association. Think about how people's incomes change as they go through adulthood. How should the association between age and income behave? First, it seems clear that younger adults who are still in school or just entering the workforce earn less than middle-aged or older adults who have been in the workforce longer (we'll ignore for now the influences of education, unemployment, job training, and other important variables). But what

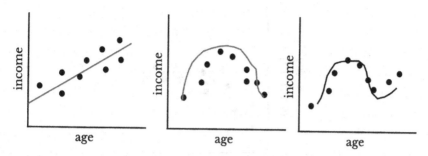

FIGURE 11.8
Examples of linear and nonlinear associations between age and income.

happens as people get older and reach their 60s? Many begin to retire and their personal income decreases. Thus, we have already departed from proposing a linear association between age and personal income; a curvilinear association is likely, with average income rising with age until it reaches a maximum point, after which it begins to decrease as people leave full-time work or retire. But we should also consider some important demographic issues concerning age. One issue involves death. Wealthier people, on average, tend to live longer than poorer people partly because they have access to better health care. As we begin to look at the age–income association at older ages, say 75 or 80, we may begin to see the effects of certain people dying earlier than others. If, on average, those with less income don't live as long, then income should appear higher among older adults. Many complex demographic phenomena are at play here that we don't have time to address, but we should consider that the age–income association is not linear and may have one or more "bends."

The graphs jointly represented in Figure 11.8 display three possible associations between age and income. The first graph shows a linear association, with income increasing at a steady pace with increasing age. The second graph exhibits what's known as a *quadratic* association, with one "bend" in the association. The third graph shows a *cubic* association, with two bends—and three distinct pieces—to represent the last part of the discussion about age and income. Once we begin considering nonlinear associations, they can quickly blossom into some very complex shapes with numerous bends and twists. In fact, any association is bound to have some curvature if we look closely enough. To manage the modeling, we should therefore always consider conceptually the associations in the set of variables. As we'll learn in Chapter 12, routinized ways to estimate regression models that include nonlinearities are available (e.g., stepwise selection), but they should be avoided if not guided by clear theoretical considerations.

When assessing associations among any set of variables, think about possible nonlinear associations. You may be surprised at how often they occur. For instance, the association between age and various outcomes is often nonlinear. The effect of explanatory variables such as education on outcomes

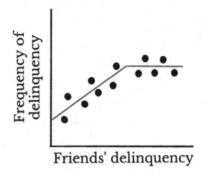

FIGURE 11.9
Hypothetical ceiling effect of the association between friends' delinquency and frequency of delinquency.

frequently have *floor* or *ceiling effects*: a negative or positive association exists up to a certain point, followed by a flat association.

Figure 11.9 provides an example of a hypothetical ceiling effect concerning the association between friends' involvement in delinquent behaviors and one's own delinquent behaviors. As the number of friends involved in these behaviors increases, one's own behaviors also tend to increase up to a certain point. High-frequency juvenile delinquents, though, are often psychologically different from others (e.g., more impulsive, less empathetic, or more callous), so having more friends who engage in illegal behaviors may not affect the frequency with which they engage in these behaviors.[14]

Testing for Nonlinearities in LRMs

Some simple tools are available for examining suspected nonlinear associations among variables in an LRM. Perhaps the most useful is the ordinary scatter plot. As we've seen previously, scatter plots are easy to construct in R and help us visualize the association between two variables. For instance, let's use the *StateData2018* dataset to construct a scatter plot with average life satisfaction on the *y*-axis and median household income on the *x*-axis. We'll add a couple of fit lines: a linear fit line and a *locally weighted regression* (lowess) fit line. The lowess function parcels the data into smaller (local) sections and fits regression lines to each section, which is useful when we're not sure

---

[14] See Margaret Kerr, Maarten Van Zalk, and Håkan Stattin (2012), "Psychopathic Traits Moderate Peer Influence on Adolescent Delinquency," *Journal of Child Psychology and Psychiatry* 53(8): 826–835.

what type of nonlinearity might occur.[15] The R code requesting these lines should be executed just after the function that creates the scatter plot.

R code for Figure 11.10
```
with(StateData2018, plot(MedHHIncome, LifeSatis,
    ylab="Life satisfaction", xlab="Median household
    income"))
abline(lm(LifeSatis ~ MedHHIncome,
        data = StateData2018)) # linear fit line
with(StateData2018, lines(lowess(LifeSatis ~
    MedHHIncome, f=.3), col="blue")
 # lowess fit line
```

The scatter plot and fit lines are furnished in Figure 11.10. The lines demonstrate a largely flat association between the median household income and life satisfaction, though with some nonlinearities. The lowess fit line suggests

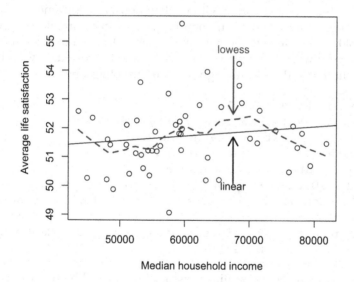

FIGURE 11.10
Scatter plot of average life satisfaction by median household income with linear and lowess fit lines.

---

[15] In R's lowess function, f is called the *smoother span* and dictates how wide or narrow each section of data that estimates the constituent model should be. The larger the number, the wider the section and the smoother the line. It typically takes some trial and error to find the value of f that reveals best the nonlinear association. The lowess method was introduced in Abraham Savitzky and Marcel J. E. Golay (1964), "Smoothing and Differentiation of Data by Simplified Least Squares Procedures," *Analytical Chemistry* 36(8): 1627–1639. See Cleveland (1993), *op. cit.*, for more information.

the association decreases then increases slightly followed by a decrease at the high end of median household income.

Scatter plots with fit lines are a good way to look for nonlinearities prior to estimating LRMs, though it's always important to keep theory and past research in mind before searching willy-nilly for them. But how can we check for nonlinearities after we've estimated an LRM, which is important for determining whether the model satisfies the linearity assumption? The main diagnostic tool remains a scatter plot but focuses on the LRM's residuals.

To see a diagnostic plot in action, let's use the same simple example: average life satisfaction is the outcome variable and median household income is the explanatory variable. After estimating LRM11.2, we'll save the studentized deleted residuals using the post-regression rstudent function (recall that these residuals are scaled using a *t*-distribution). Next, we'll plot these residuals against median household income in a partial residual plot (see Chapter 9 for examples of this plot to assess homoscedasticity[16]).

Since the residuals gauge all the influences on the average life satisfaction except for its *linear* association with median household income, we may use the plot to explore whether any nonlinear association implicating median household income affects the outcome variable. To get a better sense of the association between the explanatory variable and the residuals, we'll include a lowess fit line in the plot (though we could try other lines, too).

R code for Figure 11.11

```
LRM11.2 <- lm(LifeSatis ~ MedHHIncome,
              data = StateData2018)
StateData2018$rstudent <- rstudent(LRM11.2)
 # compute studentized residuals
with(StateData2018, plot(MedHHIncome, rstudent, ylab =
     "Studentized residuals", xlab = "Median household
     income"))
with(StateData2018, lines(lowess(rstudent ~ MedHHIncome,
     f=.3), col="blue"))
```

Figure 11.11 suggests a nonlinear association missed by the linear model. In fact, the lowess fit line looks similar to the fit line in Figure 11.10. The association is negative, slightly positive, and then negative as median household income exceeds about $70,000.

R includes a function in its car package that creates a *component-plus-residual plot* (CPRP), which is also useful for investigating nonlinear

---

[16] We utilize the raw residuals in the partial residual plots presented in Chapter 9. Here we use the studentized deleted residuals. Practically speaking, it often makes little difference which type of residual is used in the plot, but, as mentioned earlier, many experts recommend the studentized deleted residuals.

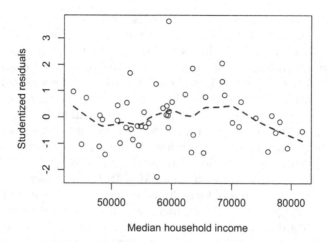

FIGURE 11.11
Scatter plot of studentized residuals by median household income with lowess fit line to assess nonlinearities from LRM11.2.

associations.[17] The organization of a CPRP is similar to a partial residual plot, with the explanatory variable on the x-axis and a form of the residuals on the y-axis. To create this plot, activate the car package and then use the following code:

R code
```
crPlots(LRM11.2)
```

The resulting graph (not shown) includes a linear fit line and a lowess fit line. Once again, look for a lowess fit line that is not straight for evidence of nonlinearities in the association. The lowess line is flatter than its analogous line in Figure 11.11, but still indicates a modest nonlinear association between the residuals and median household income. We'll leave further exploration of the nonlinearity in the relationship between average life satisfaction and median household income as an exercise.

Incorporating Nonlinear Associations in LRMs

Let's now explore an example in which we explicitly model a nonlinear association in an LRM. The *GSS2018.csv* dataset includes a variable that measures age in years. Earlier in the chapter we discussed the likelihood that age and

---

[17] The difference between the partial residual plot and the CPRP is that the former uses the residuals on the y-axis, whereas the latter uses $(\beta \times x_i) + \hat{\varepsilon}_i$, which adds the linear effects of the explanatory variable to the residuals.

income have a nonlinear association (see Figure 11.8). It's time to see if those musings are accurate. Let's examine a model that uses personal income as the outcome variable and includes a *higher-order term* for the age variable. This means we take age, square it (raise it to the power of two—age²), and determine what effect it has in the model. Including the squared value of age yields what was earlier termed its *quadratic effect*. We may also cube age (age³) and examine its *cubic effect* in the model. Remember that the quadratic effect tests whether age and the outcome variable have an association characterized by one bend in the fit line, whereas the cubic effect examines whether two bends occur in the fit line (see Figure 11.8).[18]

An important characteristic of an LRM with higher-order terms is that each lower order term should be included in the model. For instance, if age² is in the model, age should also be included. Otherwise, the interpretations are difficult to understand. R provides a straightforward way to incorporate higher-order terms in LRMs with its I function. The R code that creates LRM11.3 and its output provides an illustration.

R code

```
LRM11.3 <- lm(pincome ~ age + I(age^2), data=GSS2018)
 # note the I function to create the quadratic term
summary(LRM11.3)
confint(LRM11.3)
```

R output (abbreviated)
```
Coefficients:
              Estimate   Std. Error   t value   Pr(>|t|)
(Intercept) -7.6152445    1.2557600    -6.064   1.54e-09 ***
age          0.8870542    0.0527780    16.807   < 2e-16  ***
I(age^2)    -0.0096928    0.0005091   -19.039   < 2e-16  ***
---
Signif. codes:  0 '***' 0.001 '**' 0.01 '*' 0.05 '.' 0.1

Residual standard error: 8.144 on 2312 degrees of freedom
Multiple R-squared: 0.1721, Adjusted R-squared: 0.1714
F-statistic: 240.3 on 2 and 2312 DF, p-value: < 2.2e-16

[CIs]                 2.5%         97.5%
(Intercept)        -10.078        -5.153
age                  0.784         0.991
I(age^2)            -0.011        -0.009
```

---

[18] Including higher-order terms, such as x_i^2 and x_i^3, to an LRM is also called *polynomial regression*, where the power function is called the *degree* (h) of the polynomial. Polynomial regression takes the general form of $y_i = \alpha + \beta_1 x_1 + \beta_2 x_1^2 + \beta_3 x_1^3 + \cdots + \beta_h x_1^h + \hat{\varepsilon}_i$. LRMs in social and behavioral sciences rarely exceed $h = 3$, however.

The results show the age coefficient is positive, but the age^2 coefficient is negative. This suggests that age and personal income have a positive association until age reaches a certain point after which the association either flattens out or becomes negative.[19] In general, the signs of the original and higher-order terms indicate the particular shape of the association. Suppose, for instance, that the age and age^2 coefficients were both positive. The presumed shape of the association would be positive, with an increasingly positive slope.

A question to ask is whether this model is an improvement over a model that includes only age as the explanatory variable. We may answer this inquiry with a partial F-test or information criterion measures, AIC and BIC (see Chapter 6). You might wish to examine these for practice. Or consider that the age^2 coefficient meets common criteria for statistical significance (e.g., 95% CI: {−0.011, −0.009}), so a partial F-test is not necessary.

As a next step, consider the following inquiry: at what age do we begin to see the positive association taper off or become negative? As shown in Figure 11.12, graphing the predicted values by age provides an answer. We now see that the association flattens out somewhere in the 40- to 50-year-old age range. This may seem early but remember that the GSS data used here are cross-sectional and include respondents who have been in the work-force for varying lengths of time. One well-known phenomenon that affects many workers and leads older workers to earn less, relatively speaking, than

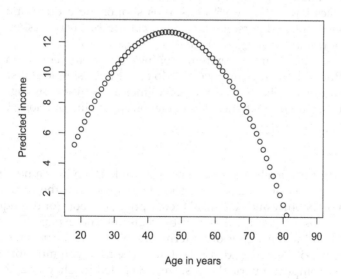

FIGURE 11.12
Plot of predicted personal income by age.

---

[19] We could extend the model to include a cubic age term by adding + I(age^3) to the LRM11.3 equation.

younger workers is wage compression. The shift after about age 40 and the diminishing income after age 50 could be an example of wage compression. For instance, full-time workers in their late 50s may earn relatively less than full-time workers in their 40s because they were hired at a lower pay when they entered the work force. The income of workers in their 40s is apt to catch up to the income of older workers after, say, those in their 40s have been in the workforce for 15–20 years.[20]

R code for Figure 11.12
```
GSS2018$predict.income <- fitted(LRM11.3)
with(GSS2018, plot(age, predict.income, ylab="Predicted
     income", xlab="Age in years", ylim=c(1,13)))
```

When using explanatory variables with higher-order terms (e.g., age and age^2) always evaluate the model for collinearity. What do the diagnostics suggest? Using vif(LRM11.3), R returns high VIFs for both age terms: 31.7 for each. The model provides strong evidence of collinearity, though this is no surprise. Another symptom of collinearity is large standardized regression coefficients (beta weights), which in most LRMs are rarely larger than 1.0. Yet, according to the lm.beta function, the beta weights are larger than one (in absolute value). Recall, though, that collinearity inflates standard errors, making it harder to claim that the inferential evidence is consistent with a presumed association in the population. Yet the coefficients for both age variables have narrow 95% CIs, thus supporting a quadratic association between age and personal income. If collinearity was masking the age effects, we would expect 95% CIs that encompass zero.

Most of us don't care for extreme collinearity in our regression models, though. But is there anything we can do about it? A simple statistical trick that diminishes collinearity or multicollinearity caused by higher-order terms is based on the statistical phenomenon shown in Equation 11.2.

$$\text{cov}\left(\left[x_i - \bar{x}\right], x\right) = 0 \tag{11.2}$$

Substituting the correlation for the covariance leads to the same result: the linear association of a variable and its deviations from the mean is zero. Books on probability and statistical theory provide proofs for this equation.[21] What it means for the LRM is that by centering the explanatory variable before computing the higher-order term and then using the new terms in the LRM, collinearity is reduced substantially. The most common form of centering is to compute a variable's z-scores. As noted in Chapter 2, R's scale

[20] For research on this topic in higher education, see James B. McDonald and Jeff Sorensen (2017), "Academic Salary Compression across Disciplines and Over Time," *Economics of Education Review* 59: 87–104.
[21] See Sheldon Ross (1994), *A First Course on Probability*, New York: Macmillan.

function provides these scores. Let's determine what happens to the model when we use the z-scores of age (see LRM11.3z).

Because the age variables are now measured in different units, the slope coefficients are different in this model than in the previous model. If we graph the predicted values by age, though, the same pattern exists as in Figure 11.12. In addition, the intercept now has a useful interpretation: it identifies the expected mean value of personal income when age is at its mean (about 40 years—but why is this the case?). Notice also that the 95% CIs for the slope coefficients do not encompass zero. More importantly, though, the VIFs are well below the thresholds indicative of collinearity ($\sqrt{1.05} < 3$). The statistical trick of centering variables is worth remembering when higher-order terms induce collinearity or multicollinearity issues in an LRM. We'll also use it later in the chapter but keep it in mind as it will come in handy throughout your career in regression modeling.

R code
```
library(car) # to use the vif function
GSS2018$z.age <- scale(GSS2018$age, center = TRUE,
                scale = TRUE)
 # compute z-scores for age
LRM11.3z <- lm(pincome ~ z.age + I(z.age^2),
              data=GSS2018)
 # use the z-scores for age in the LRM
summary(LRM11.3z)
confint(LRM11.3z)
vif(LRM11.3z)
```

R output (abbreviated)
```
Coefficients:
              Estimate   Std. Error   t value   Pr(>|t|)
(Intercept)   12.5797     0.2371       53.049   < 2e-16   ***
z.age         -1.1261     0.1734       -6.494   1.02e-10  ***
I(z.age^2)    -3.1633     0.1661      -19.039   < 2e-16   ***
---
Signif. codes: 0 '***' 0.001 '**' 0.01 '*' 0.05 '.' 0.1

Residual standard error: 8.144 on 2312 degrees of freedom
Multiple R-squared: 0.1721, Adjusted R-squared: 0.1714
F-statistic: 240.3 on 2 and 2312 DF, p-value: < 2.2e-16

[CIs]          2.5%    97.5%
(Intercept)   12.115   13.045
z.age         -1.466   -0.786
I(z.age^2)    -3.489   -2.838

[VIFs] z.age  I(z.age^2)
       1.049   1.049
```

Centering the age variable offers a convenient way to obviate the effects of collinearity but can be difficult to work with if we wish to determine at what point in the age or personal income distribution the slope flattens out or becomes negative. Therefore, the LRM with the original variables remains useful for constructing a graph to show the shape of the nonlinear association.

Nonlinear associations abound in the social and behavioral sciences (as well as in other disciplines, such as chemistry, biology, and physics). We've examined only a couple of basic examples, but when considering variables such as age, education, socioeconomic status, and many others, nonlinear shifts in their associations with myriad outcome variables may be present. Before beginning any modeling exercise, think carefully about potential nonlinear associations among the variables: do previous studies or the conceptual model guiding your research suggest any? Do visual depictions of the data, such as scatter plots with fit lines, suggest that nonlinearities exist among the variables? Once you have a sense of the likely shape of the associations, consider how to test for them in an LRM. Squaring or cubing explanatory variables scarcely uncovers the various possibilities.[22]

Interaction Effects

The models examined so far may be characterized as *additive* because, as shown in Equation 11.3, we add the products of slope coefficients and explanatory variables to predict the outcome variable.

$$\hat{y}_i = \alpha + \beta_1 x_1 + \beta_2 x_2 + \beta_3 x_3 \tag{11.3}$$

But *nonadditive* models also exist. A nonadditive term in an LRM is called an *interaction term* because it represents the association between an explanatory and outcome variable that is affected by a third variable. Another way of picturing an interaction term is that it represents one explanatory variable multiplied by another—a *multiplicative term*. Equation 11.4 furnishes this representation.

$$\hat{y}_i = \alpha + \beta_1 x_1 + \beta_2 x_2 + \beta_3 (x_1 \times x_2) \tag{11.4}$$

---

[22] In addition to some of the approaches mentioned in this section, a helpful tool for examining nonlinear associations is with the use of *splines*. Similar to a lowess fit line, a spline estimates straight-line associations in the data that are "local" or occur within specific sections of the joint distribution of x and y. The number of lines is the *degree* and the points where the lines meet are the *knots* of the spline. The R package `splines` estimates regression models with splines.

The term associated with β_3 in Equation 11.4 is listed as $x_1 \times x_2$, which indicates that the two variables are multiplied or they "interact" in some fashion to affect the outcome variable.[23]

It might be helpful to work through the algebra of the previous equation to understand how interaction effects operate, but a straightforward approach is to imagine a linear regression line and consider whether its slope is the same or different depending on another variable.

Suppose, for example, we assume that age and gender (female vs. male) independently predict self-esteem and that the association between age and self-esteem is linear. The LRM is thus estimated with Equation 11.5.

$$\text{self-esteem}_i = \alpha + \beta_1\left(\text{age}_i\right) + \beta_2\left(\text{gender}_i\right) + \hat{\varepsilon}_i \qquad (11.5)$$

Imagine that the LRM indicates positive slopes for both age and gender, so that males, on average, report higher self-esteem than females. The nonadditive model assumes that the age–self-esteem slope is the same for males and females; the only difference is in their intercepts (see Figure 11.13). If this is confusing, try plugging positive slopes for age and gender into Equation 11.5 and then compute some predicted self-esteem scores (age ranges from 15 to 25 and gender may be coded as 0 = female and 1 = male). The relative distance between the predicted values for males and females at each age point is the same. An additive model of this type is therefore also called a *different*

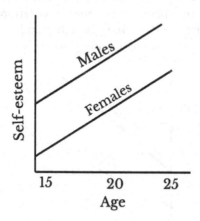

FIGURE 11.13
Illustration of different intercepts, same slope model.

---

[23] Yet another way of characterizing this model is to claim that one of the explanatory variables is a *moderator* since it "moderates" the effects of the other on the outcome variable (e.g., x_2 moderates the association between x_1 and y). For a thorough treatment of interaction effects in regression models, see Robert L. Kaufman (2019), *Interaction Effects in Linear and Generalized Linear Models*, Thousand Oaks, CA: Sage.

intercept-parallel slopes model since the slopes are the same for the two groups; just a constant distance apart. This distance is gauged by the intercept.

But perhaps you've read some research reports and suspect that self-esteem increases more among males than among females from ages 15–25: the slope is steeper for males. How can we estimate different slopes for males and females? One approach is to estimate separate LRMs for the two groups and then compare the slopes from each model. In R, the subset function is useful for splitting the data into groups and may be embedded in the lm function.[24]

To characterize this type of association in an LRM, we may also include an interaction term that multiples age by gender, as shown in Equation 11.6.

$$\text{self-esteem}_i = \alpha + \beta_1\left(\text{age}_i\right) + \beta_2\left(\text{gender}_i\right) + \beta_3\left(\text{age} \times \text{gender}\right) \qquad (11.6)$$

To determine whether the age slopes for females and males differ, examine the coefficient for β_3 and evaluate its practical relevance, *p*-value, and CI (or consider alternative inferential measures). If you conclude that β_3 is relevant, then you have evidence consistent with the notion that the association between age and self-esteem differs for females and males. This type of model is thus known as a *different intercept-different slopes* model (see Figure 11.14). As with higher-order terms such as x^2, if interaction terms appear in the model, the constituent variables should also be included for correct specification and ease of interpretation.

Let's evaluate a concrete example of an interaction term using the *GSS2018.csv* dataset by focusing on the associations among personal income, education, and gender. First, though, ponder the following issue: we know that education and income are positively associated in the U.S.,

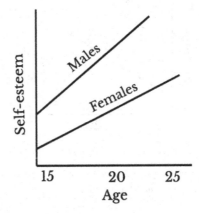

FIGURE 11.14
Illustration of different intercepts, different slopes model.

---

[24] For example: lm(esteem ~ age, data=subset(esteem.data, gender == "male")).

with people who complete more years of formal education, on average, earning higher incomes than people with fewer years of formal education (interesting exceptions are air traffic controllers and nuclear power reactor operators, which we'll ignore). Many studies also indicate that women earn less, on average, than men. Let's see if these two associations are supported by the GSS data. We'll include age as a control variable in the model since we know it is associated with both education and income (but, to simplify things, ignore the nonlinear association between age and income illustrated in Figure 11.12).

The results of LRM11.4 support both positions: statistically adjusting for the effects of age, males, on average, report higher personal income than females (β = 2.78, 95% CI: {2.1, 3.5}); and more years of education are associated with higher personal income (β = 0.83, 95% CI: {0.72, 0.94}). If we were to plot slopes for education by gender, they would appear similar to Figure 11.13: different intercepts but parallel slopes for males and females.

R code
```
LRM11.4 <- lm(pincome ~ age + female + educate,
              data=GSS2018)
summary(LRM11.4)
confint(LRM11.4)
```

R output (abbreviated)
```
Coefficients:
              Estimate  Std. Error  t value  Pr(>|t|)
(Intercept)   1.629158    0.963181    1.691  0.0909    .
age          -0.099344    0.009551  -10.402  < 2e-16  ***
femalemale    2.777799    0.346661    8.013  1.76e-15 ***
educate       0.830249    0.057977   14.320  < 2e-16  ***
---
Signif. codes: 0 '***' 0.001 '**' 0.01 '*' 0.05 '.' 0.1

Residual standard error: 8.296 on 2311 degrees of freedom
Multiple R-squared: 0.1413, Adjusted R-squared: 0.1402
F-statistic: 126.8 on 3 and 2311 DF, p-value: < 2.2e-16

[CIs]            2.5%    97.5%
(Intercept)    -0.260    3.518
age            -0.118   -0.081
femalemale      2.098    3.458
educate         0.717    0.944
```

Let's consider these variables from a different angle: what role does gender play, if any, in the association between education and income? Males may earn more, on average, than females, but does this difference depend on how much formal education people obtain? Does more education tend to equalize personal income among males and females? Or is there some type of glass ceiling that results in an enhanced income disparity among males and females at higher levels of education? Perhaps the occupational choices of highly educated men and women lead to income gaps. Or there may be no association between gender and education when it comes to personal income; they may be independently associated with this outcome. We cannot approach answers to these questions with the LRM just estimated. It provides no evidence about whether income varies by gender *and* education. Fortunately, including an interaction term in the model is useful for assembling some evidence. The R code that generates LRM11.5 demonstrates how to include a gender-by-education interaction term.[25]

R code
```
LRM11.5 <- lm(pincome ~ age + female + educate +
              female*educate, data=GSS2018)
summary(LRM11.5)
confint(LRM11.5)
```

R output (abbreviated)
```
Coefficients:
                Estimate   Std. Error   t value   Pr(>|t|)
(Intercept)     0.494563   1.220700     0.405     0.6854
age            -0.098921   0.009552   -10.356     <2e-16 ***
femalemale      5.192530   1.633960     3.178     0.0015 **
educate         0.911155   0.078878    11.551     <2e-16 ***
femalemale:
educate        -0.175891   0.116311    -1.512     0.1306
---
Signif. codes: 0 '***' 0.001 '**' 0.01 '*' 0.05 '.' 0.1

Residual standard error: 8.293 on 2310 degrees of freedom
Multiple R-squared: 0.1422, Adjusted R-squared: 0.1407
F-statistic: 95.73 on 4 and 2310 DF, p-value: < 2.2e-16

[CIs]                      2.5%   97.5%
(Intercept)              -1.899   2.888
age                      -0.118  -0.080
femalemale                1.988   8.397
educate                   0.756   1.066
femalemale:educate       -0.404   0.052
```

How should we interpret these results? First, comparing them to the results of the previous model, we see that the slope coefficient for femalemale is larger (in absolute terms). Second, the gender-by-education interaction term's

---

[25] The lm function in LRM11.5 is one way of including an interaction term. Another way is
   lm(pincome ~ age + female + educate + female:educate, data=GSS2018).

coefficient is −0.18 but its 95% CI includes zero {−0.40, 0.05}. Little evidence exists from the model to conclude that education and gender *interact* to predict personal income, or that gender *moderates* the association between education and personal income.

For pedagogical reasons, however, let's assume that we are still interested in the interaction term. How can we use it to learn about the association between education and personal income by gender? It takes experience to be able to look at the coefficients and make sense of the patterns. But here is a hint that will help: choose one of the variables involved in the interaction term to focus on and then consider what its slope coefficient *and* the slope coefficient associated with the interaction term are suggesting about its association with the outcome. Let's pick education. Its slope is 0.91. But the interaction term multiplies either a zero (to represent females) or a one (to represent males) by education.[26] For females we thus assume one slope, but when we consider males the slope is pulled in a downward direction, as indicated by the negative gender-by-education coefficient: −0.18. Similarly, as evidenced by the positive slope for gender (β(femalemale) = 5.19), males begin with an advantage in income but this is then diminished, relatively speaking, by increasing education (once again suggested by the negative gender-by-education coefficient).[27]

If this is still difficult to understand, let's predict personal income based on LRM11.5 for the following four groups: females and males with 12 and 16 years of education (we'll set age equal to 40, which is close to its overall mean) and see how the relative differences unfold (see LRM11.5a).[28]

R code
```
LRM11.5a <- data.frame(age=40, female=factor("female",
                       levels=c("female", "male")),
                       educate=c(12, 16)) # females
predict(LRM11.5, LRM11.5a) # LRM11.5 is the model
                             estimated earlier
LRM11.5b <- data.frame(age=40, female=factor("male",
                       levels=c("female", "male")),
                       educate=c(12, 16)) # males
predict(LRM11.5, LRM11.5b)
```

---

[26] The female variable consists of two character strings, but for all intents and purposes we may interpret the results with a {0, 1} coding scheme.

[27] Footnote 24 shows how to estimate LRMs with different subsets of data. Using this approach, estimate separate models for females and males. Confirm that this yields a smaller education slope coefficient for males.

[28] For a less onerous method of computing predicted values in the presence of interaction terms, examine the effect function of R's effects package. Consider, for instance, effect("gender*educate", LRM11.5). This provides predicted values for females and males at five different levels of education.

R output (abbreviated)

```
   1       2
 7.47   11.12 # predicted income for 40-year-old females
               with 12 and 16 years of education
   1       2
10.55   13.49 # predicted income for 40-year-old males
               with 12 and 16 years of education
```

What do these predicted values suggest about whether the association between education and personal income is different for females and males? Examine the differences in the two education groups. In the 12-years group, the raw difference is {10.55 – 7.47} = 3.08. In the 16-years group, the raw difference is {13.49 – 11.12} = 2.37. It appears that the personal income gap between females and males is slightly larger at 12-years of education than at 16-years of education, suggesting that the gender gap closes at higher levels of education. Comparing groups this way is often helpful but may break down in certain situations. An alternative approach is to compare relative or percentage differences within the education groups to understand the effects implied by the interaction term. The following R code uses the predicted values to demonstrate how to compute the percentage differences:

R code

```
((10.55 - 7.47) / 7.47)  * 100  # female-male difference
                                   at 12 years of
                                   education
((13.49 - 11.12) / 11.12) * 100  # female-male
                                   difference at 16
                                   years of education
```

R output (abbreviated)

```
41.2 # percentage difference at 12 years of education
21.3 # percentage difference at 16 years of education
```

The gender gap in personal income is smaller at higher levels of education (21.3%), which supports the idea of a steeper education–income slope among females than among males. To clarify this even further, compute the relative difference among females in the two education groups and compare it to the relative difference among males. This confirms that the education–income slope is steeper among females than among males.

A simpler way to assess interaction terms is to visualize their effects based on the predicted values from the model. For instance, consider the following R code and the graph it produces in Figure 11.15. Rather than using R's plot function, we use the R package ggplot2, which makes it easier to graph data from multiple groups.

R code for Figure 11.15

```
library(ggplot2)
GSS2018$pred.income <- fitted(LRM11.5)
 # request predicted values from the model 11.5 and then
    create the graph using the ggplot function
Fig11.15 <- ggplot(GSS2018, aes(x=educate, y=pred.inco
            me, color=female)) + geom_point() +
            geom_smooth(method=lm, se=FALSE,
            fullrange=TRUE)
Fig11.15 + xlab("Years of education") +
            ylab("Predicted income")
 # add x- and y-axis labels
```

Figure 11.15 includes two linear fit lines, one for females and one for males. In agreement with the percentage differences, a steeper slope occurs among females. But notice that even at 20 years of education, females still report slightly lower income than males. The income gap never closes completely

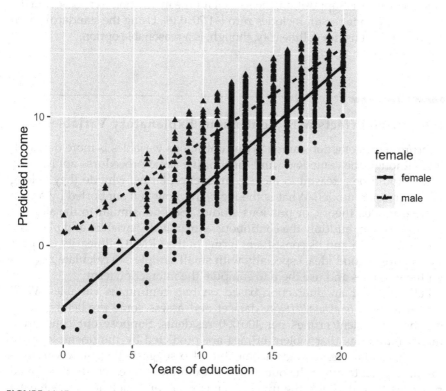

FIGURE 11.15

Scatter plot of predicted personal income by years of education with linear fit lines for males and females from LRM11.5.

(though the graph does not include confidence bands, which may be included with se=TRUE in the R code).

A problem you may have anticipated could affect our confidence in the LRM, though. Since interaction terms are created by multiplying one variable by another, collinearity between the interaction term and its constituent terms is likely. In fact, requesting collinearity diagnostics for LRM11.5 shows VIFs of 22.2 for female and 23 for female*educate. Earlier in the chapter, when computing higher-order terms for age, we learned of a solution to this problem: standardize (take the z-scores of) the constituent variables before computing the higher-order terms. It makes no sense to standardize female, though, since it is an indicator variable, but we may standardize educate, recompute the interaction term, and include it in an updated LRM along with the z-scores version of educate.

The results of this updated model (not shown) exhibit a lack of collinearity—the largest VIF is 1.85—but demonstrate the same general patterns we saw earlier. Remember that collinearity normally causes inflated standard errors (see Chapter 10), which creates, all else being equal, wider CIs. Perhaps this is why the interaction term's 95% CI in LRM11.5 included zero. But the updated model rejects this supposition: the 95% CI for the interaction term based on the z-scores of educate still includes zero {−1.20, 0.16}. Using the standardization approach to adjust for collinearity, though, is a reasonable option.

Interaction Effects with Continuous Explanatory Variables

Interpreting interactions among two continuous variables is more demanding, although the same logic and most of the same procedures apply. First, there should be a reason to include interaction terms: why do they belong in a regression model? What is the logic and how is it supported by a conceptual model, theory, or previous studies? Second, similar to the steps in the last section, multiply the continuous explanatory variables and place the interaction term and its constituent terms in the LRM. Collinearity is likely, so it's often a good idea, especially with small samples, to calculate z-scores for the variables and use them to compute the interaction term.

Let's consider an interaction based on two continuous variables. We'll return to the *StateData2018.csv* dataset and assess some predictors of the number of violent crimes per 100,000 residents. Suppose our conceptual model maintains that violent crimes are predicted by the unemployment rate and median household income, but also suggests that these variables might interact to affect the outcome. For example, a research study we read proposes that although the unemployment rate and violent crimes have a positive association, median household income serves as a moderator such that in states with wealthier households the association is stronger.

To test this highly speculative model, we'll estimate an LRM. First, though, to minimize the risk of multicollinearity, the following R code computes standardized versions of the explanatory variables (UnemployRate and MedHHIncome) and includes an interaction term that is computed from them in LRM11.6.

R code

```
# create z-score versions of the explanatory variables;
  use c( ) so the variables work with the effect
  function later
StateData2018$z.UnemployRate <- c(scale(StateData2018$
  UnemployRate, center = TRUE, scale = TRUE))
StateData2018$z.MedHHIncome <- c(scale(StateData2018$
  MedHHIncome, center = TRUE, scale = TRUE)))
LRM11.6 <- lm(ViolentCrimeRate ~ z.UnemployRate +
              z.MedHHIncome + z.UnemployRate*z.
              MedHHIncome, data=StateData2018)
summary(LRM11.6)
confint(LRM11.6)
vif(LRM11.6) # examine the variance inflation factors
              - all are below 2
```

R output (abbreviated)

```
Coefficients:
               Estimate Std. Error  t value   Pr(>|t|)
(Intercept)      349.68      16.07   21.764    <2e-16 ***
z.UnemployRate    45.15      17.18    2.628    0.0116 *
z.MedHHIncome    -30.01      17.09   -1.756    0.0857 .
z.UnemployRate:
 z.MedHHIncome    25.41      13.97    1.819    0.0755 .
---
Signif. codes: 0 '***' 0.001 '**' 0.01 '*' 0.05 '.' 0.1

Residual standard error: 113.1 on 46 degrees of freedom
Multiple R-squared: 0.2769, Adjusted R-squared: 0.2297
F-statistic: 5.87 on 3 and 46 DF, p-value: 0.00176

[CIs]                                 2.5%     97.5%
(Intercept)                        317.335   382.017
z.UnemployRate                      10.567    79.740
z.MedHHIncome                      -64.419     4.389
z.UnemployRate:z.MedHHIncome        -2.712    53.536
```

The unemployment rate is, as expected, positively associated with the number of violent crimes per 100,000. Since the unemployment rate is measured in z-scores, we may interpret its slope coefficient as "statistically adjusting for the effects of the other variables in the model, states that are one standard deviation higher in their unemployment rates have, on average, 45 more violent crimes per 100,000 residents."[29] The state's median household income is negatively associated with the number of violent crimes per 100,000, although its 95% CI is wide and includes zero. But what does the coefficient associated with the interaction term indicate? (We'll ignore its wide 95% CI.) Does median household income moderate the association between the unemployment rate and violent crimes?

One way to think about the interaction term is to first consider the slope of the unemployment rate: it's positive. But as median household income increases, this positive slope becomes more positive since the interaction term is also positive. Another way to interpret the interaction term is to focus on the slope coefficient for median household income. As already mentioned, it's negative. But as the unemployment rate increases this negative slope flattens out (or becomes less negative).

These interpretations provide general guidance for understanding the interaction effect, but many analysts find it more intuitive to compute predicted scores. The `effect` function mentioned earlier (see fn. 30) provides a simple way to calculate these predictions. We first need to decide, however, what values of the explanatory variables to use. Since each is measured using z-scores, convenient values include –1, 0, and 1, or one standard deviation below the mean (low), at the mean (medium), and one standard deviation above the mean (high). We may thus utilize the following function:

R code
```
library(effects)
effect("z.UnemployRate*z.MedHHIncome", xlevels=list(z.
  UnemployRate= seq(-1, 1, 0.5), z.MedHHIncome=seq(-1, 1,
  0.5)), LRM11.6))
```

Table 11.1 displays predicted violent crimes per 100,000 based on a portion of the output from the `effect` function. What do they indicate about the associations with the unemployment rate and median household income? Notice, first, that the lowest predicted mean, 249, occurs when a state has a low unemployment rate and a high median household income. The highest predicted mean, 399, occurs when a state has a high unemployment rate and a low median household income. But how can we determine the moderating effects of, say, median household income on the association between the unemployment rate and the number of violent crimes? Compare the predicted values down the columns. For instance, at low values of median household income, predicted violent crimes are about 11% higher (360 → 399), whereas, at high values of median household income, predicted violent crimes are 57% higher (249 → 390). In other words, the

---

[29] To be precise, since we've used z-scores this expected association occurs when the median household income variable is at its mean. Consider why this is the case.

TABLE 11.1

Predicted Number of Violent Crimes per 100,000 Population, by Unemployment Rate and Median Household Income

	Median household income		
Unemployment rate	One SD below mean	At the mean	One SD above mean
One SD below mean	360	304	249
At mean	380	350	320
One SD above mean	399	395	390

SD = standard deviation.

slope of the unemployment rate on the number of violent crimes is steeper when median household income is high. But don't forget to consider the conceptual model or hypothesis to determine whether these differences are consistent with expectations. Why, for instance, does the unemployment rate have a stronger association with violent crimes in wealthier states?

Another way to make sense of these results is to represent them with a graph in which we plot one of the slope coefficients conditional on the other explanatory variable. For instance, we might wish to examine the linear association between unemployment and violent crimes at particular values of median household income. An R package called `interplot` makes this easy to execute by plotting the estimated magnitude of a slope coefficient at different levels of another variable. Based on LRM11.6, we may use the following code:

R code for Figure 11.16

```
library(interplot)
interplot(m = LRM11.6, var1 = "z.UnemployRate", var2 =
         "z.MedHHIncome") + aes(color = "red") +
         theme(legend.position="none") + geom_
         hline(yintercept = 0, linetype = "dashed") +
         xlab("Median household income") + ylab("Slope
         for unemployment rate-violent crime rate
         association")
```

The function takes the LRM11.6 object and plots slope coefficients representing the unemployment rate (`var1`)—violent crime association by median household income (`var2`).[30] Consistent with Table 11.1, Figure 11.16 demonstrates that at higher levels of median household income (across the *x*-axis), the size of the slope coefficient representing the unemployment rate-violent

---

[30] The `effects` package used to compute the predicted values may be employed as an alternative to the `interplot` package. It also furnishes visualizations and provides other characteristics of nonlinearities and interactions from the same regression model (see Fox and Weisberg (2018), *op. cit.*).

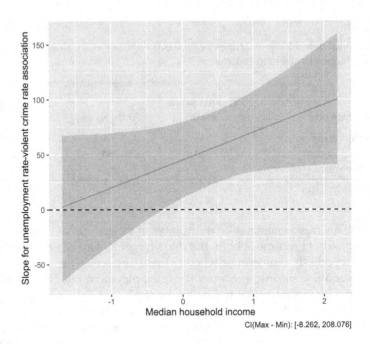

CI(Max - Min): [-8.262, 208.076]

FIGURE 11.16
Plot of the unemployment rate–violent crime rate slope by median household income with confidence bands from LRM11.6.

crimes association increases (from low to high on the y-axis). The shaded area provides the confidence band of the slope coefficients and the dotted line at $y = 0$ represents a nil slope coefficient. It appears that as we observe states with at least an average level of median household income ($x = 0$), the confidence band no longer includes zero. When median household income is about two standard deviations above the mean, the slope coefficient of the unemployment rate is expected to be about 100.[31] Compare this with the slope coefficient for the unemployment rate in the LRM11.6 output ($\beta = 45.2$). (Given the coding of the variables, at what point do we expect this slope to occur?)

As a final note: it's a good idea to examine a normal q-q plot of the residuals from this model. Interaction terms may introduce additional challenges for an LRM, such as non-normally distributed residuals. What does the plot indicate? Are you more or less confident in the results after examining the distribution of the residuals? More importantly, though, what does the literature on violent crime tell us about the model? As noted earlier, why would higher median household income, assuming it represents something about the wealth of a state, drive a stronger association between the unemployment and violent crimes?

---

[31] How would you interpret this number?

Even though the focus thus far has been on interaction terms based on two explanatory variables, LRMs are not limited to these *two-way* interactions. They can also accommodate *three-way* interaction terms, which implies multiplying three explanatory variables (e.g., $x_1 \times x_2 \times x_3$). However, this type of LRM must include all of the constituent terms as well as all the possible two-way interactions ($x_1 \times x_2$; $x_1 \times x_3$; $x_2 \times x_3$).[32] Conceivably, four-way, five-way, or even higher-order interactions are possible, but quickly become difficult to work with and are challenging to interpret.

Classification and Regression Trees (CART)

Using interaction terms in LRMs is a convenient approach for examining whether one explanatory variable moderates the association between another explanatory variable and the outcome variable. But models rarely include multiple two-way interactions or three- or four-way interactions; employing the latter, in particular, may create problems with interpretation or substantial multicollinearity issues. Yet, some alternative analytic tools are designed, explicitly or implicitly, to determine whether outcome variables are predicted best by multiple interacting effects among explanatory variables.

The rise of data analytics and machine learning has led to the growing popularity of one method of studying multiple interacting effects. Known generally as *classification and regression trees* (CART) or generically as *decision trees*, these methods may be visualized as a flowchart in which the explanatory variables are partitioned into levels that best predict the outcome variable, with the best predictor partitioned first, the next best predictor partitioned next, and so forth. If an explanatory variable does a poor job of predicting the outcome, then it may be omitted from the tree.

Suppose, for example, that we wish to predict the number of robberies per 100,000 residents across states in the U.S. The proposed explanatory variables are the unemployment rate, the percentage of males, and average household income. Figure 11.17 furnishes a simple illustration of a CART.

Each "branch" of the tree contains the predicted split of an explanatory variable and how it relates to the next explanatory variable in its ability to predict the outcome. For instance, the unemployment rate is partitioned into states that have less than 5% unemployment and states that have more than 5% unemployment. The unemployment rate and percent male interact since different combinations

---

[32] R makes this easy to do. The following code shows how to include a three-way interaction of some explanatory variables: LRM11.6a <- lm(ViolentCrimeRate ~ z.UnemployRate * z.MedHHIncome * z.DomMigRate, data=StateData2018). R automatically includes the constituent terms (e.g., z.UnemployRate * MedHHIncome).

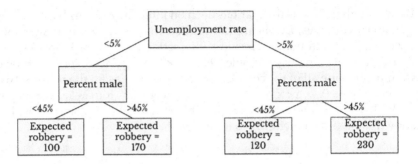

FIGURE 11.17
Classification and regression tree illustration.

predict distinct rates of robberies. Average household income is missing from the tree since it does not have sufficient predictive power in the model.

The expected number of robberies is highest—230 per 100,000 residents—when a higher unemployment rate (> 5%) combines with a higher percentage of males in a state (> 45%). The lowest expected number of robberies (100 per 100,000 residents) occurs in states with low unemployment and relatively few males.

Let's examine a similar example as in an earlier section, with personal income predicted by age, gender, and education in the *GSS2018* dataset. The R party package provides one set of routines for estimating CART models. We first create a subset of the dataset that removes the missing values with the complete.cases function and then creates a factor variable for female (which the party package prefers). We then use the ctree function to request two CART models (see LRM11.7a and LRM11.7b).

R code
```
library(party)
set.seed(6728) # for replication
# create a subset of the GSS2018 dataset with no missing
  data
sub.GSS <- GSS2018[c("age","female","educate","pincome")]
GSS.comp <- sub.GSS[complete.cases(sub.GSS), ]
# transform the variable female into a factor variable
  (the party package prefers factors)
GSS.comp$female <- factor(GSS.comp$female)
# create two CART models with party's ctree function
LRM11.7a <- ctree(pincome ~ female + educate, data=GSS
           .comp, controls = ctree_control(testtype =
           "MonteCarlo"))
LRM11.7b <- ctree(pincome ~ female + educate + age,
           data=GSS.comp, controls = ctree_
           control(testtype = "MonteCarlo"))
```

```
# plot the results of the CART models
plot (LRM11.7a)
plot (LRM11.7b)
# examine the actual and predicted income values by
   education groups from LRM11.7b
GSS.comp$pred.income <- predict(LRM11.7b)
GSS.comp$educ.cat <- cut(GSS.comp$educate, breaks=c(10,
     11, 12, 15, 16, 21), labels=c("Not HS grad", "HS
     grad", "Some college", "College grad", "Grad
     school"))
aggregate(GSS.comp[, 4:5], by=list(GSS.comp$educ.cat),
          FUN=mean)
  # request means
```

R output (abbreviated)

Group	pincome	pred.income
1 Not HS grad	5.07	5.06
2 HS grad	7.83	7.84
3 Some college	8.98	8.88
4 College grad	12.81	12.84
5 Grad school	12.95	12.91

A substantial degree of correspondence exists between personal income and the second model's predictions based on education level. In fact, CART models usually offer more precise predictions than LRMs because they reveal complex patterns in the data.[33]

The two plots requested in the R code are not shown given their large and cumbrous nature. But consider the first plot R creates based on LRM11.7a, which you may wish to resize in the plot window to make it wide. You should see boxes at the bottom of the plot labeled "Node." The nodes are groups in the data based on varying levels of personal income. Within each node is a boxplot that provides the distribution of income for each group. And the classification into these nodes is predicted by different combinations of partitions of the explanatory variables. For example, Node 7 appears to comprise those reporting the lowest level of income, close to zero on average. Following the branches that lead to this node specifies that those classified within it are females with less than 12 years of education. Node 12 consists of those who report personal incomes of about, on average, 18 or 19. Members of this node are males who report more than 15 years of education.

The second plot is more intricate than the first since it includes age, which is partitioned into many branches of the tree. One interesting pattern, though,

---

[33] This can be a blessing and a curse. For those whose aim is prediction, CART models are valuable tools. But for researchers more interested in assessing conceptual or explanatory models, CART models are limited because they tend to be data-specific and have low external validity or generalizability beyond the sample data used to estimate them.

is the skewed distributions for some of the older people in the sample (see nodes 24 and 31, for example). What does this suggest about personal income among older adults in the U.S.? It should be apparent that CART models can become quite complex even with a small number of explanatory variables since they expose a complicated set of empirical pathways—another way of representing interactions—that predict particular outcomes. These models are also employed most often as predictive tools, although if one's conceptual model is detailed enough, they might also be used to assess explanatory relationships.[34]

A Cautionary Note about Interaction Effects

Before concluding this chapter, consider a word of warning: some statisticians claim that interaction effects, whether in traditional LRMs or CART models, are challenging and sometimes misleading because they fail to disentangle the actual ordering of the variables. If one of the variables used to compute an interaction term or in a branch of a CART is accounted for by the other, then the usual interpretation of the interaction effect becomes questionable. Think about the personal income models estimated in this chapter: perhaps gender influences education in some way, or, as we say, education is *endogenous* in the model. What about the violent crime model? Median household income may affect a state's unemployment rate so the unemployment rate could be endogenous in the model.[35] Chapter 12 discusses these types of problems in more detail. In any event, the lesson is that we should always think carefully about the nature of the variables in the model when considering interaction terms.

---

[34] The field of data science and machine learning offers several other tools for predicting outcomes, some of which are suitable for exploring complex interactions among variables. In fact, some analysts consider CART models outdated and recommend methods such as random forests, support vector machines, and neural networks. R has several packages that estimate these and other predictive models (see Scott V. Burger (2018), *Introduction to Machine Learning with R*, Sebastopol, CA: O'Reilly Media).

[35] For example, states with high household incomes might be less prone to high unemployment because of their relative wealth or successful economies. On the other hand, high unemployment, which reflects a struggling economy, might lower a state's median household income over time.

Chapter Summary

This chapter has covered a lot of important material, so a summary should be helpful. Nonlinear associations and outcome variables and LRM residuals that are not normally distributed are quite common in the research world. Associations between two variables may also depend upon—or be moderated by—some third variable. Interaction terms provide a valuable way to test for moderator effects in regression models. As a review and conclusion to what we've learned, here are some suggestions:

1. Determine whether the residuals from an LRM follow a normal distribution (or are at least close to it) with a normal q-q or a kernel density plot (other plots may also be helpful). Small departures are usually acceptable, but highly skewed residuals lead to inefficient estimates and suggest that other forms of the outcome or explanatory variables be considered. Several useful transformations are available. But don't overdo it—transforming variables could lead to biased estimates.

2. Use conceptual models based on sensible theories and previous literature to guide your thinking about nonlinear associations and interactions among variables.

3. Construct scatter plots of each explanatory variable by the outcome variable. Plot nonlinear lines—such as lowess fit lines—in these scatter plots. Examine partial residual plots after estimating LRMs. When faced with nonlinear associations, consider the various transformations available, such as quadratic terms. You might be able to find one that will linearize the associations or that will better represent a nonlinear association between two variables.

4. When including higher-order (e.g., quadratic) or interaction terms in an LRM, consider using standardized versions of the explanatory variables to diminish the risk of collinearity. Graph the predicted values to figure out what the nonlinear or interaction term implies about the associations in the model.

5. Analysts can get carried away by combining quadratic terms, cubic terms, and interactions in a single regression model. Take care to avoid doing too much since models can easily become unmanageable when using these types of variables. Have a clear idea—guided by a theory, a conceptual model, and previous research—about why associations exist rather than searching for them by estimating multiple models. Remember that, by chance alone, you will probably come across what appear to be fascinating nonlinear and interactive relationships. If they were not anticipated by the conceptual model

and you cannot explain them in a reasonable way, they should be avoided.[36]

Chapter Exercises

The dataset *StateCrime.csv* consists of data from U.S. states collected in 2010. The variables include

- `State`   state identification number
- `crimes`   total crimes per 100,000 population
- `males14_24`   number of males aged 14–24, per 1,000 population
- `educate`   average years of schooling for persons aged 25 and older
- `percappolice`   per capita police expenditures, 2009 (in dollars)
- `lfp`   labor force participation per 1,000, males aged 14–24
- `population`   state population in 100,000s
- `unemploy14_24`   unemployed males aged 14–24 (per 1,000 population)
- `unemploy35_39`   unemployed males aged 35–39 (per 1,000 population)
- `wealth`   median value of goods and assets per capita (in dollars) (this is a common measure of a state's wealth)
- `inc_inequal`   income inequality index (lower values indicate low income inequality; higher values indicate high income inequality)

We shall explore these data to test assumptions concerning normality and linearity and evaluate a conceptual model that proposes an interaction effect. After importing the data into R, complete the following exercises.

---

[36] In Chapter 1, one of the "best statistical practices" is "Read results with skepticism, remembering that patterns can easily occur by chance (especially with small samples), and that unexpected results based on small sample sizes are often wrong." A wise researcher always heeds this advice.

1. Estimate an LRM that uses `crimes` as the outcome variable and the following explanatory variables: `males14_24`, `percappolice`, and `inc_inequal`.

   a. Interpret the slope coefficient, *p*-value, and 95% CI associated with the variable `inc_inequal`.

   b. Interpret the Multiple R^2 from the model.

2. Test the model estimated in exercise 1 to determine whether it satisfies the assumption of normality. Describe what test you used and include evidence of this test.

3. We suspect a nonlinear relationship between crime rates and income inequality. Criminological theory suggests that more income inequality leads to more crimes, but only up to a certain point. Create a graph that includes `crimes` on the *y*-axis and the variable that measures income inequality on the *x*-axis. The graph should include a test of whether there is a nonlinear association between these two variables. What does the graph suggest about our suspicion of a nonlinear association?

4. Estimate a linear regression model with `crimes` as the outcome variable and the following income inequality measures: a linear component and a quadratic component. Also include in the model the variable `percappolice`. What do you conclude about the association between the crimes per 100,000 population and income inequality?

5. Evaluate the model in exercise 4 for collinearity issues. What does your evaluation suggest? If you find evidence of a problem, re-estimate the model using the "statistical trick" discussed in this chapter. What does the model that adjusts for collinearity suggest about the association between crimes per 100,000 population and income inequality?

6. We also suspect because of previous research we've conducted and criminological theories regarding crime rates that police expenditures per capita and income inequality interact to affect crimes across U.S. states. In particular, when there is more police funding (which suggests a stronger police presence) and higher levels of income inequality, this represents a conflictual state climate that is reflected in more crimes than either condition alone. Extend the model estimated in exercise 1 by including an interaction term between income inequality and per capita police expenditures. What does the model suggest about our contentions regarding an interaction effect? (Suggestion: you might try graphing the interaction and its association with the outcome variable using the `interplot` function.)

7. *Challenge*: return to the LRM estimated in exercise 1. Assuming you found evidence that the normality assumption was not satisfied

(exercise 2), use a Box-Cox transformation and determine whether you can find a model that "normalizes" the residuals. What is the lambda value of this model? Provide evidence that it operates as expected by plotting the residuals using a normal q-q plot and a kernel density plot.

12

Model Specification

We've now learned about the five core assumptions of LRMs, including tests to determine if they are satisfied and adaptations to consider if they are not. But, as introduced in Chapter 4, some specific situations create issues that make it difficult to satisfy the assumptions. The first of these concerns whether or not we have constructed a suitable empirical model, or what is normally called the correct *specification* of the model. *Specification error* (also called *specification bias*) occurs when we have not set up the model correctly. For instance, in Chapter 11 we discussed one type of misstep that can occur: specifying a linear association when, actually, a nonlinear association exists between an explanatory variable and an outcome variable. Remember the figure from the last chapter that depicted a nonlinear association (reproduced in Figure 12.1)?

A second issue discussed in Chapter 11 can also lead to questionable model specification: assuming additivity when the nature of an association is nonadditive. This can occur when the true association of an explanatory variable and the outcome variable depends on—is conditional upon or moderated by—a third variable. To give a highly speculative example, suppose that the association between median household income and opioid deaths at the state level is positive in states east of the Mississippi River and negative in states west of the Mississippi, such that the *average linear association* represented in an LRM is zero. One would conclude that there is no statistical association between median household income and opioid deaths. Yet, if one includes an interaction term between median household income and eastern vs. western states, the actual association is revealed. The lessons of Chapter 11—including the need for careful conceptualization—can thus help diminish the likelihood of specification error and lead to correct model specification.

A third issue involves a type of variable discussed in Chapters 4 and 6: confounding variables. When these variables are not included in a model, analysts can misconstrue the true association between an explanatory variable and an outcome variable. If we leave confounding or other important variables out of a model, the independence assumption is not satisfied (see Chapter 8) because, as you'll recall, the error term includes all the variation of y that is not included in the model—including potentially important variables. In addition, since we rarely have control over the x variables, we must assume they behave similarly from sample to sample. If they do not, then the errors may be associated with some aspects of the x variables. When the error term includes an unobserved confounding variable or is associated

DOI: 10.1201/9781003162230-12

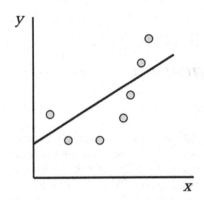

FIGURE 12.1
Illustration of a curvilinear association.

in some way with an explanatory variable, a source of systematic variation exists that is not independent of the model.[1]

After offering some advice on how one should choose the variables to include in an LRM, this chapter addresses four common types of specification error in multiple LRMs: (1) including irrelevant variables, (2) leaving out important variables, (3) misspecifying the "causal ordering" of the variables in the model; and (4) selection bias.[2] The first three are also called *overfitting*, *underfitting*, and *endogeneity* or *simultaneous equations bias*. All of these present problems for LRMs, but to different degrees.

Variable Selection

Variable selection is a central topic in regression modeling. How do we know which variables to place in a model? A large number of automated variable selection procedures are available; a few are described later in the chapter. At this point, understand that much of the work on variable selection by mathematical statisticians and data scientists emphasizes obtaining the best *prediction* equation at the expense of testing a reasonable *explanatory* model.[3] Regression models are clearly useful as predictive tools. Most of us would

---

[1] To modify the Chapter 8 notation: when the Xs and the errors are not independent, we fail to satisfy the independence assumption.

[2] As already suggested, Chapter 11 discusses two specific sources of specification error: not identifying nonlinearities or nonadditive associations.

[3] Footnote 14 in Chapter 10 makes a similar point by characterizing, perhaps a bit unfairly, data science and machine learning techniques as prediction exercises. But given the fondness the purveyors of the data science movement have with predictive accuracy, this point is not entirely unreasonable.

prefer, for instance, that medical researchers use statistical tools to predict with a high degree of certainty that a drug will cure us (or won't injure us!). Perhaps we don't care much about explaining the underlying biological mechanisms that link the drug with the cure. But, in social and behavioral sciences, explanations are the driving force at the core of research.[4] We wish to be able to explain *why* one variable is associated with another (e.g., why are neighborhood unemployment rates associated with crime rates?) rather than to say only whether one predicts another. When we select variables for a regression model, we therefore should rely on an explanatory framework—based on theory or a conceptual model, as well as logical reasoning and what previous research has taught us—to decide which variables to include in the LRM. This presupposes a solid understanding of previous studies that have examined the outcome variable and complete familiarity with the variables in the dataset, including how they are constructed, what they purport to measure, and so forth.

Overfitting—or the Case of Irrelevant Variables

Examples of overfit models in research studies are easy to find. Browse through an issue of just about any social and behavioral science journal and you will find regression models with variables that add nothing to the objective of predicting—or even explaining anything about—the outcome variable. Researchers have various reasons for including seemingly irrelevant variables in an empirical model. Perhaps most previous studies have included them. Some researchers also seek to guard against omitting potential confounding variables and so include just about any variable that might be important. Although this may violate the sacrosanct principle of parsimony and complicate models based on small samples, it normally will not present a problem for LRMs that rely on large samples. Researchers should be careful, though, about putting too many variables in a regression model.

Overfitting does not bias slope coefficients; in other words, the regression slopes on average are still accurate. The main problem with overfitting is that it leads to inefficient estimates of these coefficients. Recall from Chapter 2 that an inefficient estimator has, on average, a larger variance (less precision) than does an efficient estimator. To understand the consequences of overfitting, consider, once again, the formula for the standard errors of multiple LRM slope coefficients (see Equation 12.1).

---

[4] In fact, many physicists argue that explanation, rather than mere prediction, is also at the core of their scientific endeavors. See, for example, David Deutsch (1997), *The Fabric of Reality*, New York: Penguin Books. Explanation and prediction are not independent, though: a valid explanation should yield accurate predictions.

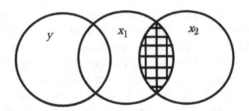

FIGURE 12.2
Illustration of an overfit LRM using overlapping circles.

$$se\left(\hat{\beta}_i\right) = \sqrt{\frac{\Sigma\left(y_i - \hat{y}_i\right)^2}{\Sigma\left(x_i - \bar{x}\right)^2\left(1 - R_i^2\right)\left(n - k - 1\right)}} \tag{12.1}$$

As noted in Chapters 4 and 10, the quantity $\left(1 - R_i^2\right)$ is called the *tolerance*, which is estimated from an auxiliary regression equation denoted as $x_{1i} = \alpha + \beta_2 x_{2i} + \ldots + \beta_k x_{ki}$. Suppose an extraneous variable (x_2) included in the model is associated with an important variable (x_1) but is not associated with the outcome variable (y). Then, all else being equal, the tolerance for x_1 is smaller (closer to zero) when x_2 is in the model and the standard error of its slope coefficient in the LRM is relatively larger. Figure 12.2 illustrates this situation.

The tolerance in the standard error computation for x_2 is represented by the overlap between the two explanatory variables. If the overlap is large, then the standard error can shift considerably, becoming large enough in some cases to affect decisions about the statistical significance of x_1's coefficient whether we use a *p*-value or CI to judge it. (Keep in mind that we want substantial variation (i.e., large circles) among the explanatory *and* outcome variables.) If a statistical association between the explanatory variables does not exist (e.g., $\text{cov}\left(x_1, x_2\right) = 0$), then the standard error associated with the important variable's (x_1) slope coefficient is unaffected. In Figure 12.2, this would be presented by non-overlapping circles between x_1 and x_2.[5]

Underfitting—or the Case of the Absent Variables

A more serious situation arises when important variables are left out of the regression model. In this situation, we may reach the wrong conclusions about our ability to predict—or explain—the outcome variable. This

---

[5] Overfitting also leads to inflated R^2 values, which can mislead us into thinking a model is particularly predictive, and often creates models that are idiosyncratic to specific samples and thus do not generalize well to other samples or to the target population (see Burger (2018), *op. cit.*, chapter 4).

problem is also called *omitted variable bias* and is why researchers should always think carefully about potential confounding variables. Recall the example in Chapter 7: there appears to be an association between fundamentalist Christian affiliation and family income until we include education in the model (see LRM7.4 and Table 7.3). In brief, the association is, to use a common term, *spurious*.

As mentioned several times, think of the error term in the LRM equation as including all of the factors, random or otherwise, that might be associated with the outcome variable but are not represented by the model's explanatory variables. The errors, thus, include information from omitted variables. We still assume, though, a zero correlation between the Xs and the errors (or their estimates—the residuals: $\text{cor}(x, \hat{\varepsilon}_i) = 0$). Suppose, as Equation 12.2 illustrates, that a regression equation includes x_1 and x_2, but omits x_3 and x_4. The variable x_3 is also associated with y and with x_2 and affects their association. In this situation, the slope coefficient affiliated with x_2 is biased. In general terms, we have not satisfied the independence assumption. Unfortunately, we cannot know for certain the direction of the bias: is the estimated slope too large or too small? Is it negative or positive? The answers to these questions depend on the joint associations among x_2, x_3, and y.

$$y_i = \alpha + \beta_1 x_1 + \beta_2 x_2 + \hat{\varepsilon}_i \text{ where } \hat{\varepsilon}_i = x_3 + x_4 + \text{random error} \qquad (12.2)$$

$$\left(\text{cov}(x_2, x_3) \neq 0 \ \& \ \text{cov}(y, x_3) \neq 0\right)$$

The underfit model in Chapter 7 is helpful for understanding this issue, but let's explore another. The *GSS2018.csv* dataset includes the variable lifesatis (see Figure 11.2), a measure of life satisfaction that is based on some questions that ask about satisfaction with one's marriage, work, and in general. Higher values of this variable indicate a greater sense of life satisfaction. Research suggests that life satisfaction is associated with education, occupational prestige, involvement with religious organizations, and several other variables. We'll examine some of these associations in an LRM.[6] First, though, let's look at a bivariate correlation matrix of the variables designed to measure each.[7]

R code

```
sub.GSS <- GSS2018[c("occprest","educate", "attend",
            "lifesatis")]
```

---

[6] This model is based on an example in William D. Berry and Stanley Feldman (1985), *Multiple Regression in Practice*, Newbury Park, CA: Sage. But it uses more recent GSS data.

[7] Values for life satisfaction are recorded for only 1,161 respondents in the *GSS2018* dataset. In order to analyze models using lifesatis, restrict the analytic sample to only those respondents who have non-missing values on this variable. As shown in the R code, the subset function is useful in this situation. Just be careful not to overwrite the original dataset; use a new descriptive name for the data frame that contains the analytic sample.

```
# create a subset of the relevant variables
sub.GSS.nomiss <- subset(sub.GSS, (!is.na(sub.
                   GSS$lifesatis)))
# remove the missing values of life satisfaction: !is
   .na = is not NA (missing)
cor(sub.GSS.nomiss) # request the correlation matrix
```

R output (abbreviated)

	occprest	educate	attend	lifesatis
occprest	1.000	0.456	0.105	0.110
educate	0.456	1.000	0.042	0.056
attend	0.105	0.042	1.000	0.121
lifesatis	0.110	0.056	0.121	1.000

The correlations suggest that all three of the proposed explanatory variables—occupational prestige, education, and religious service attendance (attend)—are positively associated with life satisfaction. Education has the smallest correlation with life satisfaction ($r = 0.056$), though. But also notice the substantial correlation between education and occupational prestige ($r = 0.46$), which should not be surprising: highly educated people tend to be employed in positions that are considered more prestigious (e.g., judges, bank presidents, and even college professors!), and occupations that require more formal education are often deemed more prestigious. Pay attention to education and occupational prestige as we consider a couple of LRMs. Unlike the approach in Chapter 7, let's begin with the full model and work backward (see LRM12.1 and LRM12.2).

R code

```
LRM12.1 <- lm(lifesatis ~ ., data=sub.GSS.nomiss)
 # when using a subset of a dataset, the lm function,
    when it finds ~., recognizes that it should use all
    the variables not listed as y as explanatory
    variables
summary(LRM12.1)
confint(LRM12.1)
```

R output (abbreviated)
Coefficients:

	Estimate	Std. Error	t value	Pr(>|t|)	
(Intercept)	69.52317	1.64085	42.370	< 2e-16	***
occprest	0.07850	0.02736	2.869	0.00419	**
educate	0.03377	0.12451	0.271	0.78623	
attend	0.44670	0.11743	3.804	0.00015	***

```
Signif. codes: 0 '***' 0.001 '**' 0.01 '*' 0.05 '.' 0.1
```

```
Residual standard error: 11.04 on 1157 degrees of freedom
Multiple R-squared: 0.02431, Adjusted R-squared: 0.02178
F-statistic: 9.61 on 3 and 1157 DF, p-value: 2.868e-06
```

[CIs]	2.5%	97.5%
(Intercept)	66.304	72.743
occprest	0.025	0.132
educate	-0.211	0.278
attend	0.216	0.677

The first thing to notice in the output is that, once we account for the association between occupational prestige and life satisfaction, the positive association between education and life satisfaction in the correlation matrix does not appear to be notable. The association between religious service attendance and life satisfaction remains, however. In this situation, we might say that the association between education and life satisfaction is spurious, perhaps confounded by occupational prestige. One way to represent this is with Figure 12.3, which suggests that although there appears to be an association between education and life satisfaction (identified by the arrow that is crossed out), occupational prestige is likely associated with both of these variables in such a way as to account completely for their presumed association.

Suppose that we omit occupational prestige from the model because we failed to read the literature on life satisfaction carefully. We might think that highly prestigious occupations demand so much of people that they can't possibly report high satisfaction with their lives. Or we simply don't care much about employment patterns because, we argue, education and religious practices are much more important.

In LRM12.2, the slope coefficient for education is much larger than in the full model (0.03 → 0.20) and its 95% CI shifts upward quite a bit. We know, however, that this regression model is misspecified because *cor*(educate, occprest) ≠ 0. The model is underfit because it does not include a variable

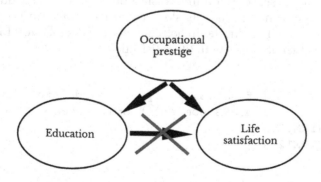

FIGURE 12.3
Illustration of an underfit LRM.

that is associated not only with the outcome variable but also with education in such a way as to account for its presumed association with life satisfaction. The education slope coefficient is inflated when we omit occupational prestige from the model.

R code
```
LRM12.2 <- lm(lifesatis ~ educate + attend, data = sub
            .GSS.nomiss)
                # omit occupational prestige
summary(LRM12.2)
confint(LRM12.2)
```

R output (abbreviated)
```
Coefficients:
              Estimate  Std. Error  t value  Pr(>|t|)
(Intercept)   70.6871     1.5949     44.321   < 2e-16  ***
educate        0.1962     0.1112      1.764   0.078    .
attend         0.4794     0.1172      4.089   4.63e-05 ***
---
Signif. codes: 0 '***' 0.001 '**' 0.01 '*' 0.05 '.' 0.1

Residual standard error: 11.07 on 1158 degrees of freedom
Multiple R-squared: 0.01737, Adjusted R-squared: 0.01567
F-statistic: 10.23 on 2 and 1158 DF, p-value: 3.929e-05

[CIs]          2.5%   97.5%
(Intercept)   67.558  73.816
educate       -0.022   0.414
attend         0.249   0.709
```

Another nested model to consider omits religious service attendance (see LRM12.3). We have seen that this variable is associated—at least in a statistically significant sense—with life satisfaction whether we include occupational prestige or not. In fact, its slope coefficient, standard error, t-value, p-value, and 95% CI are similar in both models. But does it affect the associations of the other variables with life satisfaction?

R code
```
LRM12.3 <- lm(lifesatis ~ occprest + educate, data =
              sub.GSS.nomiss) # omit attend
summary(LRM12.3)
confint(LRM12.3)
```

```
R output (abbreviated)
Coefficients:
              Estimate   Std. Error   t value     Pr(>|t|)
(Intercept)   70.46416      1.63151    43.190     < 2e-16 ***
occprest       0.08859      0.02739     3.235     0.00125 **
educate        0.03038      0.12522     0.243     0.80838
---
Signif. codes: 0 '***' 0.001 '**' 0.01 '*' 0.05 '.' 0.1

Residual standard error: 11.1 on 1158 degrees of freedom
Multiple R-squared: 0.01211, Adjusted R-squared: 0.0104
F-statistic: 7.097 on 2 and 1158 DF, p-value: 0.0008643

[CIs]           2.5%    97.5%
(Intercept)   67.263   73.665
occprest       0.035    0.142
educate       -0.215    0.276
```

The education coefficient in this model is similar to the coefficient in the full model (LRM12.1). Why do you think this is the case? Examine the correlation matrix—the bivariate correlation between educate and attend is not large ($r = 0.04$). One would not expect much of a change with attendance omitted since it has such a weak association with education. Although this last model is underfit, the consequences of specification error are less severe than when we exclude occupational prestige. One lesson to learn from these models is that specification error is a virtual certainty: we cannot include all the variables that are associated with outcome variables. The goal is to strive for models with a low or manageable amount of specification error and hope we don't reach the wrong conclusions about the associations that the models do represent. A vital task as we consider how to build LRMs is thus to identify the potential confounders.[8]

The last question to ask in this section is whether the full model, which includes all three explanatory variables, is overfit. Education is, at best, weakly associated with life satisfaction once occupational prestige is in the model, but should it be included in the model? On the one hand, including education does not change our main conclusions. On the other, it might make the other standard errors larger than they should be. At this point, the best way to check for

---

[8] Recall Chapter 2's discussion of John Stuart Mill's criteria for establishing causality (see fn. 5). The third criterion is that other explanations or factors that account for the presumed cause–effect association must be eliminated. Confounding variables are these "other" factors; therefore, if we ever hope to establish the true "causal" relationship between two variables, we need access to all potential confounders. This is a tall order and not achievable in most research projects. Statisticians have, however, developed a number of empirical tools, such as propensity score weighting and matching estimators, that purport to offer ways of establishing causality with nonexperimental data. For a thorough overview of these methods, see Imbens and Rubin (2015), *op. cit.* For applications in R, see Walter Leite (2016), *Practical Propensity Score Methods Using R*, Los Angeles, CA: Sage. See also Chapter 17 for additional information.

overfit is to estimate an LRM that omits education and inspect the coefficients. The results of such a model (not shown) indicate that the slope coefficient for occupational prestige increases slightly when education is omitted. Unless we are zealous in our desire to figure out the precise numeric association between occupational prestige and this measure of life satisfaction, though, whether or not education should be in the model is a judgment call.[9]

Endogeneity Bias

Early in Chapter 3 we observed that another name for the outcome variable is the *endogenous* variable. The term *endo* refers to something that is inside or within. An endogenous variable is, therefore, one produced within the system or within the equation. The linear regression equation, you've noticed by now, assumes that the outcome variable, but not the explanatory variables, is "produced" within the system. The explanatory variables are considered *exogenous* or produced only by factors outside of the system. A common claim is that the explanatory variables are "predetermined"; *pre* implying they are formed outside the system inferred by the model. Such an understanding of the model is so important some experts call it the *exogeneity assumption*.[10]

Endogeneity issues arise when the LRM has not been specified correctly and generally involve when explanatory variables are associated with the error term. The last section discussed how this can occur when we have omitted variable bias because $\text{cov}\left(\mathbf{x}, \hat{\varepsilon}_i\right) \neq 0$), but it also includes whether we have identified the correct ordering of the variables. For example, think about a set of variables such as education, occupational prestige, race/ethnicity, and life satisfaction. Models are often proposed that define a particular set of these variables as affecting one another in some way. In the last section, for instance, we saw that occupational prestige confounded the presumed association between education and life satisfaction. But is it feasible to imagine that occupational prestige is endogenous in the model since it is influenced by education? Probably, and this implies a different sort of relationship, with occupational prestige *mediating* the association between education and life satisfaction. Figure 12.4 represents this model.

The figure characterizes two models that are interconnected: the first uses occupational prestige as the outcome, whereas the second uses life

---

[9] One thing that might persuade us to keep education in the LRM is the fact that most other studies have used it in models designed to account for variations in life satisfaction (I'm not sure this is true, but assume it is). As a practical matter, convincing seasoned researchers that it's all right to omit a variable that has been used frequently in previous studies is difficult.

[10] Although endogeneity bias is distinguished from other specification issues in this section, omitted variable bias is actually one type of endogeneity issue. Measurement error, discussed in Chapter 13, is another type.

FIGURE 12.4
Illustration of a mediation model.

satisfaction as the outcome. When two (or more) such models are considered in a regression context, this approach is called a *simultaneous equations model*.[11] If such an association exists yet we fail to model it appropriately, we are at risk of *simultaneous equations bias*.

Similarly, consider the model in Equation 12.3 that includes race/ethnicity.

$$\text{Life satisfaction}_i = \alpha + \beta_1\left(\text{occup. prestige}_i\right) + \beta_2\left(\text{race/ethnicity}_i\right) + \hat{\varepsilon}_i \quad (12.3)$$

The model implies that occupational prestige and race/ethnicity independently combine to affect life satisfaction. But suppose that occupational prestige and race/ethnicity are not independently determined; rather, because of various historical and social factors that are associated with race/ethnicity in the U.S. and elsewhere, occupational prestige is, in part, a product of an individual's race/ethnicity. Similar to the previous example, occupational prestige is endogenous in the system specified by Equation 12.3. The problem for the LRM is that the estimated slope coefficient for occupational prestige in the equation is biased. As an exercise, estimate a model with occupational prestige as the outcome variable and a variable that measures race/ethnicity (race or ethnic) as the explanatory variable. Include education as a control variable. What do the results tell you about the possible endogeneity of occupational prestige?

A second endogeneity issue asks whether one or more explanatory variables might be affected by the presumed outcome variable. Suppose we wish to estimate a model with adolescent alcohol use as the outcome variable and friends' alcohol use as the explanatory variable. We may assume that one's friends influence one's behavior to a certain degree, thus leading to the model specification shown in Figure 12.5. But it may also be true that one's choice of friends depends on one's behavior. Perhaps youth who drink alcohol are more likely to choose friends who also drink alcohol. This issue implies that alcohol use and friends' alcohol use are involved in a *reciprocal association*.

We won't pursue these particular endogeneity issues in any more detail because they involve (1) thorny conceptual issues and (2) complex statistical issues. Endogeneity is conceptual because we need to think carefully about the models and evaluate meticulously whether one or more of the explanatory variables is potentially endogenous. When we estimate models with variables such

---

[11] These models can be complex, with multiple submodels. Books on structural equation models (SEMs) discuss simultaneous equations in depth. For information about how to estimate these models in R, see Finch and French (2015), *op. cit.*

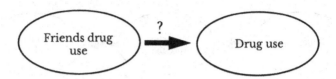

FIGURE 12.5
Illustration of questionable temporal ordering of variables.

as education, income, and life satisfaction—or one's own behavior and one's friends' behavior—we should think about the order and intra-system relationships among the variables. Are the explanatory variables truly exogenous—are they determined "outside" the model? Or could one or more "depend" on the other explanatory variables? Moreover, could the presumed outcome variable affect one or more of the explanatory variables? One should consider these questions to obtain a properly specified model.

Endogeneity is also a complex statistical issue because its solution usually requires complicated systems of equations. In addition to simultaneous equations models, Chapter 13 discusses another modeling approach, called *two-stage least squares*, that is useful for addressing endogeneity in LRMs. R also provides several other methods for addressing this topic.[12]

Selection Bias

Selection bias occurs when a researcher analyzes data among a nonrandom subsample of observations but attempts to infer the results to an intended target population. This form of bias can occur for several reasons, including using nonrandom sampling strategies to collect data or attrition of sample members in a longitudinal study.[13] As mentioned in Chapter 2, much of the edifice of inferential statistics is built on the assumption that observational data are collected randomly, so selection bias has a detrimental effect on our ability to judge LRM results.

---

[12] R even has a package called endogenous!

[13] Researchers have discussed the consequences of selection bias for many years. For instance, a debate between Karl Pearson and John Maynard Keynes concerning the former's 1910 study of parental alcoholism centered largely on whether the sample was appropriate for making inferences to a larger population of families (Stephen M. Stigler (1999), *Statistics on the Table*, Cambridge, MA: Harvard University Press, chapter 1). The term selection bias likely emerged later in the 20th century, though, as the analysis of survey data became a common method of predicting various social, economic, and political phenomena (see, for instance, Raymond Franzen and Paul F. Lazarsfeld (1945), "Mail Questionnaire as a Research Problem," *Journal of Psychology* 20(2): 293–320, and Gunner Ekland (1959), *Studies of Selection Bias in Applied Statistics*, Upsalla, SE: Almqvist & Wiksells).

A particular type of selection bias common in regression modeling occurs when some portion of the sample does not experience the outcome for systematic (nonrandom) reasons. Briefly stated, selection is not random with regard to the outcome variable, *y*. For instance, suppose a researcher wishes to estimate an LRM to predict personal income among a sample of adults. But a subset of people from the sample reports no personal income, perhaps because they are not in the workforce. The process that determines who earns no income and who earns some positive amount of income is probably not random; rather, there are systematic reasons people have no or various positive amounts of income.

A similar selection issue occurs when the outcome variable is *censored*. This arises when, say, a measure of income identifies only whether people earn $1–$15,000 per year, but not the specific amount (so it might be coded ambiguously as 7.5 (the midpoint of the category in $1,000s) or 15 (the maximum of the category in $1,000s)). But the variable does measure the amount of income if it's more than $15,000 (e.g., 16, 16.2, or 17.5). The variable is thus censored at the low end of the income distribution, which is also called *left-censoring*. Variables may also be censored at the high end (*right-censoring*) (e.g., $150,000 or more). A related phenomenon is when a variable is *truncated*. This occurs when information is collected on only a subsample of those who, for example, earn in excess of $25,000 per year and we wish to predict how much this subgroup earns based on a set of explanatory variables. The variable is thus *truncated from below*. Variables may also be *truncated from above*. In general, the sample data are missing a portion of information that theoretically exists in the population.[14]

The problem for the LRM arises if unmeasured factors affect the probability of selection, such as who reports some income or whether income levels are censored or truncated. For example, if an unmeasured factor, such as motivation to work, affects who reports any income and affects the frequency of income, the slope coefficients are biased because the selection process (*s*) and, in many situations, the *x*s are correlated with the error term $(cor(s, \hat{\varepsilon}_i) \neq 0; \ cor(\mathbf{x}, \hat{\varepsilon}_i) \neq 0)$. If the selection process is random, however, then the slope coefficients are unbiased.

When examining outcome variables affected by these issues, such as those that measure the frequency of behaviors (e.g., how much alcohol one drinks) or the amount of money or other resources earned or distributed (e.g., monetary contributions to charities), making adjustments to the regression model that account for selection is important for attaining correct model specification.[15]

---

[14] As inferred from the discussion of missing data in Appendix A, selection bias is a consequence of data *missing not at random* (MNAR).

[15] Some analysts argue that selection bias, or some forms of it, is a type of endogeneity bias because, for example, the process that leads to the "selection" of, say, personal income is a separate endogeneity issue from the process that leads to variation in the amount of personal income. Here, though, we distinguish endogeneity bias as an issue primarily for the explanatory variables and selection bias as an issue for the outcome variable, even if the distinction is not always used in practice.

How Do We Assess Specification Error and What Do We Do about It?

Specification error is a problem all careful researchers wish to avoid. But, as implied earlier, considering its many potential sources is not an easy task. The best advice is to always strive for a clear and convincing conceptual model or theory to guide the analysis. Of course, conceptual models do not spring forth fully grown like Athena from Zeus's skull. They result from examining previous studies and engaging in sensible thinking about the processes involved in creating an association between two or more variables. Proficient researchers do not even begin to estimate regression models until they have read carefully the research literature that addresses the outcome variable, including the variables likely to account for it and whether sample selection issues might affect how it is distributed. They also read studies involving the key explanatory variables. In this way, they minimize the risk of ignoring variables that are associated with the key explanatory variables used in the regression model.

Unfortunately, identifying and including all the variables that might be associated with the explanatory and outcome variables is an implausible goal. So, as suggested earlier, specification error is always lurking, we just hope to minimize it. Some limited tools are available for assessing whether certain types of specification error affect the results of the LRM. The first such tool may be understood by considering Equation 12.2 provided earlier in the chapter, reproduced as Equation 12.4.

$$y_i = \alpha + \beta_1 x_1 + \beta_2 x_2 + \hat{\varepsilon}_i \text{ where } \hat{\varepsilon}_i = x_3 + x_4 + \text{random error} \qquad (12.4)$$

When we initially saw this equation, we were interested in what happens to the regression model when x_2 and x_3 associated. Although their association might affect the conclusions, how might we test whether x_3 has an influence if we do not measure it directly? Can we figure out a way to assess it so we may, at least indirectly, examine the possible association between x_2 and x_3? As mentioned in earlier chapters, the predicted values from the model are useful for computing the residuals as an estimate of ε_i (see Equation 12.5).

$$\text{residual}_i\left(\hat{\varepsilon}_i\right) = \left(y_i - \widehat{y_i}\right) \qquad (12.5)$$

Can we then assess whether the explanatory variables included in our model are correlated with the residuals? Unfortunately, because of the way the OLS estimators are derived, a linear association between the xs and the residuals is rare. Sometimes, though, we might find a nonlinear association (see Chapter 11) and thus identify one type of specification problem.

R has some automated ways to test for specification error in LRMs. A common method that is not automated (at least as I write this), but is used often in research studies, is the *link test*. This test is based on the notion that if an LRM is properly specified, we should not be able to find any additional explanatory variables that are related to the outcome except by chance. The test creates two new variables, the variable of prediction, _hat ($\hat{y}_i$), and the variable of squared prediction, _hatsq ($\hat{y}_i^2$). The model is then re-estimated using these two variables as predictors. The first, _hat, should have a substantial effect on the outcome since it represents the predicted values. On the other hand, _hatsq should have a small to nil effect because, if the LRM is specified correctly, the squared predictions should have no explanatory power. We should thus examine the slope coefficient and other information about _hatsq (e.g., CI, p-value) to evaluate specification error.

Based on the LRM estimated earlier (LRM12.1), the following outlines the steps for conducting the link test in R.

R code
```
predict.life <- fitted(LRM12.1)
 # save the predicted values from model 12.1
new.life <- na.omit(GSS2018$lifesatis)
 # copy the original outcome variable and remove its
   missing values
LRM12.4 <- lm(new.life ~ predict.life +
           I(predict.life^2))
 # assess whether a nonlinear effect is predictive
summary(LRM12.4) # the link test
confint(LRM12.4)
```

R output (abbreviated)
Coefficients:

	Estimate	Std. Error	t value	Pr(>|t|)
(Intercept)	-316.68285	535.80239	-0.591	0.555
predict.life	9.43714	14.27137	0.661	0.509
I(predict. life^2)	-0.05617	0.09500	-0.591	0.554

The results of LRM12.4 suggest no specification problems since the _hatsq (predict.life2) coefficient does not appear large nor meaningful based, for instance, on its p-value (0.554).[16] The 95% CI for _hatsq also includes zero, and plotting the results suggests that a nonlinear association between the predicted and actual values does not exist.

---

[16] It may seem odd that the slope coefficient for _hat has such a large p-value. But this is explained by the very high VIFs in the model (check this). Once the collinearity involving the two x variables is resolved, the model provides a much smaller p-value for _hat ($p < 0.01$).

An alternative, but similar, approach is Ramsey's RESET[17] (*regression equation specification error test*), which is implemented in R using the `resettest` function that is part of the `lmtest` package. The main difference between the link test and the RESET is that the latter includes in its model the original explanatory variables. Additional higher-order predicted values, such as $\hat{y}_i^3$, may also be included. RESET uses a multiple partial (nested) *F*-test (see Chapter 6) to compare the model with all the prediction terms to a model with just the explanatory variables. Tests suggesting no difference between the two models—based on *F*-values with large *p*-values—offer little evidence of specification error. The following furnishes an example of the RESET.

R code
```
library(lmtest)
resettest(LRM12.1, power=2, type="fitted")
 # limit the powers of the explanatory variables to
   quadratic
```
R output
```
RESET test
data: LRM12.1
RESET = 0.3493, df1 = 1, df2 = 1156, p-value = 0.5546
```

This test has as its null hypothesis the absence of specification error, which is probable in this situation ($p = 0.55$), thus reaffirming the finding that this type of error is not present in the model.

A third test we won't demonstrate is called a *Durbin–Wu–Hausman test* (also known as the *augmented regression test* for endogeneity).[18] The test is more generally applicable than the other two tests for investigating endogeneity and is used for various types of regression models. Unfortunately, none of these tests is designed to determine whether the specification error, if it exists, is due to omitted explanatory variables; they are most useful for determining whether nonlinear associations are missed by the model.[19] We are often better off, though, examining partial residual and similar plots (see Chapter 11) to evaluate nonlinearities. In general, theory and a command of the previous research on the topic are the most important diagnostic tools for specification errors. Always ask: are the correct explanatory variables in the model? Does the model include the proper functional forms of the *x*s?

---

[17] Introduced in J. B. Ramsey (1969), "Tests for Specification Error in Classical Linear Least Squares Regression Analysis," *Journal of the Royal Statistical Society: Series B* 31(2): 350–371.

[18] The founding documents are James Durbin (1954), "Errors in Variables," *Review of the International Statistical Institute*, 22(1/3): 23–32; De-Min Wu (1973), "Alternative Tests of Independence between Stochastic Regressors and Disturbances," *Econometrica* 41(4): 733–750; and Jerry A. Hausman (1978), "Specification Tests in Econometrics," *Econometrica* 46(6): 1251–1271.

[19] Consider that the link test and the RESET use higher-order terms in their evaluations. This means they assess whether nonlinear associations have been omitted.

Testing for underfit or overfit is much easier when researchers have access to all (or most) of the important explanatory variables in the dataset. We obviously know that a regression model is underfit if we add variables that are associated with the outcome variable. As discussed in Chapters 5 and 6, researchers recommend using adjusted R^2s, RMSEs, and nested F-tests to compare models. If we add, say, a set of explanatory variables to an LRM and find that the adjusted R^2 increases or the RMSE decreases, then we have evidence that the initial model is underfit. If, on the other hand, we take away a set of variables (perhaps because we judge their associations with the outcome as weak) and find that the adjusted R^2 does not decrease, then we may conclude that the model is overfit.

Do not rely only on adjusted R^2s, RMSEs, or nested F-tests, though. A better method is to combine an evaluation of these goodness-of-fit statistics, as well as others, such as the AICs or BICs (see Chapter 6), with a careful review of the slope coefficients and standard errors of the remaining explanatory variables before concluding that a particular LRM is overfit or underfit. But we must also be aware of overdoing it, of testing so many models by adding and subtracting various sets of explanatory variables that we cannot decide on the appropriate model or we choose a model that emerges simply by chance.[20]

What to Do about Selection Bias?

As mentioned earlier, selection issues can lead to biased slope coefficients, especially as the factors—which might not be observed—that affect who is selected are also related to the levels or frequency of the outcome. Let's consider personal income. Imagine that some sample members do not earn any income, perhaps because they're not in the workforce, whereas others who are in the workforce earn varying amounts of income. One way to envision this situation is to consider that those who report no income have the potential to earn—their probability of earning income is not zero—but do not for various reasons. They therefore fall below a theoretical threshold that distinguishes no earnings from any earnings. Those who do report income range from low to high amounts but are above a theoretical threshold that

---

[20] An important point about estimating and comparing LRMs is that, if we test enough of them, we are bound to find one that seems to fit the data quite well, but may simply be due to chance and, thus, unique to a particular dataset. We are, after all, dealing with probabilities and as we examine more and more models, the probability that one of them appears as exceptional becomes high. This demonstrates, once again, that researchers should have a clear and logical conceptual model before any LRM is estimated and not go on "fishing expeditions" to discover highly predictive yet theoretically questionable models.

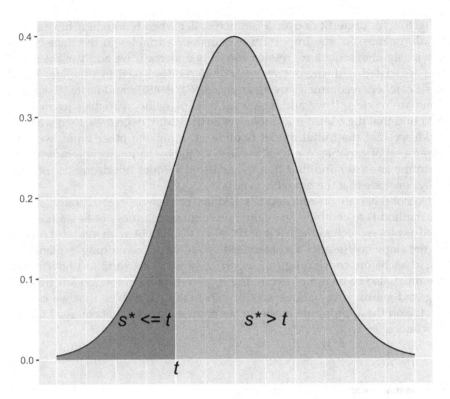

FIGURE 12.6
Assumed distribution of the outcome variable in a sample selection model.

identifies any income.[21] Figure 12.6 illustrates the underlying distribution of personal income, denoted by s^*, with t representing the theoretical threshold.

The measure of personal income in a sample is inherently comprised of a binary variable (no income vs. some income) and a continuous variable (positive amounts of income). In other words, we only observe the continuous portion of income's distribution in Figure 12.6 when $s^* > t$. To account for these two portions of personal income, we might therefore consider two regression models to estimate each. But, since the positive portion of income ($s^* > t$) is likely endogenous to the binary portion of income ($s^* \leq t$), the errors from the two regression models are apt to be correlated ($cor(v_i, \varepsilon_i) \neq 0$, where v_i signifies the errors from the binary model) and the slope coefficients biased.

One solution to this situation is to estimate a regression model for the binary portion of income with a set of explanatory variables that account for whether people report zero income or positive income. We then use this

---

[21] Some experts consider this type of distribution as censored because, theoretically, those who report no income could have an income—or the probability of positive income—we just don't observe it directly.

model to create a *selection variable* that predicts the probability of any income. The selection variable (or some suitable form of it) is included in an LRM, along with relevant explanatory variables, designed to predict the amount of income reported. Assuming the first model is specified correctly, the selection variable should control for the process that determines who reports zero vs. positive income. Equation 12.6 depicts the two models, with the selection variable part of the set of x variables in the frequency equation.

$$s_i = \alpha + \gamma z + \upsilon_i \quad \text{selection equation}$$

$$y_i = \alpha + \beta x + \varepsilon_i \quad \text{frequency equation} \tag{12.6}$$

$$s_i = \begin{pmatrix} 1 & \text{if } s_i^* > t \\ 0 & \text{if } s_i^* \le t \end{pmatrix} \quad \text{the observed binary variable}$$

Consistent with Figure 12.6, the last part of Equation 12.6 indicates that income is actually observed only if the probability exceeds the threshold t. The z and x denote explanatory variables at each stage, though there is usually some, but not complete, overlap among them. The two errors are assumed to meet the *iid* assumption (see Chapter 3, fn.7) and, if the selection process is nonrandom, have a nonzero correlation $(cor(\upsilon_i, \varepsilon_i) \neq 0)$. We also presume that the explanatory variables in each equation are independent of the error terms $(cor(z, \upsilon_i) = 0; \ cor(x, \varepsilon_i) = 0)$.

The regression model that estimates the amount of income (frequency component) may thus take the form shown in Equation 12.7.

$$(y_i \mid z, \upsilon) = \beta x + (\varepsilon_i \mid \upsilon_i) \tag{12.7}$$

You may recognize this equation as representing conditional probabilities: the outcome variable—amount of income—is conditional upon a nonzero amount of personal income reported. The selection portion of the model is estimated with a probit regression model, which is designed for binary outcome variables (see Chapter 16). The frequency portion is estimated with an LRM.

We'll use the R sampleSelection package's heckit function to estimate a *Heckman selection model* with variables from the *GSS2018.csv* dataset.[22] A key decision is what variables to include in the selection portion and in the frequency portion of the model. Specifying the correct selection model is especially important since, if one uses weak predictors, the coefficients can

---

[22] This is a commonly used type of two-stage regression model designed to adjust for selection bias and was popularized by the economist James J. Heckman (1976) ("The Common Structure of Statistical Models of Truncation, Sample Selection and Limited Dependent Variables and a Simple Estimator for Such Models," *Annals of Economic and Social Measurement* 5: 475–492).

have a higher degree of bias than in an LRM that ignores selection. In the following example, personal income (pincome) is the outcome variable. Before estimating the model, we must create a binary selection variable that is coded as 0 = zero income and 1 = positive income. This variable is labeled anyincome.[23] The selection variables include age, age^2, female, and childs (number of children living in the household). The frequency portion includes as explanatory variables educate, female, and attend. The heckit function setup is similar to the lm function, but lists the selection model first, followed by the frequency model (see LRM12.5). The selection variable computed from the first model is called the *inverse Mills's ratio*[24] and is based on a transformation of the predicted values.

R code

```
GSS2018$anyincome <- as.numeric(GSS2018$pincome >= 1)
 # create selection variable
table(GSS2018$anyincome) # check the frequency
                              distribution of anyincome
# Estimate the Heckman selection model
LRM12.5 <- heckit(anyincome ~ age + I(age^2) + female +
           childs, pincome ~ educate + female + attend,
           data=GSS2018)
summary(LRM12.5)
# Compute 95% CIs
coef.heck <- coef(LRM12.5)
se.heck <- sqrt(diag(vcov(LRM12.5)))
cbind(LL = coef.heck - qnorm(0.975) * se.heck, UL = coef
           .heck + qnorm(0.975) * se.heck)
```

R output (abbreviated and edited)

```
  0    1
891 1424 # 891 sample members report zero income

Tobit 2 model (sample selection model)
2-step Heckman / heckit estimation
2315 observations (891 censored and 1424 observed)
12 free parameters (df = 2304)
```

---

[23] In most examples that utilize income as the outcome variable, the binary selection variable is workforce participation since those with a zero value should be selected out because, if they don't participate in the workforce, they should have no personal income. But the *GSS2018* income variable could have zero values for other reasons.

[24] Named after John P. Mills (1926), who used this ratio to study areas under the normal curve ("Table of the Ratio: Area to Bounding Ordinate, for Any Portion of Normal Curve," *Biometrika*, 18(3/4): 395–400). The function that underlies it is traceable at least back to work by French mathematician Pierre-Simon Laplace (1805), however (*Traité de Mécanique Céleste, Tome IV* [*Treatise on Celestial Mechanics, Volume IV*], Paris: Chez Courcier).

```
Probit selection equation:                              95% CIs
             Estimate Std. Error t value Pr(>|t|)    2.5%   97.5%
(Intercept) -1.638      0.221      -7.397 < 0.001 *** -2.07  -1.20
age          0.114      0.009      11.619 < 0.001 ***  0.09   0.13
I(age^2)    -0.001    < 0.001     -13.816 < 0.001 *** -0.002 -0.001
femalemale   0.215      0.057       3.771 < 0.001 ***  0.10   0.33
childs      -0.046      0.019      -2.442   0.015 *   -0.08  -0.01

Outcome equation:
             Estimate Std Error  t value  Pr(>|t|)
(Intercept)  5.709      0.894      6.387 < 0.001 ***  3.96   7.46
educate      0.773      0.053     14.611 < 0.001 ***  0.67   0.88
femalemale   2.403      0.338      7.108 < 0.001 ***  1.74   3.07
attend       0.048      0.056      0.871   0.384      -0.06  0.16

   Error terms:
               Estimate Std. Error t value Pr(>|t|)
invMillsRatio -4.948       0.634     -7.799 < 0.001 *** -6.19 -3.70
sigma          6.627        NA        NA      NA
rho           -0.747        NA        NA      NA
Multiple R-Squared:0.2085,     Adjusted R-Squared:0.2062
---
Signif. codes: 0 '***' 0.001 '**' 0.01 '*' 0.05 '.' 0.1 ' ' 1
```

The estimates of the probit selection equation in the LRM12.5 output are interpretable as predicted z-score shifts in the probability of any (nonzero) income. They suggest generally that age has a quadratic (inverted U-shaped) effect on the probability of any income, whereas males are more likely than females to report any income. Having more children at home is associated with a diminished probability of any income. The outcome equation's estimates are simply LRM coefficients, though, to obtain correct standard errors, they are computed with *generalized least squares* (GLS) rather than OLS. The results indicate that more years of formal education are associated with more income and males tend to report higher earnings than females. The inverse Mills's ratio (selection variable) shown in the Error terms panel (invMillsRatio) is relatively large (−4.95, 95% CI: {−6.2, −3.7}) and implies a strong selection effect in the model. The coefficient labeled sigma (σ) is the estimated standard error of the error term of the outcome equation ($se(\varepsilon_i) = 6.6$), whereas the coefficient labeled rho (ρ) is the estimated correlation between the error terms from the two models ($cor(v_i, \varepsilon_i) = -0.75$). The large correlation coupled with the substantial inverse Mills's ratio suggests that a single equation LRM is misspecified and its slope coefficients are likely biased (estimate this model and compare the results).

Although LRM12.5 offers evidence of selection bias, a critical assumption is that the selection equation is specified correctly. Simulation studies also suggest the Heckman selection model works best with large samples and when the correlation of the error terms is high, but does not tend to overcome selection bias when the correlation of the errors is small, the normality assumption is not met, or the sample size is small. Other assumptions of LRMs, such as independence and homoscedasticity, should also be considered. Similar models that use maximum likelihood, semiparametric, and nonparametric estimation methods, as well as alternative error distributions, may surmount some of the limitations of the Heckman model.[25]

Models designed for outcome variables that are censored or truncated are also available. These include the tobit model, hurdle models, endogenous switching models, and truncated regression models. The R packages censreg, endoSwitch, mhurdle, truncreg, and VGAM (vglm function) allow the estimation of these models.

Cross-Validation

As mentioned earlier, a growing area of research is known by names such as data mining, data science, data analytics, and machine learning. Each focuses on something you are probably familiar with—*big data*. The popularity of this field of inquiry is due, in part, to the applied setting in which various types of organizations use large datasets to try to gain a better understanding of customer and client wants and needs. For example, large retail chains wish to know their customers' buying patterns so they can advertise and specialize in providing tailor-made products and sales to their different customer bases. Health-care organizations are interested in the most efficient methods for treating clients and keeping costs low. A general concern with large datasets has led to growing interest in powerful predictive models. These models include linear regression, as well as many other techniques that are designed to predict some outcome.[26]

One of the key concerns among data scientists is overfitting, which is discussed earlier in the chapter. Overfit leads not only to overly large R^2 values and inefficient slope coefficients but can also result in models that capitalize on chance and are overly complex. Alternative methods—in particular, clear conceptualization and careful diagnostics—are available to create suitable

---

[25] See Vincent Kang Fu, Christopher Winship, and Robert D. Mare (2004), "Sample Selection Bias Models," in *Handbook of Data Analysis*, M. A. Hardy and A. Bryman (Eds.), 409–430, Newbury Park, CA: Sage, and Yulia V. Marchenko and Marc G. Genton, (2012), "A Heckman Selection-*t* Model," *Journal of the American Statistical Association* 107(497): 304–317.). R's sampleSelection, SemiParSampleSel, and quantreg packages include functions to estimate these models.

[26] One class of models, CARTs, was introduced in Chapter 11. Several others are described in Burger (2018), *op. cit.*, and Paul Attewell and David Monaghan (2015), *Data Mining for the Social Sciences: An Introduction*, Oakland, CA: University of California Press.

prediction models and avoid overfitting, but in data analytics a popular technique is known as *cross-validation*.[27]

Consider the typical situation in which we have a dataset and a particular outcome we wish to predict (or explain) using a set of explanatory variables. Given what we've learned thus far, it seems as if we are limited to a single swipe at the data. Yet a particular model's findings may be due to chance or biased in some way. Cross-validation is helpful in this situation because it involves splitting the dataset randomly. One of the sub-datasets is called the *training set* and the other is called the *testing set*. The regression model is first estimated on the training dataset (the term training is used because the model is "trained" on it) and the analyst uses it to examine the model's assumptions and make corrections as needed. Once it is "trained," the model is then re-estimated on the testing dataset to determine its predictive ability. Various measures of fit are evaluated. If the model does a good job of predicting the outcome and meets other criteria—such as satisfying assumptions, low error, and generalizability—then overfit is unlikely and some data scientists claim the model is *validated*. If it does a poor job of predicting the outcome, though, then perhaps a problem with overfit or some other issue is present, such as underfit or nonlinearities the researcher has overlooked. A general advantage of cross-validation is that it furnishes information with which to judge how well a regression model generalizes from one set of data to another. This minimizes the risk that a model's finding are due to chance alone or some idiosyncratic characteristic of a sample.

Various types of cross-validation are available. The one just described is known as the *holdout method*: split the dataset into two random parts and define one part as the training set and the other as the testing set. Another type is called *k-fold cross-validation*: divide the dataset into k sets, or folds, and then use the holdout approach for each of them. The estimated errors of prediction across the comparisons are averaged so the fit can be evaluated. A third type is termed *leave-one-out cross-validation*: similar to the jackknife (see Chapter 9, fn. 18), it omits one observation at a time and predicts the outcome based on a model estimated with the remaining observations. Once again, the estimated errors of prediction and model fit are averaged across the models.

Cross-validation is a promising approach for avoiding overfit and maximizing the predictive power of models, but it does not normally consider underfit or even correct model specification. Problems such as using the wrong functional form, omitting important variables, and endogeneity biases may still exist. Too often, cross-validation is used without much attention to these issues. Yet, each is vital to developing useful models

---

[27] The point at which this method was initially created is not clear, but it dates back to at least the 1930s when there was interest in dividing samples and estimating regression models with each. For a brief history, see Mervyn Stone (1974), "Cross-Validatory Choice and Assessment of Statistical Predictions," *Journal of the Royal Statistical Society: Series B* 36(2): 111–133.

that help us understand some phenomenon. Always remember that theory and conceptualization are essential. When the goal is to maximize prediction, though, cross-validation is valuable, as are CART models (see Chapter 11).

Let's use a subset of the *GSS2018* dataset to estimate a model with sei (socioeconomic status) as the outcome variable and the following explanatory variables: female, age, educate, and marital. But we'll use the holdout method of cross-validation by first splitting the dataset into two randomly selected portions: the training data include 70% of the sample and the testing data include the remaining 30% of the sample.

R code
```
set.seed(8765) # set a seed number so we can replicate
                the results
sub.GSS <- GSS2018[c("age", "educate", "sei", "female",
            "marital")]
# create a subset of the relevant variables
sub.GSS$marital <- factor(sub.GSS$marital, levels =
                c(1,2,3,4), labels = c("married",
                "widowed", "div.sep", "never.marr"))
                # make the marital variable a factor
                variable
part.samp <- floor(0.70 * nrow(sub.GSS))
 # request 70% of the rows from the sub.GSS dataset
random.sample <- sample(seq_len(nrow(sub.GSS)), size =
                part.samp)
 # draw the random sample of 70% of the rows/
   observations
training.data <- sub.GSS[random.sample, ]
 # create the first sample - the training set to "train"
   the model
testing.data <- sub.GSS[-random.sample, ]
 # create the second sample - the testing set to "test"
   the model's fit
```

The training sample consists of 1,620 of the *GSS2018* observations. The testing sample includes the remaining 695 observations. Let's now estimate the LRM using the training dataset (see LRM12.6a).

R code
```
LRM12.6a <- lm(sei ~ female + age + educate + marital,
                data=training.data)
summary(LRM12.6a)
```

R output (abbreviated)
```
Coefficients:
```

```
                Estimate Std. Error t value Pr(>|t|)
(Intercept)     -13.83645  2.78112   -4.975  7.22e-07 ***
femalemale        0.38953  0.94058    0.414  0.678826
age               0.17738  0.03159    5.615  2.31e-08 ***
educate           4.01473  0.15087   26.611  < 2e-16  ***
maritalwidowed   -7.66141  1.85833   -4.123  3.93e-05 ***
maritaldiv.sep   -4.71858  1.26230   -3.738  0.000192 ***
maritalnever.marr -5.43197 1.22851   -4.422  1.05e-05 ***
---
Signif. codes: 0 '***' 0.001 '**' 0.01 '*' 0.05 '.' 0.1
```

```
Residual standard error: 18.53 on 1613 degrees of freedom
Multiple R-squared: 0.3366, Adjusted R-squared: 0.3342
F-statistic: 136.4 on 6 and 1613 DF, p-value: < 2.2e-1616
```

The model using the training set provides an R^2 of 0.34, which suggests that it accounts for 34% of the variability in socioeconomic status.

Next, we'll predict sei using the testing portion of the sample based on this model and examine its fit (see LRM12.6b).

R code
```
LRM12.6b <- lm(sei ~ female + age + educate + marital,
               data=testing.data))
```

The R^2 from this model is about 0.31, which is a bit smaller than in the first model. But, with the exception of age,[28] the models are similar and suggest some convergence of evidence regarding how well the explanatory variables predict socioeconomic status. In a better executed cross-validation, though, we would have examined the training model in more detail. For instance, are the assumptions of the LRM met in the training model? What should we do with the female variable—is it producing model overfit?

An easier way to conduct a holdout method analysis is to use the functions in R's caret (*classification and regression training*) package.[29] It automates some of the steps and offers additional features. LRM12.7 furnishes one approach using two subsets from the *GSS2018* dataset.

---

[28] The slope coefficient for age in the training dataset is 0.18. but only 0.09 in the testing dataset. A careful researcher would explore differences across the model. Are age and socioeconomic status associated in a nonlinear fashion? Do other peculiarities exist? Or could this difference simply be due to random variation?

[29] Additional information about using the caret package for cross-validation—as well as similar methods—in R, is available in Burger (2018), *op. cit.*, chapter 8.

R code
```
library(caret)
set.seed(67599) # set a seed number so we can replicate
                the results
# create a random 70% - 30% split and the training and
  testing datasets
in_train <- createDataPartition(sub.GSS$sei, p = 7/10,
            list = FALSE)
# request 70%/30% random split
training.data <- sub.GSS[ in_train, ] # create the
                                      training dataset
testing.data <- sub.GSS[-in_train, ] # create the
                                     testing dataset
LRM12.7 <- train(sei ~ female + age + educate + marital,
         data = training.data, method = "lm")
 # requests the linear regression model (lm) using the
   training dataset (other models are also available in
   the caret package's train function)
summary(LRM12.7)
LRM12.7 # provides the fit statistics: RMSE, R-squared &
          mean absolute error (MAE)[30]
```

R output (abbreviated)
```
Coefficients:
                   Estimate Std. Error t value Pr(>|t|)
(Intercept)        -14.9924    2.8420   -5.275 1.50e-07 ***
femalemale           1.0462    0.9270    1.129 0.259252
age                  0.1456    0.0311     .681 3.10e-06 ***
educate              4.2272    0.1566   27.000 < 2e-16  ***
maritalwidowed      -6.4032    1.8374   -3.485 0.000505 ***
maritaldiv.sep      -5.7130    1.2484   -4.576 5.09e-06 ***
maritalnever.marr   -6.6120    1.2095   -5.467 5.30e-08 ***
---
Signif. codes: 0 '***' 0.001 '**' 0.01 '*' 0.05 '.' 0.1

Residual standard error: 18.3 on 1615 degrees of freedom
Multiple R-squared: 0.3478, Adjusted R-squared: 0.3454
F-statistic: 143.5 on 6 and 1615 DF, p-value: <
2.2e-1616

1622 samples
  4 predictor
```

---

[30] The MAE is the average size of the absolute value of the residuals: $\dfrac{\sum|\hat{y}_i - y_i|}{n}$. It serves a similar purpose as the MSE or RMSE.

No pre-processing
Resampling: Bootstrapped (25 reps)

```
  RMSE  Rsquared   MAE
 18.471   0.337   15.088
```

The R^2 and RMSE values are similar to the earlier models and do little to further confirm the results. The next step is to use the results of the LRM with the training data to predict the outcome with the testing data. In other words, we use the characteristics of the model from the training dataset and apply them to the testing dataset. We then assess the correlation between sei and its predicted values based on the model.

R code
```
sei.pred <- predict(LRM12.7, testing.data)
cor(sei.pred, testing.data$sei)
```

R output (abbreviated)
```
[1] 0.535
```

The correlation is 0.54, which is about what is expected since the R^2 is 0.34 $\left(\sqrt{0.34} = 0.58\right)$. The prediction of socioeconomic status is not impressive, so the model is likely underfit.

The holdout method is the simplest approach to cross-validation. Other cross-validation methods are also available that go beyond two subsamples and estimate the model on several samples. One of the more popular was described earlier: *k-fold cross-validation*, wherein the software constructs k subsamples and one is "held-out" as the model is estimated on the remaining subsamples. The holdouts serve as the testing samples. The results of the models are then averaged. With fewer folds (subsets of data), less variability occurs across the results since fewer models are estimated, but the tradeoff is higher bias. More folds lead to lower bias but more variability across models.

The next example demonstrates a k-fold cross-validation using the same LRM designed to predict sei (see LRM12.8).

R code
```
library(caret)
set.seed(40902) # set a seed number so we can replicate
                 the results
data.control <- trainControl(method = "cv", number = 5)
 # sets up a cross-validation (cv) with 5 folds
LRM12.8 <- train(sei ~ female + age + educate +
         marital, data=sub.GSS, trControl = data.
         control, method="lm", na.action = na.pass)
          # estimates the linear regression model (lm)
            using the specifications in the command
```

```
LRM12.8                           # provides the fit
                                    statistics
LRM12.8$finalModel                # lists the slope
                                    coefficients
LRM12.8$resample                  # shows how the fit
                                    statistics vary across
                                    the five folds
sd(LRM12.8$resample$Rsquared)     # provides the variability
                                    of the R2s across the
                                    five folds
```

R output (abbreviated)
```
Linear Regression
1899 samples
   4 predictor

No pre-processing
Resampling: Cross-Validated (5 fold)
Summary of sample sizes: 1521, 1518, 1519, 1519, 1519
Resampling results:

  RMSE    Rsquared  MAE
  15.590   0.325    12.352

[Slope] Coefficients:
(Intercept)  gendermale   age    educate    maritalwidowed
  -12.897       0.703     0.154   4.029         -7.410
maritaldiv.sep maritalnever.marr
   -5.002            -6.228

    RMSE    Rsquared   MAE     Resample
1  18.871    0.315    15.544    Fold1
2  18.191    0.355    15.032    Fold2
3  18.321    0.350    15.397    Fold3
4  19.263    0.265    15.620    Fold4
5  18.211    0.342    15.168    Fold5
[1] 0.0374
```

The results are consistent with those from the earlier models, though the slope coefficients are slightly different. The R^2 varies from 0.27 to 0.36 across the folds, which suggests rather high variability, as does the variance estimate of 0.037. It might be a good idea to increase the number of folds and see what effect this has on the fit and variability measures.[31] Nonetheless, the models suggest that socioeconomic status is predicted modestly well by

---

[31] Try increasing the number of folds to 25. What do you find? Keeping in mind the tradeoff between variability and bias, do you recommend more or fewer folds?

age, education, and marital status. Yet, one should be curious about whether endogeneity or selection issues affect the results.

Variable Selection Procedures

Another method used to specify models and reduce the risk of specification error is discussed briefly in Chapter 10: *automated variable selection procedures*. They are aimed at finding regression models that predict the outcome well. Unfortunately, many, if not most, of these variable selection approaches contravene a point made earlier in this chapter: let your theory or conceptual model guide the selection of variables. Because automated procedures are discussed so often in presentations on LRMs, though, they're worth examining.

The first type we'll discuss is called *forward selection*: explanatory variables are added to the model based on their correlations with the unexplained component of the outcome variable. Yet this relies on biased estimates at each point in the selection process, so it should be avoided. A type of forward selection that you might encounter is stepwise regression wherein partial F-statistics are computed as variables are removed and put back in the model. For various reasons discussed later, however, it should also be avoided.

The second type is known as *backward selection*: the analyst begins with all the explanatory variables in a model and then allows the software to selectively remove those with slope coefficients that are not statistically significant (usually at $p > 0.05$ or 0.10, but the analyst may choose the threshold) or when their absence results in a higher AIC or BIC. Backward selection is reasonable if researchers are seeking the best predictive model and are worried about overfit, yet is overly automated and largely ignores the vital conceptual work that should guide variable selection. Nonetheless, we'll consider some steps that help make backward selection somewhat better than more arbitrary or biased approaches.

Several steps to selecting variables should be considered if one wishes to rely on backward selection.[32] First, as mentioned earlier, estimate an LRM with all the explanatory variables.[33] Second, choose a fit measure to compare models. Third, use backward selection to remove variables—this may be done one at a time or in chunks—from the model. Fourth, assess the reliability of the preferred model with a split sample procedure, such as the holdout method or k-fold cross-validation. This last step is reasonable if we

---

[32] A good description of these steps is found in Kleinbaum et al. (1998), *op. cit.*, chapter 16.

[33] Kleinbaum et al. (1998), *op. cit.*, propose that higher-order terms (e.g., x^2, x^3) and interactions should be included in the model. See Chapter 11 for a discussion of these forms of variables.

wish to verify the reliability of an LRM regardless of how we decided which variables to include.

These steps are fairly simple using statistical software, but we should say a little more about the second step: choosing a fit measure to compare LRMs. The automated procedures typically rely on p-values, nested F-tests, or information criterion measures (AIC or BIC; see Chapter 6) to choose the best fitting model. But analysts should also consider other measures, such as the adjusted R^2 and RMSE. Looking for the largest adjusted R^2 or the smallest RMSE from among a set of nested models is a reasonable step since, assuming we wish to make good predictions and find strong statistical associations, we'd like to have the most explained variance or the least amount of variation among the residuals.

Another common measure of model fit related to the others is called Mallow's C_p.[34] An important feature of this measure is that it is minimized when the LRM includes only those slope coefficients below a certain p-value threshold (e.g., $p < 0.05$). The formula for Mallow's C_p is shown in Equation 12.8.

$$C_p = \frac{\text{SSE}(p)}{\text{MSE}(k)} - \left[n - 2(p-1)\right] \qquad (12.8)$$

where p = number of explanatory variables in the restricted model.

The SSE(p) is from the restricted model and the MSE(k) is from the full model. Smaller values of Mallow's C_p designate models that fit the data better. Using this measure does not require a model with all the explanatory variables and some model nested in it. It may include any model that uses subsets of the explanatory variables as long as at least one is nested within another. Mallow's C_p is used most often, though, to compare the full model with various models nested within it.

R offers a substantial number of automated variable selection functions, with several that we haven't discussed (e.g., stepwise AIC forward selection). The following example uses one of the backward selection methods (ols _ step _ backward _ p) available in the olsrr package that removes variables if their p-values exceed a certain threshold value (the default is $p \geq 0.3$, but can be changed by specifying prem = [p-value level]). The full model uses the *StataData2018.csv* dataset and is designed to predict opioid deaths per 100,000 residents (OpioidODDeathRate) with the following explanatory variables: the domestic migration rate, the percent of residents of ages 0–18, the percent of residents of ages 65 and older, the percent of residents who are White, and the unemployment rate. The function estimates three models, with information on which variables were removed at

---

[34] Initially described in C. L. Mallows (1973), "Some Comments on CP," *Technometrics* 15(4): 661–675.

each step along with five model selection fit statistics: the R^2, adjusted R^2, Mallow's C_p, AIC, and RMSE (see LRM12.9).

R code
```
library(olsrr)
LRM12.9 <- lm(OpioidODDeathRate ~ DomMigRate +
              PerAge0_18 + PerAge65Plus + PerWhite +
              UnemployRate, data=StateData2018)
summary(LRM12.9)
ols_step_backward_p(LRM12.p) # request backward
                               selection
```

R output (abbreviated)
                Elimination Summary

Step	Variable Removed	R-Square	Adj. R-Square	C(p)	AIC	RMSE
1	DomMigRate	0.484	0.438	4.026	352.851	7.708
2	PerAge65Plus	0.475	0.441	2.796	351.718	7.690

The backward selection method suggests that the best fitting model excludes the domestic migration rate and the percent of residents of ages 65 and older. The other explanatory variables should remain in the model. The final model with three explanatory variables accounts for 48% of the variability in opioid deaths. The RMSE is 7.69. Relative importance measures (see Chapter 4) suggest that the percent of residents of ages 0–18 is the strongest predictor of opioid deaths per 100,000 residents, though the association is negative. The final model may offer good accuracy in its predictions and eliminate some issues of model overfit, but, as mentioned earlier, many statisticians advise against using automated variable selection procedures.[35]

The lesson from using these procedures is that some of them, such as backward selection, are reasonable tools if the goal is to develop a model that includes the set of explanatory variables that offers the best predictive power. But they cannot substitute for a good theoretical or conceptual model that

---

[35] Yet another variable selection procedure that is gaining adherents relies on the *false discovery rate* (FDR) (also called a *q-value* to contrast it with a *p*-value) by estimating many LRMs using a large set of explanatory variables. It determines which variables to retain in the model by adjusting the *p*-values for multiple testing, which reduces the chance of type-I errors (false-positives). The FDR is also helpful for multiple hypothesis testing (see Yoav Benjamini and Yosef Hochberg (1995), "Controlling the False Discovery Rate: A Practical and Powerful Approach to Multiple Testing," *Journal of the Royal Statistical Society: Series B (Methodological)*, 57(1): 289–300). The R function p.adjust includes an fdr option that provides *q*-values in multiple testing scenarios. The package SignifReg also has an FDR function that is useful for model selection.

not only predicts the outcome variable but, more importantly, *explains* why it is associated with the explanatory variables.[36] Automated procedures may also provide misleading results because they are designed to fit the sample data, thus overstating how precise the results appear to be when we wish to make inferences to the population from which the sample was drawn.[37] This is why many statisticians and data scientists recommend splitting the sample first and then testing the model on both sets. Better yet, use a training dataset to fine-tune the model and then validate it using a testing dataset.

Some other reasons to avoid certain automated procedures include the following:

1. The model's R^2 values are biased;
2. The coefficients' confidence intervals are biased (they are too narrow);
3. The p-values are not meaningful—if they're ever particularly meaningful (see the section on best statistical practices in Chapter 1)—without complicated corrections;
4. The procedures have problems when collinearity exists among the explanatory variables (see Chapter 10);
5. Some of the tests used to compare models (e.g., nested F-tests) are supposed to be used only with pre-specified models; and
6. They make us lazy and overconfident: we should focus on the conceptual model and not rely so much on statistical software to make decisions for us.

Chapter Summary

Specification errors are a common problem in social and behavioral sciences. Knowing which variables to include in an LRM is challenging and requires careful consideration. The fact that we rarely have control over the explanatory variables makes this problem all the more difficult to resolve. Conceptualizing well, knowing the literature on a topic, and considering different ways that variables may be related are thus vital steps; no statistical procedures can substitute for them.

---

[36] For example, why is the percentage of young people in a state negatively associated with opioid deaths? What is it about states that trend young that might offer a lower risk of these deaths? It does not seem to involve fewer elderly people since the percent of residents of ages 65 and older is not associated with opioid deaths, but it might involve other aspects of the age structure of states. I hope it's clear from this and other examples that good prediction does not equal good or even adequate explanation.

[37] For more on this point, see Fox (2016), *op. cit.*, chapter 13.

Testing for specification errors is helpful, nonetheless, since it may point toward ways that models can be improved. Many associations are not linear, for example, and, as discussed in Chapter 11, treating them using a straight line or flat surface is a type of specification error that can be avoided with careful evaluation of the model.

Selection bias is a common specification problem that occurs when the observations in the analysis are a nonrandom subsample of the sample from the target population. If, for instance, the outcome variable is observed for only a portion of the sample and whether it is observed is not random, then selection bias is likely. To obviate this form of bias, it's important to have a clear understanding of the nature of the sample and variables used in the analysis. Is the outcome variable measured in such a way that we cannot observe certain portions of its distribution (e.g., we observe personal income, monetary contributions, or the frequency of volunteer hours only once they've crossed a threshold)? If yes, then develop a model that accounts for the selection process (e.g., any volunteer work) *and* the "frequency" portion of the outcome you do observe (e.g., number of hours of volunteer work). Various models are available to estimate these two processes.

Cross-validation is a useful tool for guarding against model misspecification, especially overfit. Automated variable selection procedures should generally be avoided, unless the overarching goal is to identify the most highly predictive model. But even when this is the objective, other methods are better at reaching a high degree of predictive accuracy, such as CART and related models discussed in Chapter 11.

Chapter Exercises

The dataset called *BikeRental.csv* contains information on daily bicycle rentals from several establishments in Victoria, British Columbia, over approximately a two-year period. We are interested in examining some predictors of the number of bikes rented per day. The following variables are included in the dataset:

- `casid`   identification number—rental day
- `season`   season of the year (1 = fall, 2 = winter, 3 = spring, 4 = summer)
- `holiday`   was the day a holiday? (0 = no, 1 = yes)
- `workday`   was it a normal workday? (0 = no, 1 = yes)
- `weather`   what was the weather like on that day? (1 = sunny, 2 = cloudy, 3 = rainy)

- `temperature`   average temperature (Celsius)
- `humidity`   average humidity
- `windspeed`   average windspeed (kilometers per hour)
- `casual`   number of casual bike renters
- `registered`   number of registered bike renters
- `rentals`   total number of bike rentals

After importing the dataset into R, complete the following exercises.

1. Estimate an LRM with `rentals` as the outcome variable and the following explanatory variables: `temperature`, `humidity`, and `windspeed`.

   a. Interpret the slope coefficients and 95% CIs associated with the variables `temperature` and `windspeed`.

   b. Use the model to predict the mean number of bike rentals when the temperature is 10°, 20°, and 30°, and humidity and windspeed are set at their means.

2. Test the LRM in exercise 1 for underfit using the link test and Ramsey's RESET test. Describe what the tests imply about the LRM.

3. Given what we learned about what these tests are designed to evaluate, estimate an LRM that considers other functional forms of the explanatory variables. Assess this model with Ramsey's RESET test and determine whether underfit is still an issue (hint: it is unlikely you will be able to find a RESET value that has a p-value less than 0.05, but you should be able to reduce the value substantially).

4. Re-estimate the LRM in exercise 1, but add the following two explanatory variables: `holiday` and `workday`. Compare this model to the model in exercise 1.

   a. Which model fits the data better?

   b. Upon what do you base your judgment (hint: see Chapter 6)?

   c. Given this evidence, is one of the models underfit? Is one overfit? Explain your answer.

5. Use a backward selection procedure with the set of explanatory variables in exercise 4. What does this selection procedure suggest about which set of explanatory variables offers the best prediction of daily bike rentals?

6. *Challenge*: choose the model from exercises 1, 4, or 5 that you think offers the best predictions of bike rentals and evaluate it with a *k-fold cross-validation* (try varying the number of folds). What does this method indicate about the model?

13

Measurement Errors

How do researchers measure concepts in social and behavioral sciences? They often attempt to measure things that don't have clear or consistent definitions, or that are difficult to examine in an unambiguous fashion. Measuring concepts such as self-esteem, depression, happiness, social capital, antisocial behavior, aggression, impulsivity, political ideology, marital satisfaction, social support, and dozens of other phenomena is demanding because the available assessment instruments are crude compared to those that measure more tangible concepts such as the height and weight of Siberian Husky puppies, the amount of fertilizer applied to plants, or the quantity of a particular medication given to cancer patients. Some concepts used by social scientists, such as monthly income in dollars, official rates of crime, or years of education, are easier to measure since there are more tangible and widely accepted ways of assessing their magnitudes. But even concepts such as these may be measured poorly since we usually rely on people to report them accurately. Sometimes people do not report these things with care or sources of error outside of their control affect what they do report.[1]

Social and behavioral scientists often gauge various concepts using self-report instruments. To measure adolescent drug use, for example, they usually ask a sample of adolescents to report if and how often they've used alcohol, cigarettes, marijuana, cocaine, or other illicit drugs. To measure happiness, researchers ask people whether they are happy in particular spheres of life (e.g., job and family), or use global questions such as "Do you consider yourself a happy person?" Happiness or similar concepts such as life satisfaction could be assessed in other ways, but we generally must be content with asking people about themselves. Only rarely is there an external instrument

---

[1] Consider how official crime rates are measured in the U.S., for example. National crime statistics are compiled by the Federal Bureau of Investigation (FBI) as part of its Uniform Crime Reporting (UCR) program. But this program relies on local police agencies—there are more than 15,000 in the U.S.—to provide accurate reports of crimes and arrests in their jurisdictions. Yet some agencies do not participate in the UCR program and others, especially smaller agencies, may have few employees responsible for gathering and submitting the information to the FBI. Even when faced with official government statistics, the possible sources of reporting error are legion (see Sharon L. Lohr (2019), *Measuring Crime: Beyond the Statistics*, Boca Raton, FL: CRC Press). The following delightful and apropos quote is attributed to the early 20th-century economist and statistician Sir Josiah Charles Stamp and is thus called "Stamp's Law": "The government is very keen on amassing statistics. They collect them, add them, raise them to the nth power, take the cube root and prepare wonderful diagrams. But you must never forget that every one of these figures comes in the first instance from the village watchman, who just puts down what he damn pleases."

DOI: 10.1201/9781003162230-13

useful for verifying what people report. An interesting exception involves research on illicit drug use. Saliva and hair tests may be used to determine if people have recently used illegal substances such as marijuana or cocaine. We could, I suppose, hook respondents up to lie detector machines and ask them questions, but this would likely deter most people from participating in surveys.

The literature on measurement in social and behavioral sciences is enormous and we cannot cover all the important topics. Because these disciplines must often rely on imprecise measurement instruments and personal reports, various sources of *measurement error* afflict research projects in general and statistical models in particular. Errors may arise from various sources, including research respondents, interviewers, survey instruments, and data and cleaning methods, such as when data are incomplete, duplicated, or improperly formatted. Errors due to imprecise measurement instruments is called *method error*, whereas the fact that people sometimes misunderstand the questions researchers ask or don't answer them accurately for some reason (lack of interest or attention, exhaustion from answering lots of questions) is called *trait error*. Some people avoid providing information about personal topics. One of the most personal topics, it seems, is personal income. When researchers ask questions about a person's income or tax returns, they should be prepared for many refusals or even deceptive answers.

Another source of measurement error is *recording* or *coding errors*, such as when a data entry person forgets to hit the decimal key so an income of $1,900.32 per month becomes $190,032 per month. Normally, this type of measurement error is easy to detect during the data screening and cleaning phase. If we make it a point to always check the distributions of all the variables in the analysis using means, medians, standard deviations, ranges, q-q plots, box-and-whisker plots, and other exploratory techniques, and are familiar with the instruments used to measure the variables, we will usually catch coding errors. (In addition, don't forget to always check the missing data codes in your dataset! See Appendix A.)

But suppose we have carefully screened the data for coding errors and fixed all the apparent problems. We still need to be concerned with measurement error in the variables because method or trait error might contaminate them. Thus, we need to ask what happens to regression models when the variables are measured with error.[2] We'll discuss two situations and their consequences before offering some possible solutions: (a) the outcome variable, y, is measured with error and (b) one or more of the explanatory variables, x, is measured with error.

To understand these situations better, let's revisit some of the assumptions of LRMs (see Chapter 4) and their implications that relate to measurement

---

[2] For a brief overview, see Michael Wallace (2020), "Analysis in an Imperfect World," *Significance* 17(1): 14–19. For a thorough exposition, see Paul P. Bieder et al. (Eds.) (2004), *Measurement Errors in Surveys*, New York: Wiley.

error. First is the notion derived from the independence assumption that the X variables are fixed or under the control of the researchers (nonstochastic). Suppose researchers do have control over the explanatory variables. They may then introduce different values or quantities (such as by applying different amounts of fertilizer and water to plants and examining their growth) and determine with quite a bit of accuracy whether these variables affect the outcome. This reduces the risk of faulty measurement since researchers have control over the variables and don't have to rely on other methods, such as self-reports. Second, given that the previous ideal situation is rare in social sciences, and we usually rely on x variables that are presumed random (stochastic), another implication is the strict assumption that the Xs are uncorrelated with the error term $\left(\operatorname{cor}(\mathbf{X}, \varepsilon_i) = 0\right)$. This issue is discussed in Chapters 8 and 12, with the latter chapter providing examples.

Sufficient control over X or x variables is rarely achieved in social sciences. Experimental research—especially in psychology, but also in sociology and economics—that manipulates explanatory variables and applies them randomly to study participants is not uncommon. But important ethical guidelines limit the ability of social and behavioral researchers from controlling variables. For example, we should not—perhaps cannot—manipulate an adolescent's friends in studies of peers and illicit behaviors. Another limitation is that many interesting variables cannot be manipulated: we cannot dictate a person's ethnicity or gender, for instance. Therefore, most social science research relies on random samples to consider a variety of values of explanatory variables. Although deliberate manipulation of the xs is preferred, considering the presumed random nature of explanatory variables and observing their variability, or adjusting for their effects, in a statistical model is a convenient and commonly used substitute.

We'll learn in this chapter that some degree of measurement error almost always occurs either because the concepts are not well defined or for some other reason (e.g., the instruments are not accurate enough). In addition, measurement error is a form of specification error when it involves systematic variability that is left out of a model but is related to the explanatory variables. We thus should try to reduce this type of error to a manageable level.

The Outcome Variable Is Measured with Error

Suppose we're interested in an outcome variable, such as happiness, that, in all likelihood, is measured with error, such as trait error. For now, we'll assume that the explanatory variables, say education and income, are measured without error. We'll furthermore presume that somewhere within the measurement of happiness is a *true measure*, or what is often called the *true* *score.* Imagine, then, that we represent the outcome variable (y_i^*) in Equation 13.1 as the sum of two components—true score plus error.

$$y_i^* = y(\text{true score})_i + u(\text{error})_i \tag{13.1}$$

Using these terms, what does an LRM look like? Equation 13.2 provides one way to represent it.

$$y_i^* = \alpha + \beta_1 x_{1i} + \hat{\varepsilon}_i = y_i + u_i = \alpha + \beta_1 x_{1i} + \hat{\varepsilon}_i \tag{13.2}$$

Subtract u_i from both sides of the equation. By solving, we obtain Equation 13.3.

$$y_i = \alpha + \beta_1 x_{1i} + \hat{\varepsilon}_i + (-u_i) \tag{13.3}$$

The error in measuring y can therefore be considered another component of the error term, or simply another source of error in the LRM. As long as this source of error is not associated systematically with the explanatory variable $(\text{cov}(x_{1i}, u_i) = 0)$ —or with one or more x variables in a multiple LRM—we have met one of the assumptions of the model and our slope estimates tend to be unbiased. But if x_{1i} is correlated with the measurement error in y, then we have the specification error problems described in Chapter 12, in particular omitted variable bias.[3] Unfortunately, including this error explicitly in the regression model to solve the specification problem is challenging.

Even when no statistical association exists between x_{1i} and the error in y, the R^2 from the LRM tends to be lower than if measurement error is not present. In the presence of errors in y, the SSE is larger because of the additional variation that is not accounted for by the model. The standard errors of the slope coefficients are also inflated, thus making p-values larger and CIs wider. Recall the formula for the standard error of the slope coefficient, which is shown in Equation 13.4.

$$se(\hat{\beta}_i) = \sqrt{\frac{\Sigma(y_i - \hat{y}_i)^2}{\Sigma(x_i - \bar{x})^2(1 - R_i^2)(n - k - 1)}} \tag{13.4}$$

Notice what happens when the SSE (the numerator) is larger. All else being equal (or *ceteris paribus*, to use an urbane Latin term favored by some social

---

[3] Consider the following example. Suppose we're studying whether age predicts happiness using an LRM. The measure of happiness is accompanied by trait error that is caused by memory problems (people with poorer memories tend to overestimate their actual happiness). Assuming that memory problems are also associated with age (*cor(age, memory problems)* > 0), we have a condition where the explanatory variable is associated with the error in measuring the outcome. In this situation a strong likelihood exists that the slope coefficient associated with age is biased.

scientists), the standard error is also larger. All is not lost, though. If we take an inferential approach and find p-values below a predetermined threshold (e.g., $p < 0.05$) or CIs that don't include zero, then we have a measure of confidence that we will arrive at a similar conclusion as if y has no measurement error.

The Explanatory Variables Are Measured with Error

A scenario that is graver is when one or more of the explanatory variables is measured with error. For now, let's assume that the outcome variable is measured without error, so we only have to deal with y (*true score*). But the explanatory variable as observed is made up of a true score plus error $\{x_{1i}^* = x_{1i} + v_{1i}\}$, which is represented in Equation 13.5.

$$y_i = \alpha + \beta_1 x_{1i}^* + \hat{\varepsilon}_i = y_i = \alpha + \beta_1(x_{1i} + v_{1i}) + \hat{\varepsilon}_i \tag{13.5}$$

Distributing by β_1 leads to the regression model shown in Equation 13.6.

$$y_i = \alpha + \beta_1 x_{1i} + \beta_1 v_{1i} + \hat{\varepsilon}_i \tag{13.6}$$

The error term now has two components $(\beta_1 v_{1i}$ and $\hat{\varepsilon}_i)$ with the strong likelihood that the true score x_{1i} is correlated with at least one of them (if we're fortunate, x_{1i} is not correlated with the measurement error (v_{1i}), but this is rare).

When the explanatory variable is affected by measurement error, the estimated slope is normally biased toward zero. To illustrate, consider that both x_{1i} and v_{1i} have unique variances, which we'll identify as σ_x^2 and σ_v^2. Even if the measurement error (v_{1i}) is independent of x_{1i}, the LRM still does not provide an unbiased estimate of β_1. Instead, it yields β_1^*, which is related to β_1 in the manner shown in Equation 13.7.

$$\beta_1^* = \beta_1\left(\frac{\sigma_x^2}{\sigma_x^2 + \sigma_v^2}\right) \tag{13.7}$$

The ratio of the variance elements is also called the *reliability* of the observed variable, x_{1i}^*. Since, by definition $\sigma_x^2 / (\sigma_x^2 + \sigma_v^2) < 1$, the two slope coefficients are related in the following manner: $|\beta_1^*| \le |\beta_1|$. Thus, the LRM slope coefficient for x_{1i}^* is closer to zero than the slope coefficient for the true score, x_{1i}. This phenomenon is also called *attenuation bias.*[4]

---

[4] See Wallace (2020), *op. cit.*

The degree of attenuation bias is rarely known; it depends on how much error exists in the observed explanatory variable. The standard error of the slope is also inflated; thus the t-value tends to be smaller than it would be if there was no measurement error. Combining these two problems—a slope coefficient closer to zero and a larger standard error—you should immediately see that the p-value is larger and the CI wider when the variable is measured with error. This does not mean, however, that the explanatory and outcome variables do not have a meaningful association. It simply means such an association may not be detectable using traditional inferential criteria if there is sufficient measurement error.

Let's further imagine we have a multiple LRM and several of the explanatory variables are measured with error: the problem is compounded, sometimes greatly. To make it even worse, consider what happens when both the outcome variable and several of the explanatory variables are measured with error. Now we have a truly unfortunate situation: the R^2 is biased, the slope coefficients are biased, and the standard errors are probably incorrect. Such a model can be misleading. Yet, many variables in social and behavioral sciences are riddled with measurement errors. A key problem is we rarely know if and how much measurement error exists. Fortunately, several techniques are available to address this problem.

What Should We Do about Measurement Error?

We'll discuss four (partial) solutions to measurement error. The first is easy to state, yet slippery to implement: use good measuring instruments. This is not meant as a flippant response to an important question. The development and testing of the instruments used in surveys and experiments are substantial areas of research. Though mostly within the domain of psychology and cognitive studies, it also draws from a variety of fields such as sociology, marketing, political science, and opinion research.[5] We have neither the space nor the ability in this chapter to go over the many aspects of good measurement, including important topics such as the validity and reliability of instruments. One piece of advice, though, is to use, where possible, well-established instruments to measure phenomena rather than try to create them for a particular study. For instance, if researchers wish to measure symptoms of depression, an instrument such as the Hamilton or Beck Depression Inventory is a better choice than trying to develop a new scale.[6]

---

[5] See, for example, Jon A. Krosnick and Leandre R. Fabrigar (2005), *Questionnaire Design for Attitude Measurement in Social and Psychological Research*, New York: Oxford University Press.

[6] See Raymond W. Lam et al. (2005), *Assessment Scales in Depression, Mania and Anxiety*, New York: Taylor & Francis.

The second solution to measurement error may not be particularly useful but comes in handy in certain situations. It requires that we have some external knowledge that is normally not available: knowing the variance of the true score and the measurement error of x: σ_x^2 and σ_v^2. If we do have this information, however, we can use it to compute the reliability: the ratio of the variance of the true score to variance of the measured score. As suggested earlier, the reliability is depicted as in Equation 13.8.

$$r_{xx^*} = \frac{\sigma_x^2}{\sigma_x^2 + \sigma_v^2} \tag{13.8}$$

The r in the equation is not the Pearson's correlation coefficient (though there are statistical similarities). Recall that $\sigma_x^2 / (\sigma_x^2 + \sigma_v^2) < 1$, therefore, the reliability gets closer to one as the error in measurement decreases because, holding the true score variance constant, when less error occurs, the denominator gets smaller.

Information on reliability is sometimes available from previous research. Suppose, for example, that we are conducting a study with Mudd's Scale of Jocundity (not an actual scale) to measure happiness among teenagers. Several sophisticated studies have found that when Mudd's scale is used among this population, it yields a reliability coefficient (r_{xx^*}) of 0.8. (We won't bother trying to figure out how researchers came up with this quantity.) Since errors in measuring the explanatory variable tend to bias the slope toward zero, the observed slope coefficient is too close to zero if it is contaminated by measurement error. But we now know the degree of bias because of the reliability coefficient. We may therefore multiply the observed slope by the inverse of the reliability coefficient $\left(\frac{(\sigma_x^2 + \sigma_v^2)}{\sigma_x^2} \right)$ to obtain the unbiased slope. Let's say we estimate an LRM using happiness as the explanatory variable and high school GPA as the outcome variable. The LRM estimates an unstandardized slope coefficient of 1.45. Multiplying this coefficient by $1/0.8$ (1.25) yields an unbiased slope of $\{1.45 \times 1.25\} = 1.81$. This approach is difficult to use, though, when the model has more than one explanatory variable.

Latent Variables as a Solution to Measurement Error

The word *latent* means hidden, concealed, or present but not revealed to the senses and has been adopted in applied statistical analysis to refer to variables that exist in some indistinct sense but are not directly discernible using observable methods. In the social and behavioral sciences, one might argue, almost all variables are latent: depression, anxiety, happiness, attitudes, antisocial behavior, prosocial behavior, etc. We do not see, hear, smell, taste, or touch these variables. Perhaps we can be confident in our ability to recognize phenomena, like U.S. Supreme Court Justice Potter Stewart when he famously wrote that he could detect pornography in motion pictures because "I know it when I see it."[7] Most of us, though, would likely have a difficult time coming up with a universal way to put fixed boundaries around the many phenomena of interest to social and behavioral scientists. One solution to this problem is to measure concepts indirectly through the use of variables that we can measure well. Measuring concepts using latent variables provides the most common approach.

As the name implies, latent variables—which are mentioned in Chapter 10—are thought to exist, but are hidden from our senses (actually, we assume the variables are hidden from the direct capabilities of our measuring instruments). The accepted logic is that we can measure aspects of a latent concept. By accumulating information from these measurable aspects, we may construct a latent variable. The statistical area that permits this approach is called *multivariate analysis* because it is concerned with variance that is shared among multiple variables. Under the broad area of multivariate analysis are statistical techniques such as principal components analysis (PCA), factor analysis, latent variable analysis, log-linear analysis, partial least squares, and several other quantitative procedures.

We'll discuss briefly a frequently used technique called *factor analysis*,[8] which is designed to take *observed* or *manifest* variables and reduce them to a set of latent variables that underlie the observed variables.[9] Perhaps the simplest way to understand this statistical technique is to consider that variables that are similar share variance. If sufficient shared variance occurs among a set of observed variables, we claim that this shared variance represents the latent variable. Conceptually speaking, we presume that the latent variable predicts the set of observed variables. The diagrams in Figures 13.1 and 13.2 represent two simplified depictions of a latent variable. Figure 13.1 shows the overlap of three observed variables; the shaded area of overlap represents the latent variable, or the area of shared variability. This should remind you of

---

[7] The quote comes from Justice Stewart's concurring opinion in the case of *Jacobellis v. Ohio* (378 U.S. 184, 1964): "I shall not today attempt further to define the kinds of material I understand to be embraced within that shorthand description [*viz.* hard-core pornography]; and perhaps I could never succeed in intelligibly doing so. *But I know it when I see it*, and the motion picture involved in this case is not that" (emphasis added).

[8] Factor analysis is similar to PCA, which is discussed in Chapter 10, though they make distinct assumptions about how the observed and latent variables are related. It was introduced by British psychologist Charles Spearman in the early 1900s (see A. D. Lovie and P. Lovie (1993), "Charles Spearman, Cyril Burt, and the Origins of Factor Analysis," *Journal of the History of the Behavioral Sciences* 29(4): 308–321).

[9] Many books address factor analysis. See, for example, Paul Kline (1999), *An Easy Guide to Factor Analysis*, London: Routledge; David J. Bartholomew, Martin Knott, and Irini Moustaki (2011), *Latent Variable Models and Factor Analysis*, 3rd Ed., New York: Oxford University Press; and Bruce Thompson (2004), *Exploratory and Confirmatory Factor Analysis: Understanding Concepts and Applications*, Washington, DC: American Psychological Association.

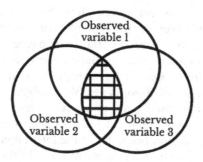

FIGURE 13.1
Latent variable represented with overlapping circles.

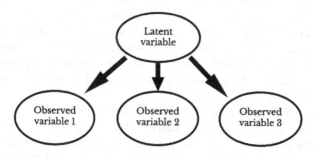

FIGURE 13.2
Latent variable represented with directional arrows.

the principal components discussed in Chapter 10's section on testing LRMs for multicollinearity. Figure 13.2 demonstrates how latent variables are often represented in research presentations. Notice the direction of the arrows. They imply that the latent variable—recall it is not directly observed—predicts the observed variables.

R has several packages that estimate factor analysis models.[10] For example, the psych package allows users to conduct various types of factor analysis. When used to measure latent variables, however, many researchers rely on a set of techniques known as *structural equation modeling (SEM)*. An advantage of this method is its ability to simultaneously execute a factor analysis that estimates latent variables and a simultaneous equations regression model to test the associations among a set of latent variables (see Chapter 12). SEM thus offers a potential solution to simultaneous equations bias, as well as measurement error. R has a couple of packages designed to estimate SEMs, including sem and lavaan.

Suppose a researcher wishes to assess the association between anxiety and self-esteem but assumes that any single question about each is afflicted with

---

[10] For an introduction to these models, see Finch and French (2015), *op. cit.* Some of the R functions and packages that estimate latent variable models are listed in Chapter 10, fn. 11.

measurement error. She therefore uses ten observed variables to measure aspects of anxiety and assumes that a latent variable—or the true measure of anxiety—accounts for their shared variance. The researcher also has 12 variables to measure self-esteem and can likewise use them to estimate a latent variable. Functions in R's sem or lavaan package allow the testing of the measurement properties of these types of variables—such as their shared variability—and assess whether there are two latent variables that account for them. They can also estimate an LRM to determine whether the latent measure of anxiety is a good predictor of the latent measure of self-esteem.

But what happened to the measurement error in the latent variable? The rationale is that the shared variance represents the true score of a latent variable, whereas the remaining unshared variability represents the error. Examine Figure 13.1 again. The shaded area where all three variables overlap denotes the true score for the latent variable and the non-shaded areas represent measurement error. Of course, we must make the assumption that the shared variance truly represents something that is noteworthy and that we've labeled it correctly. We also assume that the measurement errors for the observed variables do not covary, although this assumption may be relaxed in an SEM.

The second latent variable technique, which is based on *instrumental variables*, has been popular in econometrics. An example is the easiest way to describe this approach. Suppose we have an explanatory variable that is measured with error. We know that if we place it in an LRM, the resulting slope coefficients are biased and inefficient. However, we think we've found another set of variables that have the following properties: (a) they are highly correlated with the error-contaminated explanatory variable (the *relevance criteria*); (b) they affect the outcome variable only through their effects on the explanatory variable (the *exclusion criteria*); and (c) they are uncorrelated with the LRM's error term (which, you'll recall, represents all the other influences on the outcome that are not in the model) (the *exchangeability* or *independence criteria*). If we can find a set of instrumental variables that meet these criteria[11] we may utilize them in an LRM estimated with OLS to predict the explanatory variable, use the predicted values to estimate a latent variable as a substitute for the observed explanatory variable, and employ this latent variable in a regression model to predict the outcome variable.[12] A common estimation technique that uses instrumental variables is called *two-stage least squares* (2SLS) because it utilizes the two OLS-based regression models just

---

[11] It may be tempting to simply look for variables in the dataset that are highly correlated with the explanatory variable and use them as instruments. If only it was so easy! Instrumental variables should be justified based on theory or a clear conceptual model rather than on their mere predictive power.

[12] As with the general latent variable approach discussed earlier, if we've done a good job estimating the explanatory variable with the instrumental variables, the predicted values should reflect the true score of x and the residuals should represent measurement error.

described.[13] We presume that the latent variable is a linear combination of the instruments, thus we can estimate the association with a linear model.

The following illustrates the steps in a 2SLS in a more systematic fashion. Using the letter z to represent the instrumental variables and x to indicate the error-contaminated explanatory variable, the steps are as follows:

1. Estimate $x_i = \alpha + \beta_1 z_1 + \beta_2 z_2 + \cdots + \beta_q z_q$ with OLS.
2. Save the predicted values from this regression model, labeled $\hat{x}_{r1}$.
3. Using a second OLS-based LRM, estimate: $y = \alpha_r + \beta_{r1} \hat{x}_{r1}$, which uses the predicted values from (2).

The slope coefficient from model (3) is unbiased (assuming the outcome variable is measured without error), but is inefficient, so special methods are required to adjust the standard error. One should, therefore, use statistical software designed for 2SLS since it provides methods to compute the correct standard errors, rather than go through each step to estimate the model.

A problem that makes the use of 2SLS of limited utility involves the challenge of finding instrumental variables that meet the requirements outlined earlier. Many variables that are associated strongly with the explanatory variable are, first, also contaminated by error, and second, associated with the error term. A serious problem arises when the instrumental variables are contaminated by error or do not predict the explanatory variable well: the slope coefficients and standard errors from a 2SLS regression model may manifest even more bias than those from a single stage LRM estimated with OLS.[14]

Let's estimate a 2SLS model in R. The difficulty of finding good instrumental variables is demonstrated by the artificiality of this illustration. We'll use the *GSS2018.csv* dataset and make the following (questionable) assumptions: (a) the outcome variable, life satisfaction (lifesatis), is measured without error; (b) the true score of occupational prestige (x) predicts life satisfaction, but the observed variable (occprest) is contaminated by measurement error; (c) two instrumental variables (the zs), years of formal education (educate) and personal income (pincome), are strong predictors of occupational prestige but are uncorrelated with the error in predicting life satisfaction. In this regression model, we thus presume that higher occupational prestige is associated with higher life satisfaction, but that we should use instrumental

---

[13] The inventor of 2SLS is not certain. But most evidence points to U.S. economist Phillip G. Wright, who authored a 1928 book with an appendix that described how instrumental variables could be used in regression analyses (see James H. Stock and Francesco Trebbi (2003), "Retrospectives: Who Invented Instrumental Variable Regression?" *Journal of Economic Perspectives* 17(3): 177–194).

[14] Recall from the section on selection bias in Chapter 12 that a similar problem occurs when the model that accounts for selection uses weak predictors.

variables in a 2SLS model to reduce the effects of the measurement error in the explanatory variable.

To begin, we'll estimate the correlations and their *p*-values among the three types of variables with the psych' package's corr.test function.

R code
```
library(psych)
sub.GSS <- GSS2018[c("occprest", "educate", "pincome",
                     "lifesatis")]
                # create a subset of the data
corr.test(sub.GSS) # request the correlation matrix and
                     p-values
```

R output (abbreviated)
```
Correlation matrix
          occprest educate pincome lifesatis
occprest    1.00     0.47    0.23     0.11
educate     0.47     1.00    0.28     0.06
pincome     0.23     0.28    1.00     0.04
lifesatis   0.11     0.06    0.04     1.00

Probability values (Entries above the diagonal are
                adjusted for multiple tests)
          occprest    educate    pincome  lifesatis
occprest     0         0.00       0.00      0.00
educate      0         0.00       0.00      0.16
pincome      0         0.00       0.00      0.38
lifesatis    0         0.05       0.19      0.00
[Note: probability values listed as 0 or 0.00 are p <
  0.01]
```

The output indicates that the instrumental and explanatory variables are moderately correlated, with the largest correlation between occupational prestige and education ($r = 0.47$). At this point we'll assume that education and income are reasonable instruments for occupational prestige.

Next, assess the results of an LRM estimated with OLS that includes life satisfaction as the outcome variable and occupational prestige as the sole explanatory variable (see LRM13.1).

R code
```
LRM13.1 <- lm(lifesatis ~ occprest, data=sub.GSS)
summary(LRM13.1)
confint(LRM13.1)
```

R output (abbreviated)
```
Coefficients:
```

```
              Estimate  Std. Error  t value  Pr(>|t|)
(Intercept)   70.74823    1.13552    62.305   < 2e-16   ***
occprest       0.09162    0.02436     3.761   0.000178  ***
---
Signif. codes:  0 '***' 0.001 '**' 0.01 '*' 0.05 '.' 0.1
```

Residual standard error: 11.1 on 1159 degrees of freedom
  (1154 observations deleted due to missingness)
Multiple R-squared: 0.01206, Adjusted R-squared: 0.01121
F-statistic: 14.15 on 1 and 1159 DF, p-value: 0.0001776

```
[CIs]           2.5%    97.5%
(Intercept)   68.520   72.976
occprest       0.044    0.139
```

The R^2 for this model is 0.012, with an RMSE of 11.1. The explained variance is low but its p-value is small ($F = 14.2$, $p < .001$) (is this a concern?). The slope coefficient for occupational prestige ($\beta = 0.09$) has a small p-value and its 95% CI does not encompass zero ($p < 0.001$; CI: {0.04, 0.14}). As expected, the slope coefficient indicates that higher occupational prestige is associated with higher life satisfaction. We presumed earlier, however, that occupational prestige is measured with error. Let's see what a 2SLS regression model offers. We'll use the R package ivpack and its function ivreg to estimate the model (see LRM13.2).[15]

R code
```
library(ivpack)
LRM13.2 <- ivreg(lifesatis ~ occprest | pincome +
                    educate, data = sub.GSS)
 # the instrumental variables appear after the vertical
   bar
summary(LRM13.2, vcov = sandwich, diagnostics = TRUE)
 # show results but with robust (sandwich) standard
   errors[16]
```

R output (abbreviated)
```
Coefficients:
              Estimate  Std. Error  t value  Pr(>|t|)
(Intercept)   70.04666    2.67724    26.164   <2e-16   ***
occprest       0.10734    0.05746     1.868   0.062    .
Diagnostic tests:
```

---

[15] Another R package that estimates 2SLS models is estimatr. One advantage of its iv_robust function is that it estimates the coefficients and the robust standard errors in one line of code. The R package also includes an lm_robust function that furnishes robust (sandwich) standard errors as a corrective for heteroscedastic errors (see Chapter 9).

[16] Chapter 9 (see fn. 17) mentions that several types of robust (sandwich) estimators are available to adjust the standard errors. In this example, vcov = sandwich executes the HC0 estimator.

```
                df1   df2   statistic   p-value
Weak instruments  2   1158   156.287    <2e-16 ***
Wu-Hausman        1   1158     0.102     0.750
Sargan            1    NA      0.195     0.659
---
Signif. codes:  0 '***' 0.001 '**' 0.01 '*' 0.05 '.' 0.1
```

Residual standard error: 11.1 on 1159 degrees of freedom
Multiple R-Squared: 0.0117, Adjusted R-squared: 0.01085
Wald test: 3.489 on 1 and 1159 DF, p-value: 0.06202

Compare the occupation prestige slope coefficients from the two models: a slightly larger coefficient is found in the 2SLS model (0.11 vs. 0.09), but its standard error and *p*-value are also larger, the latter above the $p < 0.05$ threshold that many researchers use for judging statistical significance. The 95% CI (not provided in the R output) also includes zero {−0.01, 0.22}. The model does not appear to correct for the presumed measurement error in occupation prestige. An important question, moreover, is whether the instruments are suitable for the model. The Diagnostic tests in the R output provide pertinent evidence. Here's a brief description of each of these tests.

1. Weak instruments: tests the instruments at the first step using an *F*-test. The *F*-test's null hypothesis is that the instruments are weak—they do a poor job of predicting the *x* variable. Since we find evidence against the null (see the test's *p*-value), we may presume that the instruments are not weak and do an adequate job of predicting occupational prestige.

2. Wu-Hausman: though mentioned in Chapter 12,[17] in the current situation the test analyzes whether or not the OLS estimates are consistent. Recall that consistent estimators converge to the true parameter value in the population as the sample size gets larger. The null hypothesis of the test is, essentially, that the OLS and instrumental variable's estimates are analogous and, if not rejected, suggests little endogeneity in the model. The evidence provided by LRM13.2 is consistent with the null hypothesis ($p = 0.75$), so the 2SLS model may not be necessary. This implies that we may as well rely on the single stage LRM.

---

[17] Durbin has been left out of the name of the test, perhaps because his version is slightly different. If one has no access to an automated version of this test, a simple yet slightly imprecise alternative involves: (1) regressing the *x* variable on the instruments (*z*) using an OLS LRM; (2) saving the residuals from this model; and (3) regressing *y* on *x* and the residuals from (2). If the coefficient associated with the residuals has a *p*-value below 0.05, the 2SLS model is preferred to the single stage LRM. Try these steps with the model at hand. Your conclusion should be consistent with what R provides in the Wu–Hausman test.

3. Sargan: also called the Sargan *J test*, it assesses whether the instruments are genuinely exogenous (not accounted for by other variables in the model).[18] If the evidence is incompatible with the null hypothesis, then one or more of the instrumental variables is deemed invalid. The test is only applicable if there are more instrumental variables than x variables in the model. Since LRM13.2 has two instruments and one x variable assumed endogenous (occprest), the Sargan test is germane. In the model, the evidence is compatible with the null hypothesis ($p = 0.659$), which implies that the instrumental variables are exogenous.[19]

These tests offer mixed evidence that the 2SLS model is an appropriate method to account for variability in life satisfaction. The instruments satisfy the assumptions that they are not weak predictors and are exogenous, but not that the model is needed to compensate for the OLS-based model's lack of consistency. Nonetheless, theory, reasoning, and understanding the associations among variables within and outside the model should remain as the principal guide in both selecting instrumental variables and interpreting the results.

For example, we should regularly ask, are the criteria for the instrumental variables met? Are the assumptions of the model satisfied? Is the shift in the regression coefficient (0.02 units) from the OLS to the 2SLS model meaningful? Could there be other potential sources of error in either of the instruments? Suppose we learn that occupations considered prestigious are those that require more education or that pay, on average, higher salaries? Would this change our judgment of education and personal income as instrumental variables? We might decide they are not good instruments for this model. As noted earlier, finding instrumental variables is especially challenging and we will always be forced to justify our choices based on reasoning that is divorced from concrete statistical evidence.

Let's complete this section by discussing some additional issues to consider when using instrumental variables. The first involves how to estimate 2SLS models that include more than one explanatory variable. If there are multiple x variables, include *all* of them in the first stage of the regression model (e.g., $x_1 = \alpha + \beta_1 z_1 + \beta_2 z_2 + \beta_3 x_2 + \beta_4 x_3 + \cdots$). Otherwise, the intercept and slope coefficients from the first stage are biased and, thus, the prediction of the error-laden variable(s) is (are) not accurate. Since the example of occupational prestige and life satisfaction includes only one explanatory variable,

---

[18] Introduced in John D. Sargan (1958), "The Estimation of Economic Relationships Using Instrumental Variables," *Econometrica* 26(3): 393–415.

[19] For more information about these tests, see Pamela Paxton, John R. Hipp, and Sandra Marquart-Pyatt (2011), *Nonrecursive Models: Endogeneity, Reciprocal Relationships, and Feedback Loops*, Thousand Oaks, CA: Sage.

this issue doesn't apply. If there are multiple variables, however, keep this in mind.[20]

Second, 2SLS is one estimation method for instrumental variables, but several others are available, such as limited information maximum likelihood (LIML) and the continuously updated estimator (CUE). They make similar assumptions about the suitability of the instruments, but their results are not as biased when the instruments are weak.[21] Maximum likelihood estimators may not perform as well as 2SLS, though, when models are incorrectly specified.[22]

Finally, instrumental variables are also useful for endogeneity issues (see Chapter 12). To combine examples from this and the previous chapter: suppose we wish to explore whether occupational prestige is endogenous in a model that is designed to predict life satisfaction by presuming that prestige is partly due to education, race/ethnicity, and other background characteristics. In this case, we are expressly interested in the first stage of the model: the association between prestige and its predictors. Using SEM rather than 2SLS for this sort of endogeneity issue is common in sociology and psychology. Thus, if we are not content with the instrumental variables–2SLS approach, other measurement error models are available.[23]

Chapter Summary

It should be obvious by now that measurement error is an almost universal problem in social science statistics. Specification error subsumes many measurement error problems since it usually involves extra information (errors in measurement) that we hope to exclude from the LRM. Whether there will ever be tools to minimize it sufficiently so that we may convince all or most people that our models are valid is, at this time, unknown. Measurement techniques continue to improve, though, as researchers study the properties

---

[20] See Joshua D. Angrist and Jörn-Steffen Pischke (2009), *Mostly Harmless Econometrics: An Empiricist's Companion*, Princeton, NJ: Princeton University Press, chapter 4. This book also provides guidance on finding promising instrumental variables.

[21] See Neil M. Davies et al. (2015), "The Many Weak Instruments Problem and Mendelian Randomization," *Statistics in Medicine* 34(3): 454–468.

[22] See Kenneth A. Bollen et al. (2007), "Latent Variable Models under Misspecification: Two-Stage Least Squares (2SLS) and Maximum Likelihood (ML) Estimators," *Sociological Methods & Research* 36(1): 48–86.

[23] Wallace (2020), *op. cit.*, discusses several of these, including regression calibration—of which the reliability coefficient adjustment method is one type—and simulation extrapolation, which is implemented in R with the simex package.

of measuring instruments and the way that people answer survey questions or report information in general.[24]

At this point, we should focus on measuring social and behavioral phenomena accurately and reliably. Since many existing scales that measure social concepts have undergone rigorous testing, the best approach is to use them if they address some variable that one wishes to examine in an LRM or other model. Some useful tools are also available for reducing the detrimental effects of measurement error and are accessible to even novice researchers. A review of the literature on factor analysis, PCA, multivariate analysis, and SEM furnishes a guide to using these tools.

Chapter Exercises

The dataset *AdolescentAlcohol.csv* contains a subset of variables from the National Educational Longitudinal Study (NELS), a U.S. survey of high school students. The variables include

- ident   identification number
- male   0 = female, 1 = male
- latinx   0 = not Latinx, 1 = Latinx
- af_am   0 = not African American, 1 = African American
- white   0 = not White, 1 = White
- oth_eth   0 = not other ethnic status, 1 = other ethnic status
- gpa   grade point average in high school on a standard four point scale (0–4), with higher values indicating better grades
- sport   participation in high school sports (higher values indicate more participation)
- academic   participation in academic clubs (higher values indicate more participation)
- alcohol   frequency of alcohol use (higher values indicate more frequent alcohol use)

---

[24] Excellent guides to constructing survey questions are Norman Bradburn et al. (2004), *Asking Questions: The Definitive Guide to Questionnaire Design*, Rev. Ed., San Francisco: Jossey-Bass, and Krosnick and Fabrigar (2005), *op. cit.*

- ses: family socioeconomic status (standard-
                                          ized with higher values indicating
                                          higher SES)

After importing the dataset into R, complete the following exercises.

1. Estimate an LRM that uses `alcohol` as the outcome variables and
   grade point average (gpa) as the explanatory variable. Interpret the
   slope coefficient and 95% CI associated with gpa.

2. Suppose previous studies indicate that the measure of gpa used in
   the survey has a reliability coefficient (r_{xx^*}) of 0.9. Based on the results
   of the LRM in exercise 1, compute the unbiased slope coefficient
   associated with gpa.

3. Estimate a correlation matrix with the following variables: academic, sport, gpa, and ses. Assume that, because grade point
   average was self-reported and we are concerned it may be affected
   by measurement error (some adolescents don't report their grades
   accurately), we consider sport, academic, and ses as potential
   instrumental variables for gpa. Based on the correlation matrix,
   how suitable are these instruments? What other factors in addition
   to the correlations should you evaluate when considering their suitability as instruments?

4. Estimate an instrumental variables regression model that uses
   `alcohol` as the outcome variable; gpa the explanatory variable; and
   participation in sports, participation in academic clubs, and family
   socioeconomic status as instruments for grade point average.

   a. Use the tests discussed in this chapter to determine if the instru-
       mental variables approach offers improvement over the OLS
       regression model.

   b. What do you conclude about (a) the association between grade
       point average and frequency of alcohol use, and (b) whether the
       instrumental variables approach is an improvement over an OLS
       regression model?

5. *Challenge*: create a latent variable that measures school involvement
   and accounts for the joint variability of grade point average, participation in sports, and participation in academic clubs. Estimate an
   LRM with alcohol use as the outcome and the latent variable as the
   explanatory variable. Discuss whether you prefer this model to the
   models in the earlier exercises.

14

Influential Observations: Leverage Points and Outliers

Previous chapters present several examples of scatter plots. These ubiquitous plots can show a variety of patterns in relationships among variables. We've seen linear patterns, nonlinear patterns, heteroscedastic patterns, and serially correlated patterns. Nonlinear patterns, in particular, force us to question the use of a linear model to assess the association between variables (see Chapter 11). Although considering nonlinearities is important, it does not exhaust the ways that data points diverge from a linear pattern. Another common situation is to find small portions of the variables, perhaps even a single point, that deviate from the straight line. Small departures from linearity can still have considerable consequences for LRMs. Divergent data points also may affect the distribution of the errors/residuals, leading to problems satisfying the normality or homoscedasticity assumption.

This chapter focuses on *influential observations*: data points that affect LRMs to an inordinate degree. The degree of influence is not always clear but can be substantial. We are concerned with two types of influential observations that affect a regression model: *outliers* and *high leverage points*. Outliers are observations that pull the regression line (or surface) in one direction or another. Values of the y variable relatively distant from the other y values, outliers can have a strong effect on regression results, including the LRM slope coefficients and standard errors. Recall, once again, the formula for the standard error of a slope coefficient in a multiple LRM (OLS) that is shown in Equation 14.1.

$$se\left(\hat{\beta}_i\right) = \sqrt{\frac{\Sigma\left(y_i - \hat{y}_i\right)^2}{\Sigma\left(x_i - \bar{x}\right)^2\left(1 - R_i^2\right)\left(n - k - 1\right)}} \tag{14.1}$$

Can you appreciate how outliers might affect the standard errors? Look at the numerator. What effect does an outlier have on this part of the equation—the sum of squares of y?

Suppose that an outcome variable measures SAT scores from five high school seniors. We wish to predict these scores based on junior-year grades in math, science, and English (of course, we'd want more than five observations to estimate such a model, but we'll ignore this for now). Their SAT scores are {980, 1,020, 810, 900, 1,600}. The last score constitutes an outlier:

DOI: 10.1201/9781003162230-14

extreme on the y variable. You may wish to plot these observations and some fabricated math or English grades to visualize the effect of an outlier.

High leverage points are observations that are discrepant or distant from other values of the explanatory variables.[1] They may or may not also be outliers.[2] For example, if we wish to use the aforementioned SAT scores to predict college GPA, then the score of 1,600 might constitute a high leverage point. The observation could also be an outlier if, say, the student who scored 1,600 also obtained a 1.5 GPA in college whereas the other students obtained GPAs close to 3.0.

Examine the standard error formula in Equation 14.1 and you'll notice that even high leverage points that are not outliers affect the standard error of the slope coefficients. The key component affected by leverage points involves the x_i values in the denominator. Suppose that, all else being equal, we add an extreme leverage point to the equation. What happens? A larger squared value in the denominator results: the denominator increases and the standard error decreases. Think about what this implies for tests of statistical significance.

Figure 14.1 provides three graphs that show an outlier, a high leverage point, and a combination of the two. They also include estimated linear fit lines. The outlier pulls what should be a relatively steep, positive fit line upward at the low end of the x distribution, thus inducing a flatter slope than if the outlier is not part of the set of observations. The high leverage point falls on the fit line, so its slope is not influenced but, as mentioned earlier, the standard error of the slope coefficient is affected. The third graph shows a common situation: the observation is extreme on both the x and y variables. The point has a relatively strong influence on the slope of the fit line and the standard error of the slope coefficient.

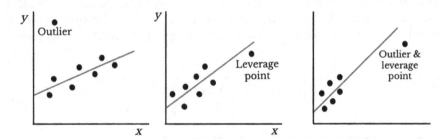

FIGURE 14.1
Illustrations of outliers and high leverage points.

---

[1] The term leverage refers to the distance of an observation's value on the explanatory variable(s) from the average value of the explanatory variable(s). Leverages can denote points that are very close to or very far away from the average(s).

[2] Rand Wilcox (2016) refers to high leverage values that are also outliers as "bad leverage points" (*Introduction to Robust Estimation and Hypothesis Testing*, 4th Ed., San Diego, CA: Academic Press, p.521).

Influential observations result from a number of sources. A common cause is a coding error, thus they may be due to errors in measurement (see Chapter 13). When entering numeric data, hitting the wrong entry key or forgetting a numeral or a decimal place is a common problem (though it has been obviated by automated forms of data entry). A good practice, therefore, is to always check the data for coding errors. Influential observations might also occur as a routine part of data collection exercises: some people do earn a lot more money than others; some adolescents do drink alcohol or smoke marijuana much more often than other youth. If the nature of a variable leads to extreme values, a common solution is to pull in these values by taking the square root or natural logarithm of the variable (see the normality assumption section of Chapter 11). Before estimating a model, it's a good idea to employ exploratory data analysis,[3] such as with q-q plots, box plots, kernel density plots, or stem-and-leaf plots to visualize the distributions and assess whether there appear to be high leverage points and outliers. Use scatter plots to examine the association between the proposed explanatory variables and the outcome variable. We can often identify unusual observations before estimating the LRM and come up with solutions early in the investigative process.[4]

Detecting Influential Observations

In a multiple LRM context, several diagnostic methods are useful for unearthing influential observations. A number of books and articles cover these in detail.[5] The following is a brief overview of some of the more frequently

---

[3] Perhaps the most highly regarded statistics book of the last 50 years is John W. Tukey's (1977) *Exploratory Data Analysis* (Reading, MA: Addison-Wesley). His advice regarding how to carefully examine a set of variables remains timely and indispensable. Heeding Tukey's instructions, whether or not one plans to estimate an LRM, is essential for data-driven research projects.

[4] Researchers too frequently view influential observations as an annoyance or a trouble spot in the data. But an important and relevant contrary perspective was expressed by geneticist W. Bateson (1908) more than a century ago: "Treasure your exceptions! When there are none, the work gets so dull that no one cares to carry it further. Keep them always uncovered and in sight. Exceptions are like the rough brickwork of a growing building which tells that there is more to come and shows where the next construction is to be" (*The Methods and Scope of Genetics*, Cambridge, UK: Cambridge University Press, p.22).

[5] For example, see David A. Belsley, Edwin Kuh, and Roy E. Welsch (2004), *Regression Diagnostics: Identifying Influential Data and Sources of Collinearity*, Hoboken, NJ: Wiley; Herman Aguinis, Ryan K. Gottfredson, and Harry Joo (2013), "Best-Practice Recommendations for Defining, Identifying, and Handling Outliers," *Organizational Research Methods* 16(2): 270–301; A. A. M. Nurunnabi, N. Nasser, and A. H. M. R. Imon (2016), "Identification and Classification of Multiple Outliers, High Leverage Points and Influential Observations in Linear Regression," *Journal of Applied Statistics* 43(3): 509–525; and Wilcox (2016), *op. cit.*

utilized techniques. We begin with some numeric methods and then discuss graphical methods.

Assuming an LRM has multiple explanatory variables, assessing leverage points requires an evaluation of the spatial patterning among the set of x values. If one observation is considerably discrepant from the others, it should show up as far from the others in a measure of spatial distribution. To understand how to develop a numeric approach for gauging the distances of observations, recall that we may represent multiple LRMs using matrix notation (see Chapter 4). The X matrix in Equation 14.2 depicts the explanatory variables with the entries representing specific observations.

$$X = \begin{bmatrix} x_{11} & x_{12} & \cdots & x_{1k} \\ x_{21} & x_{22} & \cdots & x_{2k} \\ \vdots & \vdots & \cdots & \vdots \\ x_{n1} & x_{n2} & \cdots & x_{nk} \end{bmatrix} \tag{14.2}$$

$$H = X(X'X)^{-1}X' = h_{ij} \tag{14.3}$$

The *hat matrix* based on the X matrix is defined in Equation 14.3. Recall that the matrix X' is the transpose of X, whereas X^{-1} is the inverse of X (see Chapter 4). H is called the hat matrix because $\hat{Y} = HY$: the predicted values of y are derived from the vector of y values premultiplied by H. The diagonals of the hat matrix (h_{ii}) lie between zero and one. We may compute the mean of these values to come up with an overall mean for the set of x values used in a multiple LRM. The points off the diagonals represent distance measures from this joint mean.

Many statistical software packages, including R, provide leverage values (also called *hat values*) based on the hat matrix. As already mentioned, larger values are farther from the overall mean of the xs and, thus, are more likely to influence the results of the LRM. Leverage values range from zero to $\{n - 1\}/n$, with the theoretical maximum getting closer to one as the sample size increases. A standard criterion states that any leverage value equal to or exceeding $2(k + 1)/n$ should be scrutinized as an influential observation. For example, in a model with three explanatory variables and a sample size of 50, leverage points of $2(3 + 1)/50 = 0.16$ or more should be inspected. R includes a hatvalues function that furnishes leverages following estimation of an LRM.

A numeric method for detecting outliers is to examine the studentized deleted residuals, which we've seen in previous chapters (e.g., Chapter 8). Other names for these are *jackknife residuals* or the less precise *studentized residuals*. The formula for computing them is shown in Equation 14.4.

$$t_i = \frac{\hat{\varepsilon}_i}{RMSE_{(-1)}\sqrt{1-h_i}} \tag{14.4}$$

The $\hat{\varepsilon}_i$ value is the unstandardized residual and the h_i value is from the hat matrix. Recall that deleted residuals are computed based on LRMs that remove observations one at a time, each time re-estimating the model on the other $n-1$ observations. The RMSE$_{(-1)}$ in the equation is thus based on an LRM when a particular observation is removed. The residual is called studentized because it follows Student's t distribution, which, like standardized variables, have a mean of zero and, in large samples, a standard deviation of one. We thus expect about 5% of these residuals to fall two or more units from zero. To detect outliers, look for studentized deleted residuals that are substantially greater than two or substantially less than negative two. Recall that we may request these residuals after an LRM with R's rstudent function.

Statisticians have also developed a number of measures designed to identify influential observations by combining information about residuals and leverage values. The two most prominent are Cook's D or Distance[6] and DFFITS.[7] Equation 14.5 shows how to compute Cook's D values.

$$D_i = \frac{\left(y_i - \hat{y}_i\right)^2}{(k+1) \times \text{MSE}} \times \frac{h_i}{\left(1 - h_i\right)^2} \tag{14.5}$$

k: number of explanatory variables

Larger values of Cook's D indicate more influence on the regression results. Two recommended criteria for Cook's D are (a) $D \geq 1.0$ and (b) $D \geq 4/(n - k - 1)$. Data points that meet or exceed these thresholds should be scrutinized as influential observations. In a model with three explanatory variables and a sample size of 50, for example, any data point with a Cook's D value greater than or equal to $4/(50 - 3 - 1) = 0.09$ should be inspected. R provides Cook's D values with the cooks.distance function.

DFFITS serve a similar purpose as Cook's D and are computed with Equation 14.6.

$$\text{DFFITS}_i = \frac{\hat{y}_i - \hat{y}_{i-1}}{\text{RMSE}_i \sqrt{h_i}} \tag{14.6}$$

The numerator is the difference between the predicted values from the LRM using the entire sample and the LRM that excludes the observation. A general criterion is to consider any DFFITS with an absolute value greater than or equal to two as indicative of an influential observation. (DFFITS, like studentized residuals, can be positive or negative.) A different benchmark that considers the sample size, however, is $2\sqrt{k+1/n}$; hence, if we have an LRM

---

[6] Introduced in R. Dennis Cook (1977), "Detection of Influential Observation in Linear Regression," *Technometrics* 19(1): 15–18.

[7] The earliest description of this measure appears to be in Roy E. Welsch and Edwin Kuh (1977), "Linear Regression Diagnostics," Working Paper No. 173, Cambridge, MA: National Bureau of Economic Research.

with three explanatory variables and a sample size of 50, any data point with a DFFITS value greater than or equal to $2\sqrt{3+1/50} = |0.57|$ should be evaluated as an influential observation. R includes a `dffits` function. A good practice is to compute these various measures after estimating an LRM. Calculate the threshold values for each and then identify potential influential observations in data.

Several types of graphs are also helpful for identifying influential observations and assessing their effects in an LRM, though they tend to be more suitable with small to modest sized samples. In large datasets, detecting patterns of observations is difficult. A simple way to visualize influential observations is with *added-variable plots* that display the association between each explanatory variable and the outcome variable after they have been adjusted for the other explanatory variables. We may then check for observations that are discrepant from others. An example in the next section demonstrates how to request these plots with the `avPlots` function that is part of R's `car` package.

Another graphical approach is to construct a scatter plot with leverage values on the x-axis and studentized deleted residuals on the y-axis, which may then be used to determine if one or more observations is an outlier, a high leverage point, or both. A scatter plot of the Cook's D values (or the absolute value of the DFFITS) by the leverage values or studentized deleted residuals is also useful for determining specific types of influential observations.

An Example of Using Diagnostic Methods to Identify Influential Observations

Let's utilize the *StateData2018.csv* dataset to explore a few predictors of the percent of state residents living in poverty (`PerPoverty`) and then examine some post-model diagnostics to identify influential observations. The LRM includes the following explanatory variables: median household income (`MedHHIncome`), the unemployment rate (`UnemployRate`), and gross state product (`GSP`; `GrossStateProduct`). Median household income and GSP are re-scaled to make their slope coefficients easier to interpret. After estimating the model in LRM14.1, we'll ask R to compute the following diagnostic measures: studentized deleted residuals, leverage values, Cook's Ds, and DFFITS.

R code
```
StateData2018$MedHHInc <- StateData2018$MedHHIncome/1000
 # rescale median household income into $1,000s
StateData2018$GSP <- StateData2018$GrossStateProduct/
                       100000
```

```
# rescale gross state product into $100,000s
LRM14.1 <- lm(PerPoverty ~ MedHHInc + UnemployRate +
              GSP, data=StateData2018)
summary(LRM14.1)
confint(LRM14.1)
# compute the studentized deleted residuals, leverage
  values, Cook's distances and DFFITS
LRM14.1.stddr <- rstudent(LRM14.1)
LRM14.1.lev <- hatvalues(LRM14.1)
LRM14.1.cook <- cooks.distance(LRM14.1)
LRM14.1.dffits <- dffits(LRM14.1)
```

R output (abbreviated)
Residuals:
```
  Min      1Q     Median   3Q      Max
-2.2279 -0.7015 -0.2631 0.7093 3.0265
```

Coefficients:
```
                Estimate Std. Error t value Pr(>|t|)
(Intercept)     21.79218   1.43694   15.166  < 2e-16  ***
MedHHInc        -0.23456   0.01803  -13.011  < 2e-16  ***
UnemployRate     1.34869   0.21991    6.133  1.83e-07 ***
GSP              0.09013   0.03614    2.494  0.0163     *
---
```
Signif. codes: 0 '***' 0.001 '**' 0.01 '*' 0.05 '.' 0.1

Residual standard error: 1.21 on 46 degrees of freedom
Multiple R-squared: 0.8361, Adjusted R-squared: 0.8254
F-statistic: 78.22 on 3 and 46 DF, p-value: < 2.2e-16

```
[CIs]             2.5%    97.5%
(Intercept)      18.900   24.685
MedHHInc         -0.271   -0.198
UnemployRate      0.907    1.791
GSP               0.017    0.162
```

The results suggest that all three explanatory variables are associated with poverty, though the lower bound of the GSP's coefficient's 95% CI is close to zero. The model accounts for about 84% of the variability in the percent living in poverty. But are there any influential observations that might affect the model inordinately?[8] The first step is to compute the threshold levels for each influence measure, which are furnished in Table 14.1.

Let's examine some summary statistics for the measures by creating a data frame using the as.data.frame and cbind functions that combines

---

[8] The first part of the R output lists Residuals, which might be inspected for especially large or small values, but we shall rely on the other diagnostic measures in this example.

TABLE 14.1

Influential Observation Diagnostics

Measure	Threshold	Formula
Studentized deleted residuals	$t_i > \lvert 2.0 \rvert$	Look for relatively large values
Leverage values	$l_i > 0.16$	$\dfrac{2(3+1)}{50}$
Cook's D	$D_i > 0.09$	$\dfrac{4}{(50-3-1)}$
DFFITS	$DF_i > \lvert 0.57 \rvert$	$2\sqrt{3+1/50}$

Note: the thresholds and formulas are based on LRM13.1.

the information.[9] The results include the mean, median, and first and third quartiles of each diagnostic measure, but we are most interested in the minimum and maximum values. Beginning with the studentized deleted residuals, observe that one observation has a residual of about 2.95, which might be worth inspecting.

R code
```
diag.measures <- as.data.frame(cbind(LRM14.1.stddr,
                 LRM14.1.lev, LRM14.1.cook,
                 LRM14.1.dffits))
summary(diag.measures)
```

R output (abbreviated; bold added)

LRM14.1.stddr		**LRM14.1.lev**		**LRM14.1.cook**		**LRM14.1.dffits**	
Min.	**-1.932**	Min.	0.022	Min.	0.000	Min.	**-1.329**
1st Qu.	-0.592	1st Qu.	0.040	1st Qu.	0.002	1st Qu.	-0.145
Median	-0.224	Median	0.052	Median	0.007	Median	-0.062
Mean	0.010	Mean	0.080	Mean	0.029	Mean	0.022
3rd Qu.	0.597	3rd Qu.	0.080	3rd Qu.	0.016	3rd Qu.	0.193
Max.	**2.947**	Max.	**0.510**	Max.	**0.426**	Max.	**1.282**

What about high leverage values? The largest leverage is 0.51, which is above the threshold of 0.16 shown in Table 14.1. We should examine the observation associated with this leverage value, though there might also be others that exceed the threshold. The largest Cook's D (0.43) and DFFIT ($\lvert 1.33 \rvert$) are

---

[9] If we simply ask R to cbind the measures, it creates an *atomic vector*: a vector that holds data of only a single data type. For some purposes it doesn't matter, but in many situations, requesting data frames offers the analyst more options.

also above their recommended thresholds.[10] These results suggest additional observations that might be worth assessing.

As a next step, let's create a couple of graphs to conduct a further evaluation of outliers and high leverage values. We'll build added-variable plots using the `car` package's `avPlots` function and an outlier-leverage plot available with the post-lm `plot` function.

The most disparate observation in the added-variable plots appears in the gross state product (GSP|others) graph (see Figure 14.2). Observation 5 is quite a bit larger on the *x*-axis than the others, with a value that exceeds 20 whereas the next largest value is only about 12 or 13. The Residuals vs. Leverage plot in Figure 14.3 also shows a couple of high leverage values, including one that is quite far from the others (leverage > 0.5). The plot

Added-Variable Plots

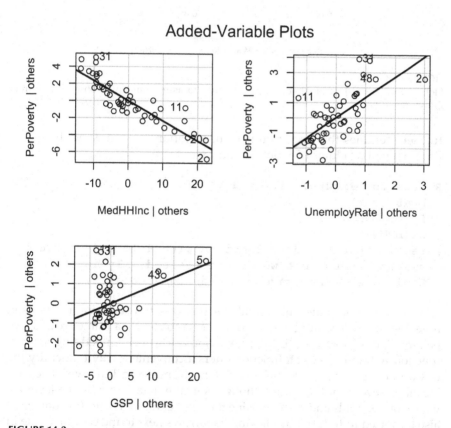

FIGURE 14.2
Added-variable plots to assess influential observations from LRM 14.1.

---

[10] R lists all the leverage values, DFFITS, and Cook's *D* values from an LRM, along with two related quantities (*dfbetas* and *covratio*), with the `influence.measures` function: `influence.measures(LRM14.1)`.

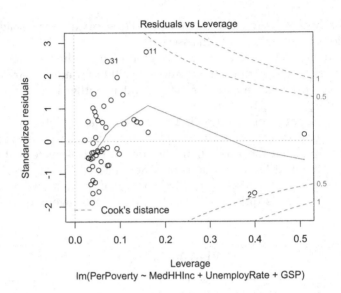

FIGURE 14.3
Plot of standardized residuals by leverage values to assess influential observations from LRM14.1.

further identifies the largest outliers, observations 11 and 31, with studentized deleted residuals of about 2.9 and 2.6.

R code for Figures 14.2 and 14.3
```
library(car)
avPlots(LRM14.1)
plot(LRM14.1, 5)
# recall that the plot function with a lm object creates
  several plots, but we are interested only in the
  Residuals vs Leverage plot (designated as 5)
```

We may explore other influential observations by examining the plots in more detail or by scrutinizing the diag.measures data frame created earlier. In a large data frame this can be tedious, however. Embedding the max function within the which function offers a convenient way to identify the row number of the largest value of a diagnostic measure. If we wish to examine all or several of the values, though, sort the data frame by the leverage values or residuals and then examine the most extreme values (the data may also be sorted in RStudio by clicking the arrows next to the column name in the Viewer window).

R code
```
diag.sort <- diag.measures[rev(order(LRM14.1.lev)),]
  # sorts from high to low based on the leverage values
```

```
View(diag.sort) # to see the entire sorted data frame in
the viewer window
head(diag.sort) # to see the first few rows
which(LRM14.1.lev == max(LRM14.1.lev)) # returns the row
                                         number with the
                                         largest
                                         leverage value
diag.sort.out <- diag.measures[rev(order(LRM14.1.
                  stddr)),]
 # sorts from high to low based on the studentized
   deleted residuals
View(diag.sort.out)
head(diag.sort.out) # to see the first few rows
tail(diag.sort.out) # to see the last few rows
which(LRM14.1.stddr == max(abs(LRM14.1.stddr)))
# returns the row number with the most extreme residual
   (absolute value)
```

The largest leverage value is associated with observation or row 5, which represents California. But what about it is extreme? One of the added-variable plots in Figure 14.2 provides a clue: California is extreme along the x-axis on GSP. GSP measures the economic output of a state but is not adjusted for population size. We should not be surprised that California, the state with the largest population in the U.S. and an economy that's bigger than all but about five or six countries, should show up as extreme on this variable.

The largest residuals, observations 11 and 31, are associated with Hawaii and New Mexico (they also show up among the maximum Cook's D and DFFITS values). Neither state has a particularly high percent poverty, so why they appear as more extreme than other observations is not clear. But examine their values on the explanatory variables: Hawaii has a low unemployment rate (2.2) and both are relatively low on GSP (Hawaii = 0.88, New Mexico = 0.94). Perhaps their predicted poverty values are discrepant from their actual poverty values due to some intricate association involving these two explanatory variables (see Figure 14.2).

What to Do about Influential Observations

We've already discussed a couple of the better solutions for influential observations: make sure the variables are coded correctly and consider whether their distributions are implicated. Influential observations might be due to coding errors or non-normally distributed variables. Carefully exploring the variables before estimating an LRM through the various exploratory techniques discussed in earlier chapters will go a long way toward alleviating problems that stem from influential observations.

Though rarely recommended, another solution is to delete an influential observation. If we cannot figure out why California is affecting the percent poverty LRM, for instance, we might simply remove it from consideration. Let's try this. The following R code estimates LRM14.2, which omits California's data.

R code
```
LRM14.2 <- lm(PerPoverty ~ MedHHInc + UnemployRate +
              GSP, data = StateData2018[-5, ])
              # omits row 5 - California
summary(LRM14.2)
confint(LRM14.2)
```

R output (abbreviated)
```
Residuals:
   Min     1Q    Median    3Q     Max
-2.2431 -0.7060 -0.2871 0.7275 3.0163
```

Coefficients:

	Estimate	Std. Error	t value	Pr(>|t|)	
(Intercept)	21.81375	1.45658	14.976	< 2e-16	***
MedHHInc	-0.23466	0.01823	-12.874	< 2e-16	***
UnemployRate	1.34963	0.22230	6.071	2.45e-07	***
GSP	0.08337	0.05080	1.641	0.108	

```
---
Signif. codes: 0 '***' 0.001 '**' 0.01 '*' 0.05 '.' 0.1

Residual standard error: 1.223 on 45 degrees of freedom
Multiple R-squared: 0.8362, Adjusted R-squared: 0.8253
F-statistic: 76.58 on 3 and 45 DF, p-value: < 2.2e-16
```

[CIs]	2.5%	97.5%
(Intercept)	18.880	24.747
MedHHInc	-0.271	-0.198
UnemployRate	0.902	1.797
GSP	-0.020	0.186

The results don't change much. The slope coefficient for GSP barely shifts $(0.09 \to 0.08)$ and the 95% CIs overlap almost completely, even though GSP's 95% CI in the second model includes zero {−0.02, 0.19} because the standard error is larger. In this situation, it may be best to ignore the extreme leverage value. In large samples, especially, ignoring the value and keeping the observation in the LRM is a reasonable option. In smaller samples, however, high leverage values and outliers can have considerable effects on the results.

The best solution, though, is to try to understand influential observations.[11] For example, what is it about California's data that produces such a large leverage value? Let's explore this some more by considering the explanatory variable—GSP—that is implicated in California's leverage value (see Figure 14.2). Recall that GSP assesses a state's economic productivity. As already suggested, California is the U.S.'s most populous state and has a huge economy based on its production of agricultural goods, large shipping ports, a world-renowned entertainment industry, and some of the most valuable real estate in the world. Given this, conduct an exploratory analysis of GSP, such as by constructing a normal q-q plot or a kernel density plot. Notice the right skew that is affected by California's value (GSP = 28; the next largest value belongs to Texas, with a GSP = 16.5). What can we do with a variable that manifests such skewness? Perhaps the distribution of GSP would benefit from taking its square root or natural logarithm (recall from Chapter 11 that these transformations are useful for normalizing right-skewed variables).

Including the logged version of an explanatory or outcome variable in R is simple since the log function may be embedded in the lm function. Let's estimate an LRM with the natural log of GSP (see LRM14.3).

R code

```
LRM14.3 <- lm(PerPoverty ~ MedHHInc + UnemployRate +
              log(GSP), data=StateData2018)
```

R output (abbreviated)
```
Coefficients:
              Estimate  Std. Error  t value   Pr(>|t|)
(Intercept) 21.59873    1.45159      14.879   < 2e-16   ***
MedHHInc    -0.23147    0.01814     -12.762   < 2e-16   ***
UnemployRate 1.36597    0.22282       6.130   1.85e-07  ***
log(GSP)     0.36408    0.16896       2.155   0.0364    *
---
Signif. codes: 0 '***' 0.001 '**' 0.01 '*' 0.05 '.' 0.1

Residual standard error: 1.229 on 46 degrees of freedom
Multiple R-squared: 0.831, Adjusted R-squared: 0.82
F-statistic: 75.4 on 3 and 46 DF, p-value: < 2.2e-16

[CIs]          2.5%    97.5%
(Intercept)  18.677   24.521
MedHHInc     -0.268   -0.195
UnemployRate  0.917    1.814
log(GSP)      0.024    0.704
```

---

[11] Remember Bateson's advice: "Treasure your exceptions!" (see fn. 4).

The results are consistent with LRM14.1, though the slope coefficient for GSP is larger in this model because its scale is now in log-units. The diagnostic measures indicate that the largest leverage value is no longer California (which state is it?). The largest Cook's D and DFFITS values also differ. But the general conclusions about the factors that predict percent poverty at the state level do not change.

Other solutions when faced with influential observations involve the use of regression models and estimation techniques that are not so sensitive to extreme values. We'll discuss four of these. First, recall that the median is known as a robust measure of central tendency because it is less sensitive than the mean to extreme values (see Chapter 2). Since the LRM estimated with least squares is based on means, an alternative regression technique based on medians should be influenced less by influential observations, especially outliers. This reasoning is the basis for two similar regression models: *median* regression and *least median of squares* (LMS) regression. Rather than minimizing the sum of squared errors (SSE), these techniques minimize the *sum of the absolute residuals* ($|\varepsilon_i|$) or the *median of the squared residuals* ($med(\varepsilon_i^2)$).

Let's examine a median regression model. In R, this is a type of *quantile regression*[12] because, as you'll recall, the median is the 50th percentile. Median regression is available in the quantreg package by using the function rq (see LRM14.4).

R code
```
library(quantreg)
LRM14.4 <- rq(PerPoverty ~ MedHHInc + UnemployRate +
              GSP, data=StateData2018, tau=0.5)
 # tau=0.5 indicates the 50th percentile; it is the
   default option
summary(LRM14.4)
```

R output
Coefficients:

	coefficients	lower bd	upper bd
(Intercept)	21.77729	18.38094	24.73351
MedHHInc	-0.24470	-0.28205	-0.20137
UnemployRate	1.46610	1.07759	2.06527
GSP	0.10513	-0.00568	0.14776

---

[12] The roots of quantile regression actually predate OLS to 18th-century work on the ellipticity of the earth by Ragusan polymath Rudjer Boscovich. The method was given new life by Irish economist Francis Y. Edgeworth (1888) who proposed a "plural median" as a competitor to the method of least squares ("On a New Method of Reducing Observations Relating to Several Quantities," *Philosophical Magazine and Journal of Science* 25(154): 184–191; see also Roger Koenker (2000), "Galton, Edgeworth, Frisch, and Prospects for Quantile Regression in Econometrics," *Journal of Econometrics* 95(2): 347–374).

The rq output is sparse compared to the lm output, with just coefficients and confidence bounds (lower bd and upper bd), which are similar to 95% CIs. The coefficients from the median regression model and the LRM differ somewhat. Comparing them directly is risky, however, since their interpretations differ. The interpretation in median regression involves differences in the expected medians rather than means. One evidentiary distinction between the original LRM and the median regression is that GSP's confidence bound includes zero in the latter model.[13]

A second type of regression model that is designed to mitigate the effects of outliers is estimated with *iteratively reweighted least squares* (IRLS). This method begins with a standard LRM but calculates weights based on the absolute value of the residuals. The weights are then applied to a second model and new weights are computed. It goes through this process repeatedly (*iteratively*) until the change in the weights drops below a certain point. In essence, it down-weights observations associated with the largest residuals or those that have the most influence on the estimated slope coefficient(s). An IRLS regression model may be estimated with the RobustRegBS function in the robustreg package (see LRM14.5). The results of the OLS and IRLS estimated LRMs are similar, which is not surprising since the former model did not appear to be influenced much by outliers (compare the output from LRM14.1 and LRM14.5).

R code
```
library(robustreg)
LRM14.5 <- robustRegBS(PerPoverty ~ MedHHInc +
                    UnemployRate + GSP, data =
                    StateData2018, anova.table =
                    TRUE)
```
R output (abbreviated)
```
Coefficients:
              estimate     std error    t value  p value
(Intercept)   21.81537484  1.50022287   14.54    0.00000
MedHHInc      -0.23892119  0.01923388  -12.42    0.00000
UnemployRate   1.38612961  0.23676837    5.85    0.00000
GSP            0.09945129  0.03694313    2.69    0.00987

Signif. codes: 0 '***' 0.001 '**' 0.01 '*' 0.05 '.' 0.1
```

A third approach is based on the fact that influential observations affect the standard errors of the slope coefficients. Similar to the corrective measures discussed for heteroscedastic errors and instrumental variables regression models in Chapters 9 and 13, we may use robust (sandwich) estimators to

---

[13] The confidence bounds using the quantreg's rq function are based on rank tests and tend to be wider than CIs from OLS models (for more information, see Roger Koenke (2005), *Quantile Regression*, New York: Cambridge University Press, chapter 3). Bootstrapped confidence bounds are also available in the rq function.

adjust the standard errors. Influential observations often mimic heteroscedasticity, so adjusting the standard errors with a robust estimator is fitting.
As in Chapter 9, we'll use the `coeftest` function that is built from R's sandwich and `lmtest` packages to obtain robust standard errors (see LRM14.6).

R code
```
library(sandwich)
library(lmtest)
LRM14.6 <- lm(PerPoverty ~ MedHHInc + UnemployRate +
              GSP, data=StateData2018)
coeftest(LRM14.6, vcovHC(LRM14.5, "HC1"))[14]
```

R output (abbreviated)
```
               Estimate   Std. Error  t value    Pr(>|t|)
(Intercept)   21.792179    1.340992   16.2508  < 2.2e-16  ***
MedHHInc      -0.234562    0.020158  -11.6360  2.652e-15  ***
UnemployRate   1.348690    0.269928    4.9965  8.895e-06  ***
GSP            0.090132    0.023756    3.7940  0.0004312  ***
---
Signif. codes: 0 '***' 0.001 '**' 0.01 '*' 0.05 '.' 0.1
```

The difference between this model and the original model (LRM14.1) is in
the standard errors, which are larger for median household income and the
unemployment rate coefficients, but smaller for GSP. (Why is this? Review
the formula for and discussion of the multiple LRM standard error early in
this chapter.)

A fourth way to diminish the effects of influential observations is through
the use of the bootstrap. As we learned in Chapter 9, bootstrapping is a resampling method that draws random subsamples with replacement from the
dataset and estimates an LRM with each subsample. The averages of the estimates from the subsamples tend to provide more robust results. Chapter 11
also demonstrates that this approach is useful when the normality assumption is not satisfied and the residuals do not follow a Gaussian (normal)
distribution.

Think about how bootstrapping works if a dataset of, say, 50 observations includes a few influential observations. If we repeatedly sample with
replacement from the dataset, some of the subsamples will not include the
influential observations, so they will have less influence on the model and its
constituent estimates once the results are averaged.

LRM14.7 offers an example of bootstrapping that uses the `car` package's
Boot function to re-estimate the model that predicts the percent of residents
living in poverty.

---

[14] For Stata users, the HC1 estimator in R produces the same standard errors as the robust
option used with the `regress` command. But, as noted in Chapter 9, other estimators are
also available (Zeileis (2006), *op. cit.*).

R code
```
library(car)
set.seed(86473) # for replication
LRM14.7 <- lm(PerPoverty ~ MedHHInc + UnemployRate +
              GSP, data=StateData2018)
LRM14.7.boot <- Boot(LRM14.7, R=500) # requests 500
                                       samples
summary(LRM14.7.boot)
Confint(LRM14.7.boot, level=.95)
```

R output (abbreviated)
```
Number of bootstrap replications R = 500
              original      bootBias      bootSE      bootMed
(Intercept)   21.792179   -0.23508670   1.519561   21.803808
MedHHInc      -0.234562    0.00047041   0.021134   -0.235062
UnemployRate   1.348690    0.06301479   0.285370    1.388456
GSP            0.090132   -0.00758592   0.039529    0.087699

Bootstrap bca confidence intervals
              Estimate     2.5%       97.5%
(Intercept)   21.792     18.507     24.356
MedHHInc      -0.235     -0.273     -0.189
UnemployRate   1.349      0.803      1.893
GSP            0.090      0.012      0.145
```

As discussed in Chapter 9, the R output furnishes the following information:

1. The slope coefficients for the entire sample (original).
2. The bootstrap bias (bootBias): the difference between the slope coefficients for the full sample and the mean slope coefficient from the bootstrapped samples.
3. The bootstrapped standard errors (bootSE).
4. The median slope coefficient from the bootstrapped samples (boot-Med). The median is frequently listed in research presentations because of its insensitivity to the effects of extreme estimates that might emerge from one of the subsamples.

The Confint function furnishes bootstrapped 95% CIs (bca: bias corrected and accelerated).[15]

As displayed in Table 14.2, the results of these various models indicate a consistent positive association between median household income or the

---

[15] For more information about these and alternative CIs, see Thomas J. DiCiccio and Bradley Efron (1996), "Bootstrap Confidence Intervals," *Statistical Science* 11(3): 189–212, and Tim Hesterberg (2011), "Bootstrap," *Wiley Interdisciplinary Reviews: Computational Statistics* 3(6): 497–526.

TABLE 14.2

Slope Coefficients and 95% Confidence Intervals, by Type of Estimator

Variable	OLS	OLS omit CA	Median regression[a]	Robust IRLS	OLS robust SEs	Bootstrap
Median household income	−0.23 {−0.27, −0.20}	−0.23 {−0.27, −0.20}	−0.23 {−0.28, −0.20}	−0.24 {−0.28, −0.19}	−0.23 {−0.27, −0.20}	−0.23 {−0.27, −0.19}
Unemployment rate	1.35 {0.91, 1.80}	1.35 {0.90, 1.80}	1.47 {1.08, 2.07}	1.39 {0.92, 1.85}	1.35 {0.82, 1.88}	1.35 {0.80, 1.89}
Gross state product	0.09 {0.02, 0.16}	0.08 {−0.02, 0.19}	0.11 {−0.01, 0.15}	0.09 {0.03, 0.17}	0.09 {0.04, 0.14}	0.09 {0.01, 0.15}

[a] Because this model involves medians rather than means, its slope coefficients are not directly comparable to the others in the table.

unemployment rate and the percent of a state's residents living in poverty. Influential observations do exist in these associations, with a couple of large residuals, but a number of estimation methods suggest that the effects of these observations would not compel us to change our general conclusions. The main difference across the models involves the significance tests for the GSP coefficients: in four models the CIs/bounds do not include zero, whereas in two they do (the LRM that omits California and the median regression model). But we should also not forget one correction method—not shown in the table—was to take the natural logarithm of GSP (see LRM14.3). Transforming GSP may offer a good solution in this situation, especially since several of the regression models are most appropriate when outliers, rather than high leverage points, are the source of problems in a model.[16]

Chapter Summary

We've only touched the surface of the abundant information about influential observations and the many robust regression techniques available in R and elsewhere.[17] The main point is that influential observations are a common occurrence in regression models. Though frequently caused by coding errors or high skewness, they are also a routine result of data collection and

---

[16] As an exercise, conduct some analyses of the model with the logged values of GSP by (a) creating a partial residual plot with the studentized deleted residuals on the *y*-axis and log(GSP) on the *x*-axis; (b) a plot of the leverage values by the studentized deleted residuals; and (c) a plot to check the normality of the residuals (e.g., normal *q*-*q* or kernel density plot). What do the plots indicate?

[17] More information is available in Jana Jurečková, Jan Picek, and Martin Schindler (2019), *Robust Statistical Methods with R, 2nd Edition*, Boca Raton, FL: CRC Press, and Peter J. Rousseeuw and Annick M. Leroy (2003), *Robust Regression and Outlier Detection*, New York: Wiley.

variable construction. Modeling issues such as heteroscedastic errors or non-linearities may also be the result of influential observations. Exploring the data carefully before estimating a model and assessing influence measures after estimating a model should, therefore, be a regular part of any project that involves linear regression. A final word of warning: rarely, if ever, is it a good idea to simply omit a data point because it meets the criteria for an outlier or high leverage point. Instead, try to figure out why it occurs and then decide on a course of action. Often, as with the example of California's GSP, the value is legitimate and a suitable corrective step is available.

Chapter Exercises

The dataset *Wages.csv* includes a set of variables from 753 two-parent families from the U.S. The variables include:

- lfp   husband is in the paid labor force (0 = no, 1 = yes)
- kids5   number of children of ages 5 or younger
- kids618   number of children of ages 6–18
- wage   wife's age in years
- wcoll   wife's college attendance (0 = did not attend, 1 = attended college)
- hcoll   husband's college attendance (0 = did not attend, 1 = attended college)
- faminc   annual family income in $1,000s
- wifewage   wife's wages per hour

After importing the dataset into R, complete the following exercises.

1. Estimate an LRM with annual family income in $1,000s as the outcome variable and the following explanatory variables: number of children ages 5 or younger, number of children ages 6–18, wife's age in years, wife's college attendance, and husband's college attendance. Interpret the slope coefficients and 95% CIs associated with wife's college attendance and husband's college attendance.

2. Assess the LRM estimated in exercise 1 for influential observations. Use the threshold levels discussed earlier in the chapter to answer the following inquiries:

   a. How many outliers does the model contain?

   b. How many high leverage points?

   c. How many influential observations overall?

    Support your answers by specifying the thresholds for these tests and by identifying the most extreme values of each by their values and row numbers.

3. Use R's plot function after estimating the LRM in exercise 1. Examine the Residuals vs. Leverage plot. What does the plot suggest about high leverage values and outliers that might be affecting the model?

4. Examine the added-variable plots for the LRM in exercise 1. What do they suggest about high leverage values and outliers?

5. As you have noticed, the model includes substantial outliers that might be affecting the results. Use two of the alternative models discussed in this chapter to try to reduce the effects of the outliers. Describe in what way these models differ from the original LRM and furnish evidence that one or both reduced the number of outliers (hint: perfection is unattainable).

6. *Challenge*: another name for median regression is *least absolute deviations* (LAD) regression. LAD is more efficient than OLS when the errors follow a skewed distribution. Find a R package that includes a lad function and estimate a LAD regression model with family income in $1,000s as the outcome variable and wife's age in years, wife's college attendance, and husband's college attendance as explanatory variables. Estimate the same model using LMS regression (hint: see R's MASS package). Compare the results of these two regression models to an LRM estimated with OLS. What similarities and differences do you find? Which model do you prefer? Assess the distribution of the residuals from the LRM. Does this affect your decision regarding which model is more suitable?

15

Multilevel Linear Regression Models

A number of scientific disciplines utilize data that are quantitative and hierarchical. Although the term quantitative is clearly understood, the term hierarchical is appreciated less often. *Hierarchical data* refer to information that is collected at multiple units of analysis. Units of analysis can include entities such as individuals, informal social groups, schools, organizations, neighborhoods, counties, and nation-states. Hierarchical data also involve individuals who are followed over time in a longitudinal design, which is described in Chapter 8. In this situation, time is nested within different individuals, so there are two levels. Some additional examples should help clarify hierarchical data structures.

First, many sociologists and education researchers collect information from students, teachers, and school administrators. This type of data collection is usually expensive and time-consuming, so researchers often collect data from samples of students in particular classrooms, the teachers who supervise the classrooms, and the school administrators to whom the teachers report. Students are thus nested within classrooms and teachers are nested within schools (see Figure 8.1 for a simple example). The researchers end up with rich data on individual characteristics from students, classroom characteristics from teachers, and school characteristics from administrators.

Second, microeconomic and labor relations scholars are often interested in the experiences of employees who work for different companies or in diverse types of industries. They collect data from employees about issues such as salary, benefits, job satisfaction, and workplace autonomy. But they also gather information about companies or industries, such as collective bargaining agreements, union membership, hiring practices, stock options, employee turnover, promotion policies, and so forth. In this type of data collection effort, employees are nested within companies or industries. How might such data be used? Researchers can investigate whether company-level variables, such as the collective bargaining agreements, and employee-level information, such as workplace autonomy, affect job satisfaction or performance.

Third, clinical psychologists frequently conduct research on the effectiveness of different therapies to treat mental health disorders such as depression, anxiety, or phobias. Data are collected on patients undergoing specific types of therapies so that these regimens may be compared. To make studies efficient, the patients are often drawn from lists of specific therapists, some of whom provide each type of therapeutic method. Consider how this is similar to the data collection effort described in the last two paragraphs: patients are nested within types of therapies, but also by specific therapists. Any one therapist may, for example,

DOI: 10.1201/9781003162230-15

TABLE 15.1

Multilevel Data Example

idcomm	id	income
25	10001	2
25	10004	4
25	10005	3
25	10008	2
25	10009	8
30	20001	2
30	20003	4
30	20005	4
30	20008	6
30	20009	2
...	...	...

Note: idcomm is the town identifier and
id is the individual identifier.

treat several of the patients participating in a study. Yet these therapists also have particular characteristics (perhaps some are more compassionate than others) that are likely to affect treatment and, ultimately, the outcome the researchers wish to study (e.g., recovery, relapse, or length of time in treatment).

To provide a concrete example of hierarchical or *multilevel data*, consider Table 15.1, which shows a small selection of data from the *MultiLevel.csv* dataset. The dataset, introduced in Chapter 8, is based on a study that collected information from samples of residents of 99 towns in the U.S. The variable idcomm references the particular towns in the dataset, whereas the id variable indicates individuals who live in these towns. The individuals are thus nested within each town. For example, the people represented by idcomm 10001 and 10004 are from town 25. In the parlance of research that utilizes multilevel data, the town level is referred to as *level-2* and the individual level is referred to as *level-1*. If we also collected data on samples of towns from several states, we could also have a *level-3*: the state level.

As we know from Chapter 8, an important assumption of regression models is that the errors of prediction are statistically independent. In other words, information left out of the model should not be shared in some systematic way by individuals who live in the same towns. Yet individuals nested in towns may influence one another and share characteristics that affect various outcomes. Using a standard LRM to determine whether an individual-level characteristic, such as educational attainment, affects an outcome, such as life satisfaction, is inappropriate because individuals who live in the same town might have all sorts of characteristics in common that are not observed in a study and thus contribute to the errors of prediction. The result of using a statistical model that does not account for the lack of independence tends to be incorrect inferences about the effects of the explanatory variables on the outcome variable. Thus, any conclusions drawn from this type of statistical exercise might be misleading.

Chapter 8 presents a simple approach to consider when data are hierarchical and the analyst does not care about variability at the group level or whether level-2 variables (e.g., town-level economic conditions) affect outcomes. The approach first identifies the variable that indicates the nesting structure, such as the idcomm variable in the *MultiLevel* dataset. This variable is then used to notify the software that the standard errors in a regression model require modification. Recall that, in R, the level-2 or group-level identifier is used along with the svyglm function from the survey package to adjust the standard errors for nonindependence (see, for example, LRM8.2 in Chapter 8).

But many studies are designed to examine not only characteristics of the level-1 observations, or the effects of individual level variables on one another, but also how these characteristics and effects might vary across groups defined by towns, schools, organizations, or other aggregates. Many researchers are also interested in the effects of group-level variables, such as town-level economic conditions, on individual-level outcomes, such as life satisfaction. Fortunately, multilevel regression models are available that allow researchers to capitalize on hierarchical data structures.

The Basics of Multilevel Regression Models

Multilevel models—also known as *hierarchical models* and *mixed-effects models*—stem from developments in agricultural statistics, educational statistics, biometry, and econometrics. Long ago, agricultural researchers studying crop growth developed methods for analyzing plants in multiple shared environments or plots of land. Yet if one wished to extrapolate findings to the broader "population" of plants, one needed to be able to consider that the relationships found in an analysis might be "random." Models that view these relationships as different, or random, rather than fixed across environments or plots, are called *random-effects models*. Not only do the characteristics of a plant vary (since there are many plants to examine), but the relationships between these characteristics (e.g., how much water the plant receives) and some outcome of interest (e.g., plant growth) vary across environments or plots of land.[1]

---

[1] The historical record is not clear about when these models were introduced, but the notion that there's random variation among individual units within and across aggregate units—and that it can be modeled—may date back to the mid-1800s. Some attribute this idea to the British astronomer George Biddell Airy as he assessed day-to-day fluctuations in the measurement of planetary data in the 1860s. Others suggest it was developed by Ronald A. Fisher in the 1910–1920s as he studied family variation in the heights of offspring, agricultural statistics, and ANOVA models (see Marc Nerlove (2014), "Individual Heterogeneity and State Dependence: From George Biddell Airy to James Joseph Heckman," *Œconomia* 4(3): 281–320).

Similar to ANOVA models, multilevel models are designed to separate the variability of the outcome variable into two components: the variability that is due to differences across individual (level-1) units (e.g., plants or people) and the variability that is due to differences across groups (level-2) (e.g., plots of land, schools, or towns).

Let's examine the *MultiLevel* dataset, which includes individual respondents nested within towns. We'll focus on the individual level variable income, which measures household income in the previous year. The variable is measured in categories, but we'll treat it as continuous. Its mean is 4.36, with a standard deviation of 2.11, and a variance of 4.45. Think about the towns the respondents reside in, however. Each town may have a different mean, reflecting higher or lower average incomes among its sample members. Whereas overall variability among the individuals in the dataset occurs, variability across towns also exists. Another way of stating this is that, although the overall fixed mean of income is 4.36, the 99 town-level means are likely to differ and thus contribute to the variability of income. The term *random-effects* reflects this variation.

Examine the results in the following R output that lists the means of the income variable for first few towns in the dataset. Town 25 has a mean of 3.74 and town 30 has a mean of 3.88, but town 35 has a mean of about 4.75. At least upon first glance, it appears the variation in means that occurs across towns should be addressed in some manner.

R code

```
aggregate(income ~ idcomm, MultiLevel[1:949, ], mean)
  # the aggregate function is useful for obtaining
    statistics for level-2 units; the data are
    subsetted so the means for the first 10 towns
    (rows 1-949) are shown
```

R output (abbreviated)

	idcomm	(mean) income
1	25	3.740
2	30	3.878
3	35	4.747
4	45	4.340
5	50	4.375
6	55	4.359
7	80	3.412
8	105	5.294
9	135	4.107
10	225	5.541

Identifying the variability across individuals and towns is analogous to conducting a one-way analysis of variance (ANOVA) model. Just as we wish to compare variances within units and across units in a one-way ANOVA, for the income variable we may compute and compare the variance that is due to individual-level variation to that due to town-level variation. We may also set up a regression model that captures this concern using the multilevel Equation (15.1).

$$y_{ij} = \alpha_j + \hat{\varepsilon}_{ij} \tag{15.1}$$

You may recognize Equation 15.1 as representing an *intercept-only model*, with the y subscripted to indicate the multiple observations (i). Recall that in an intercept-only model, α represents the mean of y. But, unlike the conventional LRM equation, the multilevel equation acknowledges the multiple groups; hence, both y and α include the subscript j. In other words, multilevel data include means for each of the j groups. In a multilevel regression model, we consider three statistics: the overall mean (α_0, or *grand mean*, as it is often called), the between-group variance (σ^2_w), and the within-group variance (σ^2_e).[2] In R, one way to estimate these quantities is with the lmer (*linear mixed-effects regression*) function of the lme4 package. The function requires two variables: an outcome and a level-2 identifier that indicates how the level-1 units (individuals in the *MultiLevel* dataset) are grouped. The variable idcomm serves as the level-2 identifier in the *MultiLevel* dataset since it groups the individuals into towns.

LRM15.1 produces a one-way ANOVA—or intercept only model—for the variable income. The 1|idcomm portion in the R code separates the estimated income variance into a fixed-effects component and a random-effects component (this is why some researchers use the term *mixed-effects* model). The random effects are allowed to vary across towns that are designated by the idcomm variable. Since the random effects in the model address the intercept—mean income—this is also called a *random-intercept model*.

R code
```
library(lme4)
LRM15.1 <- lmer(income ~ (1|idcomm), data = MultiLevel)
summary(LRM15.1)
confint(LRM15.1)
```

R output (abbreviated)
```
Linear mixed model fit by REML ['lmerMod']
Formula: income ~ (1 | idcomm)
   Data: MultiLevel
Random effects:
 Groups   Name          Variance    Std. Dev.
 idcomm   (Intercept)    0.1719      0.4146
 Residual                4.2792      2.0686
Number of obs: 9859, groups: idcomm, 99
```

---

[2] You may also find these referred to as *between-group* and *between-individual* variability. The terms used depend on the type of multilevel data, however.

```
Fixed effects:
             Estimate Std. Error t value
(Intercept) 4.36197     0.04663    93.55

[CIs]           2.5%      97.5%
.sig01         0.346      0.492
.sigma         2.040      2.098
(Intercept)    4.270      4.454
```

The first line of the output indicates the model is not estimated by OLS, but rather by REML (*restricted maximum likelihood*), which is based on a different estimation algorithm. The output also indicates 9,859 individual observations are nested within 99 groups (towns) identified by idcomm.

The Random effects panel of the output furnishes the estimated variability of income between towns (the row with idcomm (intercept), also represented as σ^2_u, and within towns (between individuals—the row with Residual), also represented as σ^2_e. The Fixed effects panel shows the overall average or grand mean of income ($\bar{y} = 4.3$), its standard error, and *t*-value.[3]

We are initially interested in the two variance measures in the random-effects panel because they give some indication of whether more variability in income occurs between towns or between individuals. In this model, the variance between individuals ($\sigma^2_e = 4.28$) is substantially larger than the variance between towns ($\sigma^2_u = 0.17$), which is often the case in community studies since variability in attitudes, behaviors, and so forth is typically larger across individuals than across groups. One way to characterize the relative degree of variability is with a ratio measure of the group-level variance to the total variance. For LRM15.1, the computation is provided in Equation 15.2.

$$ICC(\rho) = \frac{\sigma^2_u}{\sigma^2_u + \sigma^2_\varepsilon} = \frac{0.17}{0.17 + 4.28} = 0.038 \tag{15.2}$$

Called the *intraclass correlation*, or ICC,[4] it measures the proportion of variability in the outcome that is between groups. Thus, about 3.8% of the variability in income is between towns; most of the variability occurs within towns (or between individuals). The ICC is also called the *cluster effect*.

The ICC is useful because if the nesting or clustering of observations has little to no effect on results of a regression model, then researchers may simply use a model that does not account for it (e.g., an LRM estimated with OLS). If the ICC

---

[3] The 95% CIs include those for the standard deviations of the two random effects, .sig01 and .sigma (Intercept and Residual), and the single fixed effect (Intercept). The latter is simply the 95% CI for the mean of income {4.27, 4.45}. Because of the way the multilevel model is estimated, however, this CI is not the same as the CI for the mean of income that does not take into account the multilevel nature of the data (95% CI: {4.32, 4.40}).

[4] The Greek letter *rho* is often used to symbolize the ICC. The R package performance has an icc function that computes this quantity following estimation of a multilevel model (e.g., icc(LRM15.1)).

is a substantial number, however, then one should take into account the nesting structure since the observations—and errors of prediction—are probably not independent. No widely accepted threshold for the size of the ICC currently exists, but some experts recommend that the ICC be used to compute the *design effect* (d^2). Chapter 8 defined d^2 as the ratio of the sampling variance for a statistic computed using the multilevel design divided by the sampling variance of the statistic assuming a simple random sample. A design effect of two or more is usually indicative of clustering substantial enough to warrant a multilevel model (or some other statistical adjustment). If an analyst is interested in the effects of the level-2 explanatory variables on the outcome, though, then a multilevel model may be useful even if d^2 is less than two.

One way to compute the design effect is to utilize the ICC and the average group size, as shown in Equations 15.3 and 15.4.

$$d^2 = 1 + \left(\text{ave. group size} - 1\right) \times \text{ICC} \tag{15.3}$$

$$\text{LRM15.1}: 1 + \left(99.6 - 1\right) \times 0.038 = 4.75 \tag{15.4}$$

The 99.6 portion of Equation 15.4 is the average number of individuals per town (9,859/99). An interesting characteristic is that, all else being equal, as the average group size increases, the design effect becomes larger. Since the design effect in LRM15.1 is substantially greater than two, it is best to rely on a multilevel model to assess the outcome variable `income`.[5]

An issue often raised with multilevel models is the appropriate number of level-1 observations per level-2 units, or the average subsample sizes of the level-2 units. Suppose, for instance, that a dataset includes only about four or five individuals per school or town. Can we still use a multilevel model? The answer to this question depends on several issues, such as how many level-2 units are available, the ICC, the number of explanatory variables in the model and their interrelationships, and whether the analyst is interested chiefly in individual- or group-level associations. Several studies indicate that having a sufficient number of level-2 units is especially important.[6] Power analy-

---

[5] A simple and precise way to obtain the design effect for a particular variable is with the `deff` function of the Hmisc package (e.g., `with(MultiLevel, deff(income, idcomm))`).

[6] But as one expert notes, "Rules of thumb such as only doing multilevel modelling with 15 or 30 or 50 level-2 units can be found and are often personal opinions based on personal experience and varying reasons" (William J. Browne, "Sample Sizes for Multilevel Models," retrieved from www.bristol.ac.uk/cmm/learning/multilevel-models/samples.html). Research by Cora J. M. Maas and Joop J. Hox (2005) indicates that even with level-1 samples averaging only five observations, multilevel LRMs yield little bias in slope coefficients and standard errors (though the number of level-2 units in their research did not fall below 30). The main problem exists for level-2 standard errors, where considerable bias occurs when level-2 sample sizes are, on average, 50 or less. ("Sufficient Sample Sizes for Multilevel Modeling," *Methodology* 1(3): 86–92).

sis procedures for multilevel models also provide guidance regarding the proper balance of level-1 and level-2 units.[7]

The Multilevel LRM

Let's consider whether income is associated with another variable in the *MultiLevel* dataset. Reminiscent of a model from Chapter 8, we'll assess whether males report higher incomes than females across the 99 towns using the binary variable male. We already know from LRM15.1 that mean income varies across towns, but our concern shifts to whether the slope coefficients indicating the association between male and income also vary across the towns. This concern motivates a *random-slopes* model.

One way to visualize the associations is with a graph. The next set of R code produces a way to envisage the linear associations for four towns that are identified by the following idcomms: 700, 1,500, 2,075, and 4480. The code first extracts the rows of the data frame that include information from these towns. It then places four linear fit lines in a graph (see Figure 15.1) that represent the associations between the male and income variables.

R code for Figure 15.1[8]

```
new.multi <- MultiLevel[c(1939:2025, 3890:3997,
                          5493:5588, 9153:9252), ]
  # extracts the rows corresponding to four towns
library(ggplot2) # offers a convenient way to plot the
                 slopes
library(lme4)
new.multi$male <- factor(new.multi$male) # change male
                                         into a factor
                                         variable
multislopes <- lmer(income ~ male + (male | idcomm),
                    data=new.multi)
  # estimate a multilevel regression model
coef(multislopes)$idcomm
  # save the intercepts and slopes for each of the four
    towns from the multilevel regression model
intercepts <- coef(multislopes)$idcomm[, 1] # extracts the
                                              first column
```

---

[7] See, for example, the free software available at http://www.bristol.ac.uk/cmm/software/ mlpowsim. See also Joop J. Hox, Mirjam Moerbeek, and Rens van de Schoot (2017), *Multilevel Analysis: Techniques and Applications*, 3rd Ed., New York: Routledge, chapter 12.

[8] This code is not an efficient way to build the graph, but it has the advantage of showing each step in the process.

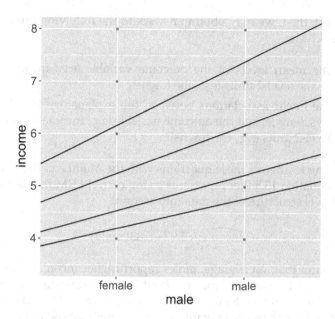

FIGURE 15.1
Plot of linear fit lines for four towns among females and males.

```
slopes <- coef(multislopes)$idcomm[, 2]
  # extracts the second column
ggplot(new.multi, aes(x=male, y=income, color = as.fact
      or(idcomm))) + geom_point(shape=20) + geom_
      abline(slope=slopes, intercept=intercepts) +
      ylim(3.5,8)
  # creates a graph with the four linear fit lines based
    on the slopes and intercepts
```

Figure 15.1 shows how different towns can have different associations between the two variables. The top fit line appears to be slightly steeper than the others. It also has a higher income on the left side of the graph (about 5.5). What does this represent? Females in this town have higher average incomes than females in the other towns.[9] The magnitudes of the slopes suggest that the difference in average income between males and females is slightly larger or smaller depending on the town.[10]

---

[9] The points where the lines meet the left side of the graph are not, however, the actual intercepts for females. Consider that the predicted values for females and males are represented where the slopes cross the vertical lines labeled `female` and `male`. For instance, in the town represented by the uppermost fit line, predicted average income among females is about six. A nuance of the way the graph is constructed is to extend the frame beyond the data.

[10] Figure 15.1 is a poor example of a graph since `male` is an indicator variable that takes on only two values. For instance, the slope at `male = 0.5` is meaningless.

In general, then, we may obtain answers to the following questions with multilevel LRMs:

1. Do the mean levels of the outcome variable (e.g., income) differ across the level-2 units (e.g., the towns)?
2. Do the linear associations between the explanatory variable (e.g., male vs. female) and the outcome variable (e.g., income) differ across the level-2 units (e.g., the towns)?

Before trying to answer these questions with the *MultiLevel* dataset, though, consider Equation 15.5 that represents a single-level LRM that does not take the multilevel structure of the data into account.

$$\widehat{income} = 3.99 + 0.83(male) \tag{15.5}$$

We may claim that, on average, males report higher incomes than females. Females report average incomes of about 3.99 units, whereas males report average incomes about 0.83 units higher ($\bar{x}_{males} = 4.82$). But, again, this is an overall effect that does not take into account differences across the towns.

Let's use R's lmer function to illustrate an LRM that simply takes into account the clustering of individuals within towns to estimate the variability within and across groups. As noted earlier, this is commonly called a random-intercept model since the intercept is allowed to vary randomly across the groups. But the model now includes male as an explanatory variable (see LRM15.2).

R code
```
LRM15.2 <- lmer(income ~ male + (1 | idcomm), data =
                MultiLevel)
summary(LRM15.2)
confint(LRM15.2))
```

R output (abbreviated)
```
Random effects:
 Groups   Name          Variance Std.Dev.
 idcomm   (Intercept) 0.1617    0.4021
 Residual              4.1179    2.0293
Number of obs: 9859, groups: idcomm, 99

Fixed effects:
            Estimate Std. Error t value
(Intercept) 3.99876   0.04892    81.74
malemale    0.81477   0.04131    19.72

Correlation of Fixed Effects:
          (Intr)
malemale -0.376
```

```
[CIs]            2.5%    97.5%
.sig01          0.335    0.478
.sigma          2.001    2.058
(Intercept)     3.903    4.095
malemale        0.734    0.896
```

Consider first that the average intercept—or predicted income for females—is about 4.0. Males, on average, report about 0.8 units more income than do females. As shown by the variance of the intercept in the Random effects panel of the output, however, the intercept—average income among females—varies across the towns ($\sigma^2_u = 0.16$). The ICC (*rho*) computed from the multilevel model in LRM15.2 is displayed in Equation 15.6.

$$\text{ICC}(\rho) = \frac{0.16}{(0.16 + 4.12)} = 0.037 \tag{15.6}$$

The model estimates only whether variability exists in the intercepts across the towns, however. The slope coefficient is a *fixed effect*, which means that, in this model, it does not vary across the level-2 units. Another way to interpret this is that the average income difference between females and males is presumed to be constant across the towns. A more interesting model, though, examines whether the slopes also vary randomly. Thus, LRM15.3 estimates a random-slope model or a *random-coefficients model* since both the intercept and slope coefficient are presumed to vary across towns.

R code
```
LRM15.3 <- lmer(income ~ male + (1 + male | idcomm),
                data = MultiLevel)
summary(LRM15.3)
confint(LRM15.3)
```

R output (abbreviated)
```
Random effects:
 Groups     Name         Variance Std.Dev.  Corr
 idcomm   (Intercept)    0.16107   0.4013
          malemale       0.01849   0.1360  -0.06

 Residual                4.11334   2.0281
Number of obs: 9859, groups: idcomm, 99

Fixed effects:
             Estimate Std. Error t value
(Intercept)  3.99863   0.04885    81.85
malemale     0.81432   0.04351    18.71
```

```
Correlation of Fixed Effects:
          (Intr)
malemale -0.374

[CIs]          2.5%   97.5%
.sig01         0.322  0.489
.sig02        -1.000  1.000
.sig03         0.000  0.281
.sigma         1.200  2.057
(Intercept)    3.903  4.095
malemale       0.729  0.900
```

```
# the 95% CIs for the intercept and male are for the
  fixed-effects only; the .sig's and .sigma are for the
  random-effects
```

The male | idcomm portion of the R code allows the male coefficient to vary across the towns. The results suggest that the male slope coefficient ($\beta = 0.81$) is variable ($\sigma^2_u = 0.02$, the malemale variance in the Random effects panel). A helpful way of representing this variability is by computing a 95% confidence band for the slope coefficient using the standard formula, as shown in Equation 15.7.

$$95\% \text{ confidence band} = 0.814 \pm (1.96 \times 0.136) = \{0.55, 1.08\} \qquad (15.7)$$

The first term is the fixed-effect slope coefficient for male and the standard deviation term is taken from the random-effects panel of the output (malemale, $\sigma_u = 0.136$). The result of Equation 15.7 suggests that about 95% of town-specific slope coefficients are expected to fall in interval that ranges from 0.55 to 1.08. In some towns the association between male and income—the average income difference between males and females—is twice as large as in other towns. But the slope coefficient is still positive across almost all of the towns represented in the dataset.[11]

Multilevel models can be made more complex by including several random-slope coefficients, though some models are difficult to estimate when the number of level-1 or level-2 observations is small (see fn. 6). The best plan is to use a conceptual model or theory to guide which slope coefficients are estimated as random and which are estimated as fixed across the level-2 units.

---

[11] Don't confuse the 95% CIs in the last part of the output with the expected range of the slope coefficients across the towns. The former estimates are interpreted in the same way as 95% CIs in a single-level LRM and are usually used in an inferential sense to make claims about the target population. For instance, the confidence interval for malemale implies that we are 95% confident that the population slope coefficient falls in the interval {0.73, 0.90}.

Examining Assumptions of the Model

Multilevel LRMs share the same assumptions as other LRMs, though they are designed to adjust for nonindependent errors of prediction. Since the model now has two levels, however, we should test the assumptions at each. To test the normality and the homoscedasticity assumptions at the individual level (level-1), for example, we may utilize some familiar diagnostic tools:

R code for Figures 15.2 and 15.3
```
plot(density(rstudent(LRM15.3)), main="")
plot(LRM15.3) # plots the Pearson residuals by the
                predicted values
```

Figure 15.2 provides a kernel density plot of the studentized deleted residuals and displays a bit of bumpiness and some degree of non-normality, but much of this is likely due to the categorical nature of the income variable. It might be a good idea to assess the distribution with a *q-q* plot to see if any other peculiarities emerge.

Figure 15.3 furnishes a residuals-by-predicted values plot useful for assessing homoscedasticity. The pattern does not show the randomness we might wish to see, but there is also no clear pattern indicative of heteroscedasticity. Again, the configuration of observations appears at least partly due to the measurement scale of the income variable.

To assess linearity, we may build partial residual plots in the customary manner (e.g., plot(rstudent(LRM15.3), *explanatory variable*)),

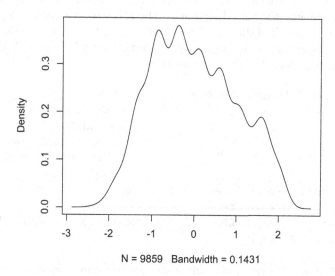

N = 9859 Bandwidth = 0.1431

FIGURE 15.2
Kernel density plot of studentized residuals from LRM15.3.

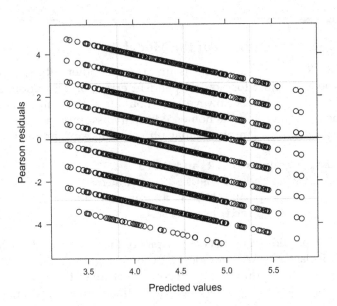

FIGURE 15.3
Plot of Pearson residuals-by-predicted values from LRM15.3.

although this wouldn't be appropriate given that, in LRM15.3, male is the only explanatory variable in the model and it consists of only two categories.

We may also examine information from level-2 to determine whether the model meets certain assumptions. But the model now includes two random effects or sources of variation at the group level: a random intercept and a random-slope coefficient. We should create plots for the residuals from each to determine how they are distributed. The R function ranef computes one type of residual called the *conditional modes*.[12] The following R code provides a sample of the conditional modes, followed by kernel density plots of them for the intercepts and slope coefficients.

R code for Figures 15.4 and 15.5
```
head(ranef(LRM15.3)[["idcomm"]]) # examine the
                                   conditional modes for
                                   the first few
                                   intercepts and slope
                                   coefficients
plot(density(ranef(LRM15.3)$idcomm[, 1] ), main="",
     xlab="Intercepts")
plot(density(ranef(LRM15.3)$idcomm[, 2] ), main="",
     xlab="Slopes")
```

---

[12] The conditional modes represent the difference between the average predicted values given the fixed effects and the predicted values of the particular individuals represented in the dataset.

```
R output (abbreviated)
   (Intercept)      male
25    -0.419       0.023
30    -0.378      -0.012
35     0.319      -0.049
45    -0.011       0.055
50    -0.039       0.002
55     0.016      -0.032
```

The conditional modes of the intercepts appear close to being normally distributed (Figure 15.4). Those for the slope coefficients, though, have a long right tail (Figure 15.5). A normal q-q plot suggests an extreme value at the upper end of the distribution.

Finally, we may wish to examine other diagnostic tests. For instance, looking for outliers is simple at the individual level since, as shown earlier, requesting residuals is straightforward. Some post-function code following the lmer function may be used to compute Cook's D and leverage values (the latter are available through the hatvalues function). At level-2 computing diagnostic variables is more complex since, as shown by the conditional modes, there are random effects from two sources: the intercept and the slope coefficient. A useful approach, though, is to plot the intercepts by the slopes across the level-2 units and look for any unusual patterns or extreme values that might have an undue influence on the model. Consider the following R code that creates Figure 15.6.

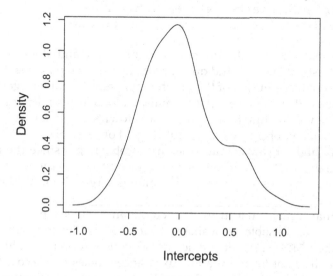

FIGURE 15.4

Kernel density plot of conditional modes (intercepts) from LRM 15.3.

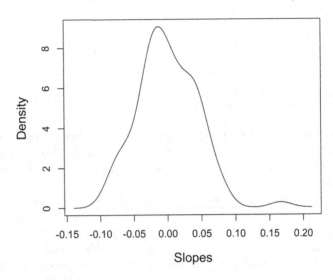

FIGURE 15.5
Kernel density plot of slopes from LRM 15.3.

```
R code for Figure 15.6
intercepts <- coef(LRM15.3)$idcomm[, 1]
 # level-2 predicted intercepts
slopes <- coef(LRM15.3)$idcomm[, 2]
 # level-2 predicted slopes
plot(slopes, intercepts, cex.lab=1.25)
abline(lm(intercepts ~ slopes), col="red")
```

Figure 15.6 suggests that most of the intercepts and slope coefficients fall in a reasonable range and don't show any unusual patterns. But notice the slope on the right side of the graph that is substantially larger than the other slopes (also identified in a normal *q-q* plot of the conditional modes). It might be worthwhile to investigate this town's features to determine if there's anything else unusual about them. For example, does it have an extremely high or low average income? Probably not since the intercept is comparable to the other towns. But it does appear as though the difference between men's and women's income is larger than in other towns. Why is this?

A similar graph that uses the conditional modes of the intercept and the male variable is available by executing the following R code: plot(ranef(LRM15.3))). It provides analogous information, although the scales are different for the *x*- and *y*-axes since it takes into account the random-effects information from the model. Consistent with Figure 15.6, one male conditional mode is substantially larger than the others.

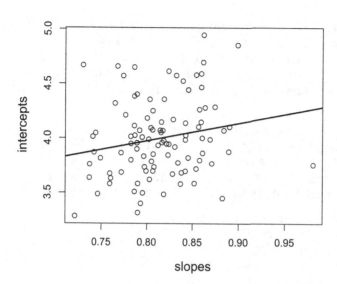

FIGURE 15.6
Scatter plot of intercepts by slopes from LRM 15.3.

Group-Level Variables and Cross-Level Interactions

A key advantage of multilevel models is that they permit not only individual-level variables, such as sex and marital status, but also group-level explanatory variables. The *MultiLevel* dataset, for example, includes variables that measure the towns' populations and levels of economic disadvantage. We might be interested in whether, say, household income is lower, on average, in larger towns or in more disadvantaged areas. LRM15.4 illustrates a model that examines these inquiries. As shown by the lmer function, we simply add group-level variables to the fixed portion of the model.

R code
```
MultiLevel$pop2020resc <- MultiLevel$pop2020 / 1000
 # the original R output issued a warning that
   recommended rescaling this variable, so it is now
   town population in 1,000s
LRM15.4 <- lmer(income ~ male + pop2020resc +
                disadvantage + (1 + male | idcomm),
                data = MultiLevel)
summary(LRM15.4)
confint(LRM15.4)
```

R output (abbreviated)
```
Random effects:
```

```
Groups   Name            Variance   Std.Dev.  Corr
idcomm   (Intercept)  0.14027      0.3745
         malemale     0.01984      0.1409   -0.20
 Residual             4.11366      2.0282
Number of obs: 9859, groups: idcomm, 99

Fixed effects:
               Estimate     Std. Error    t value
(Intercept)     3.85484      0.06134      62.844
malemale        0.81499      0.04366      18.665
pop2020resc     0.07676      0.02125       3.612
disadvantage    0.09307      0.04269       2.180

Correlation of Fixed Effects:
              (Intr)   maleml   pp2020
malemale      -0.324
pop2020resc   -0.649   0.002
disadvantag   -0.115  -0.007    0.181

[CIs]           2.5%    97.5%
.sig01         0.291    0.455
.sig02        -1.000    1.000
.sig03         0.000    0.284
.sigma         2.000    2.057
(Intercept)    3.734    3.974
malemale       0.729    0.901
pop2020resc    0.035    0.119
disadvantage   0.009    0.176
```

The Fixed effects panel indicates that towns with larger populations tend to have residents who report higher household incomes. Surprisingly, however, those areas that have more economic disadvantage tend to have residents who report, on average, higher household incomes. Although unexpected, this is the sort of association that can emerge in a multilevel analysis and is worth considering in further detail.

Researchers often wish to present some measure of the proportion of variance accounted for by a regression model. Recall, for example, the well-known R^2 statistic that is usually presented for an LRM (see Chapter 5). Analogous measures have been developed for multilevel models. Perhaps the most common is the variance accounted at the lowest level of aggregation (level-1—individuals in these models). This requires the within-group variances (σ_e^2) from the one-way ANOVA (intercept only) model (LRM15.1) and the full model (LRM15.4). The computation for an analogous R^2 measure is shown in Equation 15.8.

$$\frac{\sigma_e^2(\text{ANOVA}) - \sigma_e^2(\text{hypothesized model})}{\sigma_e^2(\text{ANOVA})} = \frac{4.28 - 4.11}{4.28} = 0.04 \quad (15.8)$$

The equation indicates that about 4% of the variance in income within the towns, and hence between individuals, is accounted for by the model that includes sex, community-level disadvantage, and population size.[13]

We may also estimate the proportion of variance accounted for between groups (towns) using information from the between-group variances (σ_u^2) in the one-way ANOVA model and the full model, which are found in the Random effects panels of the output in the idcomm (Intercept) row. Based on LRM15.1 and LRM15.4, this goodness-of-fit measure is shown in Equation 15.9.

$$\frac{0.172-0.140}{0.172} = 0.19 \tag{15.9}$$

The model accounts for about 19% of the variability in income between towns. Given that the male slope coefficient also has a variance component, we could also determine the proportion of variance explained between groups due to the random-slope component.

Another advantage of multilevel models is their ability to examine questions about contextual effects on individual-level associations. Utilizing what are termed *cross-level interactions*—which serve a similar purpose as the interaction terms discussed in Chapter 11—we may determine, for instance, whether the association between sex and income or between age and income differs depending on characteristics of the town. LRM15.5 assesses if the association between age and income depends on whether a town is high or low on the measure of economic disadvantage. The age slope coefficient is designated as random, or allowed to vary across towns, since we are assessing whether its association with income differs based on a group-level variable. By definition, the disadvantage slope coefficient is fixed since it does not vary across towns.

R code
```
MultiLevel$zage <- scale(MultiLevel$age)
 # a model with age measured in years would not converge,
    but converting age to z-scores yields a solution
LRM15.5 <- lmer(income ~ zage * disadvantage + (1 +
              zage | idcomm), data = MultiLevel)
summary(LRM15.5)
confint(LRM15.5))
```

R output (abbreviated)
```
Random effects:
```

---

[13] Given this discussion of ANOVA models, we would be remiss if we didn't mention that R's anova function is also useful for comparing nested multilevel models. For example, since LRM15.1 is nested within LRM15.4, the R function anova(LRM15.1, LRM15.4) provides a test of whether the full model fits the data better than a nested model. Rather than a partial (nested) *F*-test (which is appropriate for LRMs estimated with OLS), R provides a *likelihood ratio chi-square test*, which serves the same purpose. AICs and BICs are also useful for comparing the fit of full and nested multilevel models.

```
Groups      Name        Variance  Std.Dev. Corr
idcomm   (Intercept) 0.109838  0.33142
            zage        0.006541  0.08088 -0.80
 Residual              3.782047  1.94475
Number of obs: 9859, groups: idcomm, 99
```

Fixed effects:

	Estimate	Std. Error	t value
(Intercept)	4.35760	0.03869	112.637
zage	-0.72099	0.02139	-33.700
disadvantage	0.06580	0.03851	1.709
zage:disadvantage	-0.05016	0.02181	-2.299

Correlation of Fixed Effects:

	(Intr)	zage	dsdvnt
zage	-0.264		
disadvantag	0.000	0.004	
zag:dsdvntg	0.004	0.019	-0.277

[CIs]	2.5%	97.5%
.sig01	0.270	0.396
.sig02	-1.000	-0.204
.sig03	0.019	0.145
.sigma	1.918	1.973
(Intercept)	4.282	4.433
zage	-0.764	-0.679
disadvantage	-0.010	0.141
zage:disadvantage	-0.093	-0.007

The results suggest that age is negatively associated with income (perhaps due to the older average age of this sample, $\bar{x} = 57$), but this negative association depends partly on the town's level of economic disadvantage (β(zage × disadvantage) = −0.05; 95% CI: {−0.09, −0.01}).

An easier way to understand the association between age, economic disadvantage, and income than by decomposing the coefficients is to graph the predicted association between age and income by disadvantage. The following graphing exercise provides one approach by taking the subsample of towns with economic disadvantage scores in the lower 25th percentile (< −0.57) and comparing them to those with scores in the upper 25th percentile (> 0.57). The result is Figure 15.7. The difference in the age–income slopes is modest, with a slightly steeper association in high disadvantage towns.[14]

---

[14] This exercise offers a good example of the difference between relying on statistical significance and practical significance or effect sizes to judge the importance of a model's results. The interaction term is below the common threshold of $p < 0.05$ ($t = -2.3$, $p \cong 0.022$) with a 95% CI that does not encompass zero. Yet, the story of Figure 15.7 is that the linear association between age and income differs little by economic disadvantage.

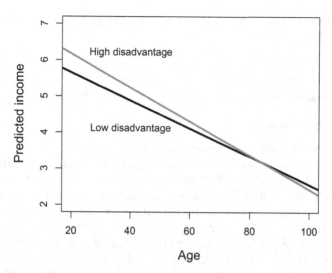

FIGURE 15.7
The association between age, economic disadvantage, and predicted income from LRM15.5.

R code for Figure 15.7

```
MultiLevel$predicted <- fitted(LRM15.5)
 # request fitted (predicted) values from model 14.5
plot(1, type="n", xlab="Age", ylab="Predicted income",
    xlim=c(20, 100), ylim=c(2, 7), cex.lab=1.25)
    # creates a graph frame on which to overlay the
        linear fit lines
abline(lm(predicted ~ age, data=subset(MultiLevel,
        disadvantage < -0.57)), col="blue", lwd=3)
        # low disadvantage
abline(lm(predicted ~ age, data=subset(MultiLevel,
        disadvantage > 0.57)), col="red", lwd=3)
        # high disadvantage
text(40, 5.8, label="High disadvantage", col="red")
text(40, 4.45, label="Low disadvantage", col="blue")
```

Finally, multilevel models are useful for estimating regression models with longitudinal data. Recall in Chapter 8 we estimated GEE models with a dataset that included seven years of information from a sample of youth (*Esteem.csv*). An advantage of the GEE model is that it allows analysts to examine different correlation structures among the estimated errors, rather than assuming, as in an LRM, that the estimated errors over time are independent. Although R's lmer function may also be used to estimate longitudinal models, the gee function offers more flexibility.[15]

Chapter Summary

Multilevel models offer substantial flexibility for testing interesting and relevant social science models and hypotheses. Since social, behavioral, education, health, labor force, and other researchers are often interested in what goes on not only among individuals but also among groups and organizations, they regularly collect multilevel data. These data and the models they motivate must be handled appropriately, however, given the high likelihood that the errors of prediction are not independent and sources of variation occur at multiple levels. Fortunately, R and other statistical software offer straightforward tools for estimating regression models with multilevel data. The chapter describes just a few of these models, but numerous resources are available for those who wish to learn more.[16]

Chapter Exercises

The dataset called *Neighborhoods.csv* includes information from 2,310 people residing in 524 neighborhoods in a large European city. Our objective is to examine predictors of family wealth using a multilevel model. The variables in the dataset include

- Neighed   Neighborhood identification number
- male   female or male
- dadeduc   Father graduated from college? (no or yes)
- momeduc   Mother graduated from college? (no or yes)

---

[15] If you wish to practice using the lmer function with longitudinal data, consider the following model that uses the *Esteem.csv* dataset: LRM15.6 <- lmer(esteem ~ male + (1 + male | newid), data = Esteem). To make it more interesting, introduce age and age² to the model (after converting age to z-scores to minimize the risk of collinearity).

[16] Two exceptional books that explain how to utilize R for multilevel modeling are W. Holmes Finch, Jocelyn E. Bolin, and Ken Kelley (2016), *Multilevel Modeling Using R*, Boca Raton, FL: CRC Press, and Andrew Gelman and Jennifer Hill (2007), *Data Analysis Using Regression and Multilevel/Hierarchical Models*, New York: Cambridge University Press.

- socialcapital   Family social capital score based on books in the home, cultural and educational events attended, and so forth (measured in z-scores)
- familywealth   Family income in previous year, properties owned, savings, and so forth (measured in z-scores)
- deprive   Neighborhood deprivation scale (measured in z-scores—higher values suggest a more economically deprived neighborhood) (level-2 variable)

After importing the dataset into R, complete the following exercises.

1. Estimate an intercept-only model with familywealth as the outcome variable.

   a. Interpret the intercept.

   b. Compute the ICC and the design effect from this model. What do these indicate about the necessity of a multilevel model?

2. Estimate a multilevel model with familywealth as the outcome variable and momeduc as the explanatory variable. Interpret the fixed-effect slope coefficient associated with the explanatory variable.

3. Estimate a multilevel model with familywealth as the outcome variable and the following two explanatory variables: momeduc and dadeduc. Request that the model estimate a random-slope coefficient for momeduc.

   a. Interpret the fixed-effect slope coefficient for dadeduc.

   b. Compute and interpret the 95% CI that gauges the estimated variability of the momeduc slope coefficient across the neighborhoods.

4. Assess whether the multilevel model in exercise 3 satisfies the assumptions of normality and homoscedastic errors at level-1. What does your assessment suggest about these assumptions?

5. Based on the multilevel model estimated in exercise 3, list the estimated slope coefficients for momeduc in the neighborhoods with the following identification numbers: 26, 29, and 31.

6. Plot the random intercepts (y-axis) by the random-slope coefficients (x-axis) based on the multilevel model in exercise 3. Include a linear fit line in the plot. What does the plot suggest about the association between the intercepts and slope coefficients across the neighborhoods? Try to describe the association in practical terms.

7. *Challenge*: recall the dataset *Esteem.csv* that was used in Chapter 8 to illustrate longitudinal data. Use these data to estimate a multilevel model of the association between self-esteem (the outcome variable), gender, and age (include a nonlinear association with age). Discuss what it suggests about the association between age and self-esteem. (Hint: plotting the predicted association between age and self-esteem with the `predictorEffect` function of the `effects` package will depict their relationship nicely. See the R code that produces Figure 4.1 in Chapter 4 for guidance.)

16

A Brief Introduction to Logistic Regression

Previous chapters emphasized outcome variables that are continuous, yet many studies in social and behavioral sciences examine other types of variables. A frequent situation, for example, is to assess outcome variables that are measured as categories, such as low, medium, and high, or yes or no. Another is to assess outcome variables that count something, such as "how many times did you see a physician in the past year?" The topic of categorical and count outcome variables requires a full book in order to examine the many modeling possibilities.[1] Rather than attempting even a modest overview, this chapter introduces and briefly discusses perhaps the most popular of the categorical variable regression methods: *logistic regression*. To be precise, we examine *binary logistic regression* because of its concern with binary outcome variables.[2]

Social, behavioral, and health sciences are replete with examples of binary variables (also called *dichotomous* variables). Chapter 7 discusses indicator variables, for example, though these are used as explanatory variables. Yet, research using binary indicators as outcome variables is common across the scientific community. Social and behavioral researchers examine issues such as support for the death penalty, graduation from high school, whether companies test for illegal drugs, if people vote in presidential elections, or if adolescents drink alcohol. All of these variables lend themselves, some better than others, to yes or no responses.

Regression models with binary outcome variables, rarely, if ever, satisfy the key assumptions of multiple LRMs. In Chapters 4 and 11 we learned that the normality assumption states that the errors of prediction should be normally distributed. The homoscedasticity assumption notes that these errors ought to have constant variance. And the linearity assumption declares that the mean value of Y for each specific combination of the Xs is presumed to be a linear function of the Xs. But imagine an LRM that uses a binary outcome variable; will it meet these assumptions? Usually not. In order to gather evidence to this effect, consider one of the variables that appears in

---

[1] See Hoffmann (2016), *op. cit.*, for an overview. For a thorough discussion of categorical data visualization and modeling in R, see Michael Friendly and David Meyer (2016), *Discrete Data Analysis with R: Visualization and Modeling Techniques for Categorical and Count Data*, Boca Raton, FL: CRC Press.

[2] A detailed history of the origins of logistic regression and similar models is furnished in J. S. Cramer (2003), *Logit Models from Economics and Other Fields*, New York: Cambridge University Press.

DOI: 10.1201/9781003162230-16

the dataset *LifeSat.csv*. The variable `satisfied` measures respondents' self-reported satisfaction with life using two categories, low and high. One might prefer a continuous measure of life satisfaction but suppose this is all that is available. Can we assess predictors life satisfaction with an LRM? Does a continuous variable such as age predict life satisfaction?

Let's first construct a scatter plot with life satisfaction on the *y*-axis and age in years (age) on the *x*-axis and overlay a linear fit line.

R code for Figure 16.1[3]

```
LifeSat$satisfied <- factor(LifeSat$satisfied)
  # change satisfied from a character to a factor
    variable
with(LifeSat, plot(age, satisfied, xlab="Age in
      years", ylab="Life satisfaction: low vs. high",
      cex.lab=1.25))
with(LifeSat, abline(lm(as.numeric(satisfied) ~ age),
      lwd=2, col="red"))))
  # change the outcome to a numeric variable in the lm
    function
```

Figure 16.1 provides the graph that R produces. If we look only at the linear fit line, age and life satisfaction clearly have a negative association. Notice, though, how peculiar the scatter plot appears. All the life satisfaction scores fall in line with the values of one and two. The pattern is not irregular,

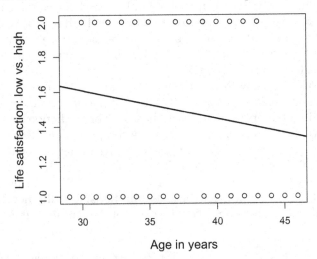

FIGURE 16.1
Scatter plot of life satisfaction (low vs. high) by age with a linear fit line.

---

[3] R's `with` function, illustrated here, provides another way of specifying the dataset in a function and is useful for those R functions that don't permit the `data = "dataset"` convention.

though, since respondents can take only these two values. But try to imagine vertical lines from the points to the fit line. Can you picture the systematic pattern to the residuals? Is this indicative of any problems we've discussed in previous chapters? Recall that we wish to have a random pattern to the residuals, yet they follow a systematic pattern, which might be indicative of heteroscedasticity or some other issue.

Let's use R to estimate an LRM with life satisfaction as the outcome variable and age in years as the explanatory variable (see LRM16.1). We'll then examine the distribution of the studentized deleted residuals from the model in a kernel density plot.

R code
```
LRM16.1 <- lm(as.numeric(satisfied) ~ age,
              data=LifeSat)
summary(LRM16.1)
confint(LRM16.1)
plot(density(rstudent(LRM16.1)), main ="")
```
R output (abbreviated)
```
Coefficients:
              Estimate Std. Error t value  Pr(>|t|)
(Intercept)   2.09417    0.40551    5.164  1.22e-06 ***
age          -0.01621    0.01072   -1.512   0.134
---
Signif. codes: 0 '***' 0.001 '**' 0.01 '*' 0.05 '.' 0.1

Residual standard error: 0.4991 on 101 degrees of freedom
Multiple R-squared: 0.02214, Adjusted R-squared: 0.01246
F-statistic: 2.287 on 1 and 101 DF, p-value: 0.1336

[CIs]         2.5%  97.5%
(Intercept)  1.290  2.899
age         -0.037  0.005
```

The output for LRM16.1 furnishes a slope coefficient of −0.016. The usual interpretation is that each one-year difference in age is associated with a −0.016 difference in life satisfaction. But does this interpretation make sense? Can life satisfaction decrease by 0.016 units? This illustrates a key problem with using an LRM to predict a binary outcome variable. Examine the kernel density plot of the residuals in Figure 16.2. Claiming the residuals follow a normal distribution constitutes a leap in interpretation no sincere researcher would be willing to make. (Yet the plot does show a bimodal distribution.) Another problem is that, even though binary outcome variables take on only two values, LRMs can produce predicted values that are outside this range. We might predict, for instance, that a person's life satisfaction score is a meaningless −0.3 or 2.75.

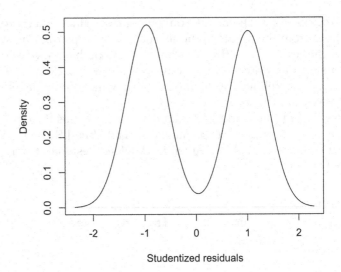

FIGURE 16.2
Kernel density plot of studentized residuals from LRM16.1.

An Alternative to the LRM: Logistic Regression

One assertion that has appeared in the statistical literature is that, as long as the binary outcome variable is distributed somewhere close to 50% in one category and 50% in the other (some claim that anything between 30% and 70% is acceptable), then the LRM—which is called a *linear probability model* when it includes a binary outcome variable—yields acceptable results. But there is little reason to rely on this model since several alternatives are available. We'll discuss the most popular alternative used in social, behavioral, and health research: binary logistic regression. As background, consider that, when interested in a non-normal outcome variable, one of the tools used is a transformation. A variable with a long right tail, for example, might be transformed to normality using the natural logarithm or square root function (see Chapter 11). Though not as apparent from a mathematical point of view, we may also transform a binary outcome variable using a specific function. Before seeing this function, let's revisit some information from elementary statistics.

Remember when you were asked in some secondary school mathematics course to flip a coin and estimate elementary probabilities? You were probably told that, given a fair coin, the probability of a heads was 0.50 and the probability of a tails was also 0.50. From a frequentist view of statistics, we thus expect that if we flip a coin numerous times about half the flips will come up as heads and the other half will come up as tails. The outcomes are usually represented as $P(T)$ or $P(H) = 0.50$, with P shorthand notation for probability (see Chapter 2). The development of logistic regression—as well as other techniques for binary outcome variables—was predicated on the notion that a binary variable could be represented as a probability. Returning to our earlier example, we might wish to know the probability that a person reports high or low life satisfaction. Using our shorthand method, this is depicted as P(high life satisfaction) = 0.49 (50/103). In practical terms, about 49% of the sample respondents report high life satisfaction.[4] A fundamental rule of probabilities is that they must add up to one. With only two possibilities, we know that the probability of low life satisfaction {P(low life satisfaction)} is 1 − 0.49 = 0.51.

Once we shift our interest from estimating the mean of the outcome variable to estimating the probability that it takes on one of the two possible choices, we may utilize a regression model designed to predict probabilities rather than means. The principal aim of a binary logistic regression model is to do this—estimate the probability that a two-category outcome variable takes on one value, such as "yes" or one, rather than another, such as "no" or zero. Understanding the mechanics of this model is simpler if we think about the outcome in a similar manner as the indicator variables introduced in Chapter 6: zero equals the comparison category and one represents the category of interest. As in the LRM, the logistic model includes explanatory variables—which may be continuous or indicators—to predict or account for the probability that a particular outcome occurs, such as high life satisfaction.

Binary logistic regression accomplishes this feat by transforming the regression equation through the use of the *logistic function*, represented in Equation 16.1.[5]

$$p(y=1) = \frac{1}{1 + \exp\left[-(\alpha + \beta_x x_1 + \cdots + \beta_k x_k)\right]} \tag{16.1}$$

The part of the denominator in the equation in parentheses is similar to the LRM we've used in earlier chapters but is transformed in a specific way. The advantage of this function is that it guarantees that the predicted values range from zero to one, just as probabilities are supposed to do. To illustrate, Table 16.1 shows several negative and positive predicted values from an LRM and their values after running them through the function. If we placed more extreme values in the function it would return numbers closer to zero or one, but, consistent with probabilities, they would never fall outside the boundary of [0, 1].

Logistic regression models do not use OLS or some similar estimation routine. Rather they employ *maximum likelihood estimation* (MLE), which is

---

[4] The simplest way to determine these probabilities in R is to request frequencies for the sat- isfied variable: table(LifeSat$satisfied). See the other uses of the table function later in the chapter.

[5] Belgian mathematician Pierre François Verhulst introduced this function in the mid-1800s to assist with early demographic studies of population growth. But it wasn't until about 100 years later that physician and biostatistician Joseph Berkson (1944) developed the "logit" model, which became logistic regression ("Application of the Logistic Function to Bio-Assay," *Journal of the American Statistical Association* 39(227): 357–365. See also J. S. Cramer (2003), *op. cit.*, chapter 9).

TABLE 16.1

Logistic Function Transformation

Initial value	Value transformed with logistic function
−10.0	0.000045
−5.0	0.006693
−0.5	0.487503
0.0	0.500000
0.5	0.622459
5.0	0.993307
10.0	0.999955

mentioned in Chapters 3 and 4. One type, called restricted MLE (RMLE), is also used by R's lmer function, as shown in Chapter 15. MLE is a common statistical method but its particulars require more statistical knowledge than we currently have at our disposal. Suffice it to say that MLE determines the most likely value of a statistic—whether the value of interest is a mean, a slope coefficient, or a standard error—given a particular set of data. Detailed information on MLE is available in books on statistical inference.[6]

A peculiar feature of many research studies is that logistic regression models are used infrequently to estimate predicted probabilities, at least directly. Rather, most researchers who use logistic regression employ *odds* and *odds ratios* (ORs) to understand the results of the model.

Odds are used regularly in games of chance.[7] Slot machines and roulette wheels, for example, are typically "handicapped" by being accompanied by the odds of winning any particular play. Horse races are also supplemented by odds: "The odds that Felix the Thoroughbred will win the Apple Downey Half-Miler are four to one." Odds are calculated by taking probabilities and placing them in Equation 16.2.

$$\frac{p}{1-p} \tag{16.2}$$

Imagine that Felix the Thoroughbred has been in ten half-mile races and won two of them. The probability that she wins a half-miler is, therefore, 0.20. This translates into an odds of winning of $0.20/(1 - 0.20) = 0.25$. Another interpretation is that the odds of winning are one to four (1:4), or for each race she wins she loses four of them.

---

[6] The contemporary form of MLE was specified by Ronald A. Fisher, but it traces its roots to mathematical work by Johan Heinrich Lambert, Daniel Bernoulli, Carl Friedrich Gauss, and Francis Y. Edgeworth (see Anders Hald (1999), "On the History of Maximum Likelihood in Relation to Inverse Probability and Least Squares," *Statistical Science* 14(2): 214–222). MLE is explained in Scott R. Eliason (1993), *Maximum Likelihood Estimation: Logic and Practice*, Newbury Park, CA: Sage.

[7] The origin of the concept of odds is obscure but dates back to at least the early 17th century. Some evidence suggests it predates probabilities (see James Franklin (2015), *The Science of Conjecture: Evidence and Probability before Pascal*, Baltimore, MD: Johns Hopkins University Press).

An example closer to the interests of the research community involves the following scenario. Suppose we conduct a survey of adolescents and find that 25% report using marijuana and 75% report not using marijuana in the past year. The probability of marijuana use {P(marijuana use)} among adolescents in the sample is 0.25. What are the odds of marijuana use?

$$\text{Odds}_{mj} = \frac{p(\text{marijuana use})}{1 - p(\text{marijuana use})} = \frac{0.25}{1 - 0.25} = 0.33 \text{ or } \frac{1}{3} \quad (16.3)$$

Equation 16.3 implies that for every adolescent who uses marijuana, we expect three adolescents who do not use marijuana or, stated differently, three times as many adolescents did not use marijuana as did use marijuana in the past year.

Restricting our attention to only one variable is seldom interesting. Let's extend the hypothetical to two variables. We'll treat adolescent past-year marijuana use as the outcome variable. To keep it simple, we'll use an indicator variable that is coded 0 = female and 1 = male. We now have four outcomes, with males and females who used or did not use marijuana. If we're interested only in probabilities, we may compare the two: P(male marijuana use) and P(female marijuana use) and see if one is higher. But keeping our interest on the odds, let's compare the odds of marijuana use among male and female adolescents.

Imagine some survey data indicate that 40% of male adolescents and 20% of female adolescents report past-year marijuana use. The odds of marijuana use among males is thus 0.40/0.60 = 0.667 or 2:3 and the odds of marijuana use among females is 0.20/0.80 = 0.25 or 1:4. Remember: these are not probabilities; odds are different. Another interpretation is that for every two males who used marijuana, three did not, and for every female who used marijuana, four did not.

An odds ratio (OR)—a key measure of association in logistic regression models—is simply the ratio of two odds. Given the odds of marijuana use in the two groups, we may now determine the OR. First, though, we should choose a focal group and a comparison group. Let's use males as the focal group since they are more likely to report use. The OR of male to female marijuana use is provided in Equation 16.4.

$$OR(mj, \text{ male } v. \text{ female}) = \frac{\text{Odds, males}}{\text{Odds, females}}$$

$$= \frac{p(\text{males})}{1 - p(\text{males})} \Big/ \frac{p(\text{females})}{1 - p(\text{females})} \quad (16.4)$$

$$= \frac{0.667}{0.25} = 2.67$$

The interpretation of the OR is that the odds of past-year marijuana use among males are 2.67 times the odds of past-year marijuana use among females. Some analysts use a shorthand phrase that males are 2.67 times *as likely* as females to use marijuana. This might mislead some people, however, since it can imply probabilities rather than odds. The ratio of the probabilities is 0.40/0.20 = 2.0—clearly not the same as the OR.[8]

The range of ORs is zero to infinity (Can you confirm this using the formula?) A number greater than one implies a "positive" association between an explanatory and outcome variable and a number between zero and one indicates a "negative" association. But keep in mind that the choice of the focal and comparison categories for the explanatory *and* outcome variables drives the interpretation. If there is no difference in the odds, then the OR is one. Suppose, for instance, that one-third of all the males *and* females report past-year marijuana use, then the OR is 1.0 (0.5/0.5).

A logistic regression model provides a way to compute predicted odds and ORs for binary outcome variables. A useful equation that lends itself to these computations is shown in Equation 16.5.

$$\ln\left[\frac{p(y=1)}{1-p(y=1)}\right] = \alpha + \beta_1 x_1 + \cdots + \beta_k x_k \tag{16.5}$$

Exponentiating both sides of 16.5 yields Equation 16.6:

$$\frac{p(y=1)}{1-p(y=1)} = \exp(\alpha + \beta_1 x_1 + \cdots + \beta_k x_k) \tag{16.6}$$

The quantity $\ln\left[\frac{p(y=1)}{1-p(y=1)}\right]$ in Equation 16.5 is also known as the *log-odds*. The term *ln* indicates the natural logarithm. Note that logistic function that was specified earlier is similar to this equation. The term *log-odds* also suggests one solution to computing the odds: take the exponential of this quantity to transform it from a log-odds to an odds (or, as we shall see, an OR).

Let's disentangle the discussion by returning to a real example of a binary outcome variable. The dataset used earlier, *LifeSat*, includes the variable female that distinguishes males and females. We'll treat female as the explanatory variable and life satisfaction as the outcome variable. Beginning with a cross-tabulation allows the computation of probabilities, odds, and the OR.

---

[8] But should this ratio of probabilities be interpreted as "male adolescents are two times as likely as females to use marijuana"? Perhaps, but no widely accepted rules are available about this approach to interpretation.

R code

```
with(LifeSat, addmargins(table(female, satisfied)))
  # a basic cross-tabulation
with(LifeSat, prop.table(table(female, satisfied), 1))
  # 1 requests row percentages that sum to 1; change it
    to 2 for column percentages
```

R output (abbreviated)

```
          satisfied
female   no yes Sum
female   48  37  85
male      5  13  18
Sum      53  50 103

         satisfied
female     no     yes
female   0.565   0.435
male     0.278   0.722
```

What are some of the relevant probabilities and odds from this output? First, the probability of high life satisfaction overall is $(37 + 13)/(48 + 5 + 37 + 13) = (50/103) = 0.485$. The odds of high life satisfaction are $(37 + 13)/(48+5) = 50/53 = 0.94$ (observe that we do not need to compute probabilities to compute odds). The overall probabilities and odds might be revealing, but they're not very informative because they don't represent differences among males and females. Let's take this into consideration: the probabilities of high life satisfaction among males and females are $13/18 = 0.72$ and $37/85 = 0.44$ (see the second part of the R output for confirmation). The respective odds are $37/48 = 0.77$ among females and $13/5 = 2.6$ among males. Males tend to report higher life satisfaction than females.

The OR provides a summary measure of the association between `female` and `satisfied`. Since we have the odds for females and males, computing the OR is simple: $OR_{\text{males vs. females}} = 2.6/0.77 = 3.4$. The interpretation of the OR is "the odds of high life satisfaction among males are expected to be 3.4 times the odds of high life satisfaction among females." Since males comprise the focal group in this exercise, we might claim a positive association between this measure of biological sex and high life satisfaction (without, of course, making any qualitative judgments about the two variables).

Computing ORs with two groups is straightforward, yet we may also wish to (1) gather evidence with which to infer something about the association in the target population and (2) add additional explanatory variables to a model. A binary logistic regression model comes in handy for both objectives. Like an LRM, this model estimates standard errors, *p*-values, 95% CIs, and other evidence, and it allows more than one explanatory variable.

R has a number of functions that estimate logistic regression models. The most general is the `glm` (*generalized linear model*) function, which estimates not only binary logistic models but several other types appropriate for categorical outcome variables. The analyst specifies the distribution of the outcome variable using the family option. To request a binary logistic regression model from the glm function, for example, the family is specified as binomial. The output's Estimate column entries are expressed in terms of log-ORs but, as shown with Logit16.1, may be exponentiated to provide ORs.

R code
```
Logit16.1 <- glm(satisfied ~ female, data=LifeSat,
family = binomial)
summary(Logit16.1)
exp(coef(Logit16.1))    # exponentiate the coefficients
                          to obtain the odds ratio
exp(confint(Logit16.1)) # exponentiate the confidence
                          intervals to obtain the 95%
                          CIs for the odds ratio
```

R output (abbreviated)
```
Coefficients:
              Estimate  Std. Error  z value  Pr(>|z|)
(Intercept)   -0.2603     0.2188     -1.190   0.2341
femalemale     1.2158     0.5699      2.133   0.0329 *
---
Signif. codes: 0 '***' 0.001 '**' 0.01 '*' 0.05 '.' 0.1

 (Dispersion parameter for binomial family taken to be 1)

 Null deviance: 142.70 on 102 degrees of freedom
 Residual deviance: 137.68 on 101 degrees of freedom
 AIC: 141.68

[exponentiated coefficient estimates]
(Intercept)  femalemale
   0.771        3.373

[CIs]        2.5%   97.5%
(Intercept)  0.499  1.181
femalemale   1.160  11.291
```

The appearance of the results is similar to those obtained from an LRM, with a few of important exceptions. First, as noted earlier, the numbers under the Estimate column of the Coefficients panel are log-ORs. Second, the model furnishes deviance values (Null and Residual) and an AIC (see Chapter 6) rather than sums of squares, R^2 values and *F*-statistics. They may be used to judge the fit of the model. Finally, the R code includes a request for exponentiated coefficients, which are the ORs (exp(coef(Logit16.1)).

Examine the OR for `femalemale`. It should look familiar because it is almost identical to the OR computed from the cross-tabulation: 3.4 (but remember this works only because `female` is the comparison group).[9] Assuming we wish to infer something about the population from which the sample was drawn, we may also use the *p*-values and 95% CIs in the same manner as in the LRM. Using a threshold value of 0.05, for instance, we may claim that the OR for `femalemale` is inconsistent with a hypothesis that the odds of life satisfaction are the same for females and males. The 95% CI for `femalemale` {1.2, 11.3} offers additional evidence. We could be more precise and formulate similar interpretations of the *p*-value (0.033) and the 95% CI as used in earlier chapters, but this is left as an exercise in rhetoric.

The interpretation of the OR is the same as it was earlier: the odds of high life satisfaction among males are expected to be 3.4 times the odds of high life satisfaction among females. We use the term "expected to be" rather than something more determinate to remind ourselves that we often wish to infer from a sample to a population. We should lack confidence, given a single sample, that the odds among males *are* 3.4 times higher. But we have a piece of evidence to infer that the odds are higher among males.

Suppose we next wish to figure out the specific odds of high life satisfaction among males and females rather than just the summary OR. The log-OR may be used to estimate predicted values. For example, the log-odds for females is [−0.26 + {1.216 × 0}] = −0.26 and the log-odds for males is [−0.26 + {1.216 × 1}] = 0.956. But recall the equation presented earlier in the chapter, reproduced here as Equation 16.7.

$$\ln\left[\frac{p(y=1)}{1-p(y=1)}\right] = \alpha + \beta_1 x_1 + \cdots + \beta_k x_k \qquad (16.7)$$

The left-hand side of Equation 16.7 is, as mentioned earlier, the logit. We may use the results of the model to estimate the logit, or more commonly, the log-odds of the predicted outcome based on one-unit differences in particular values of the explanatory variables. To transform the log-odds into odds, we simply take the exponential (inverse of the natural logarithm) of these values, which is what R did to the log-ORs when we requested `exp(coef(Logit16.1))`. The predicted odds for females are thus *exp*(−0.26) = 0.77 and for males *exp*(0.956) = 2.6. Compare these to the odds computed with the cross-tabulation shown earlier.

Some—perhaps many—researchers are discomforted with odds and ORs and prefer probabilities. The coefficients from a logistic regression model may be transformed into probabilities by utilizing the logistic function reproduced in Equation 16.8.

---

[9] Examine the OR listed for the intercept (0.77). Where have we seen this number before? Hint: it's actually an odds not an OR.

$$p(y=1) = \frac{1}{1+\exp\left[-\left(\alpha + \beta_x x_1 + \cdots + \beta_k x_k\right)\right]} \qquad (16.8)$$

We simply substitute the part of the denominator in parentheses with the logistic regression coefficients. However, we must choose specific groups to represent. With the previous model, only two groups exist, females and males, and we may predict their probabilities by including the coefficients in the equation. For females, we find

$$1 + \left[\exp\left(-\left[-0.26 + \{1.216 * 0\}\right)\right]\right]^{-1} = 0.435 \qquad (16.9)$$

Whereas, for males we compute

$$1 + \left[\exp\left(-\left[-0.26 + \{1.216 * 1\}\right)\right]\right]^{-1} = 0.722 \qquad (16.10)$$

The predicted probabilities are identical to the probabilities computed from the cross-tabulation. R also provides probabilities for each observation by requesting the fitted values after estimating the logistic regression model.

R code
```
LifeSat$probs <- Logit16.1$fitted.values
summary(LifeSat$probs)
table(LifeSat$probs))
```

We may then examine the probabilities for the different groups represented in the model. For example, you'll find only two probs values since we estimated probabilities only for females and males. These match those computed earlier: 0.485 and 0.722.

Additional options that furnish predictions, residuals, and influence statistics are available and worth exploring and understanding for those who wish to realize the capabilities and limitations of logistic regression models. Books on logistic regression provide details about these and other issues.[10]

Extending the Logistic Regression Model

We have not reached the limit of the binary logistic regression model. As with the LRM, it may be extended to include multiple explanatory variables—whether indicator or continuous—and interaction terms. Let's continue to use life satisfaction as the outcome variable and include the following explanatory variables: female, age, and IQ (intelligence quotient). We'll treat the latter two as continuous variables (see Logit16.2).

---

[10] An especially lucid guide to these models is furnished in Joseph M. Hilbe (2015), *Practical Guide to Logistic Regression*, Boca Raton, FL: CRC Press.

R code (abbreviated)

```
Logit16.2 <- glm(satisfied ~ female + age + IQ, data =
                 LifeSat, family = binomial)
summary(Logit16.2)
exp(coef(Logit16.2))
exp(confint(Logit16.2))
```

R output
  Coefficients:

	Estimate	Std. Error	z value	Pr(>|z|)	
(Intercept)	8.24667	5.39053	1.530	0.1261	
femalemale	1.22158	0.57882	2.110	0.0348	*
age	-0.09460	0.05070	-1.866	0.0621	.
IQ	-0.05383	0.04630	-1.163	0.2450	

```
---
Signif. codes: 0 '***' 0.001 '**' 0.01 '*' 0.05 '.' 0.1
```

```
 (Dispersion parameter for binomial family taken to be 1)
Null deviance: 142.70 on 102 degrees of freedom
Residual deviance: 133.82 on 99 degrees of freedom
AIC: 141.82
```

[exponentiated coefficient estimates]

(Intercept)	femalemale	age	IQ
3814.915	3.393	0.910	0.948

[CIs]	2.5%	97.5%
(Intercept)	0.133	2.574+08
femalemale	1.145	11.154
age	0.820	1.002
IQ	0.862	1.036

The model indicates that, statistically adjusting for the effects of age and intelligence scores, the odds of high life satisfaction among males are expected to be 3.4 times the odds of high life satisfaction among females. The measure is also called an *adjusted OR* since it is estimated adjusting for the presumed effects of other variables in the model. It appears, though, that considering age and IQ have little effect on the association between female and satisfied.[11]

What about interpretations for continuous explanatory variables? Consider the exponentiated coefficient for age: the OR is 0.91. Since age is measured in year increments, the interpretation is

---

[11] Because of some peculiarities with the way logistic regression models are estimated, making direct judgments about the shift in an OR from a nested to a full model is not appropriate. We should not, for instance, claim that the shift in the female variable's OR from Logit16.1 to Logit16.2 is meaningful. Making such claims requires different forms of the coefficients. For details see Kristian Bernt Karlson, Anders Holm, and Richard Breen (2012), "Comparing Regression Coefficients between Same-Sample Nested Models Using Logit and Probit: A New Method," *Sociological Methodology* 42(1): 286–313.

Statistically adjusting for female-male differences and the association with IQ, each one-year difference (or increase) in age is associated with a 0.91 times difference in the odds of high life satisfaction. More generally, older people tended to report a lower odds of high life satisfaction, after statistically adjusting for age and IQ.

The specific interpretation is not satisfactory to many because of the challenge of understanding negative associations in logistic regression models. A property of regression models that utilize natural logarithms to transform outcome variables is useful in this situation, however.[12] That is, we may use the *percentage difference or change formula* provided in Equation 16.11 to interpret coefficients from logistic regression models.

$$\{\exp(\beta)-1\}\times100 \tag{16.11}$$

The formula uses the exponentiated coefficient (OR) minus one to evaluate the difference or change in the odds associated with a one-unit difference or change in the explanatory variable. Using the age coefficient and Equation 16.11, we find $\{[\exp(-0.095)] - 1 \times 100\} = -9.1\%$. The interpretation is

Statistically adjusting for female-male differences and the association with IQ, each one-year difference (or increase) in age is associated with a 9.1% *decrease* in the odds of high life satisfaction.

We may also employ multiple logistic regression models to estimate predicted odds or probabilities. This is complicated by the presence of continuous variables, however. As with multiple LRMs, we should choose particular values of the explanatory variables to compare. Some researchers prefer to use minimum and maximum values, but any plausible values, as long as they are represented by a sufficient number of observations in the dataset, work reasonably well. Let's utilize the logistic function to compare the probabilities of high life satisfaction for males and females at the median values of age (39) and IQ (92) (see Equations 16.12 and 16.13).

$$p(\text{females}) = \frac{1}{1+\exp[-\left(8.25+\{1.22\times0\}+\{-0.095\times39\right)+\{-0.054\times92]} \tag{16.12}$$

$$= \mathbf{0.396}$$

$$p(\text{males}) = \frac{1}{1+\exp[-\left(8.25+\{1.22\times1\}+\{-0.095\times39\right)+\{-0.054\times92]} \tag{16.13}$$

$$= \mathbf{0.689}$$

---

[12] This approach may also be used when the outcome variable in an LRM has been natural log transformed to normalize its distribution.

Another way of thinking about these predicted probabilities is to infer that approximately 40% of females in the median categories of age and IQ are expected to report high life satisfaction, whereas about 69% of males in these categories are expected to report high life satisfaction.[13]

Finally, tests of fit for logistic regression models that are analogous to fit statistics for LRMs are available. The tests are not computed in the same manner, nor should they be interpreted in the same way, as F-values, RMSEs, or R^2 values. The tests are, however, similarly useful for comparing models. For example, R provides `Deviance` statistics in the logistic regression output. The residual deviance may be employed to compare nested models (but not non-nested models). Values closer to zero tend to indicate better fitting models, although additional information is required before one may reach this conclusion.

Let's compare the original model that included only one explanatory variable with the full model that included three explanatory variables. The residual deviance of the nested model is 137.68, whereas the residual deviance of the full model is 133.82. To compare these models, subtract one from the other. The result is distributed χ^2 with degrees of freedom equal to the difference in the number of explanatory variables of the two models. After taking the difference between the deviances ($137.68 - 133.82 = 3.86$), compare it to a χ^2 value with two degrees of freedom (since the full model has two additional explanatory variables). The p-value associated with this χ^2 value is 0.145; thus, difference between the two models is compatible with a hypothesis that the model fits are similar. Another interpretation is that the model's predictive capabilities do not appear to improve by including age and IQ. In the spirit of parsimony, we should conclude that the simpler model provides a better fit to the data.

A simple way to perform this test in R is with its anova function, which is employed in the same manner as in Chapter 6:

R code
```
anova(Logit16.1, Logit16.2)
```

Another way of understanding this test is that we've asked R to jointly test whether the age and IQ coefficients (log-odds) are equal to zero. The results are identical to the earlier paragraph's: a difference in deviances of 3.86. R does not automatically provide the p-value associated with this number, but allows easy computation:

R code
```
pchisq(3.86, df=2, lower.tail=FALSE)
```

---

[13] R has methods to automate this process, but we'll leave the application as an exercise.

R returns 0.145. Again, this result suggests that the restricted model fits the data at least as well as the full model. It would be unwise to reject the hypothesis that the log-odds of age and IQ are both equal to zero in the population.

Another method for comparing logistic regression models is with information criterions (ICs). As introduced in Chapter 6, the two most widely used ICs are the AIC and the BIC. R provides the AIC as part of its default summary output of the glm function. The AICs for the two models are 141.68 (nested model) and 141.82 (full model). Lower AIC scores suggest better fitting models. We therefore seem to gain little in predictive ability by adding the two extra explanatory variables: the first model with just female as an explanatory variable provides a better, more parsimonious fit to the data.

Chapter Summary

Binary outcome variables are common in the various scientific disciplines that utilize statistical modeling and provide important information about many relevant phenomena. This chapter includes only a brief introduction to one model, binary logistic regression, but we did not address its twin method: probit regression.[14] This model provides similar results in general (although the coefficients are scaled differently), but is based on the assumption that underlying a binary variable is a continuous variable that is represented along a continuum from a probability of zero to a probability of one. Either model is suitable, moreover, when an outcome variable measures a proportion, which is bounded by zero and one. For example, a researcher may wish to predict or account for the proportion of residents of towns who test positive for COVID-19 or collect unemployment insurance in the previous month.

Some analysts point out, however, that logistic regression may not be the best choice when (1) the probability of the outcome is close to 0.50 or (2) is near zero or one or (3) when a proportion is not represented well by a binomial distribution. For (1), some researchers recommend a *log-binomial*[15] or

---

[14] Introduced about 10 years before logistic regression by Chester I. Bliss (1934) ("The Method of Probits," *Science* 79(2037): 38–39). Similar to logistic regression, the glm function allows estimation of this model (e.g., glm(satisfied ~ female, family=binomial(link="probit"), data=LifeSat)).

[15] Developed by Sholom Wachholder (1986) ("Binomial Regression in GLIM: Estimating Risk Ratios and Risk Differences," *American Journal of Epidemiology* 123(1): 174–184). The log-binomial model is especially promising if you find yourself befuddled by odds and ORs and would simply prefer probabilities. Available in the R package logbin, the model furnishes coefficients as log-probability ratios, which may be exponentiated to reveal the ratio of probabilities for two or more groups.

a *robust Poisson regression model.*[16] For (2), alternatives include *exact logistic regression*[17] and the *Firth logit model.*[18] When modeling proportions, beta or beta-binomial regression is generally appropriate.[19] We'll conclude with an invitation to explore these models on your own.

Chapter Exercises

The dataset called *PATH.csv* includes information from the 2018 Population Assessment of Tobacco and Health (PATH) study. We'll use this dataset to examine some predictors of e-cigarette use, commonly known as *vaping*. The dataset includes the following variables:

- `ID`   Identification number
- `Smoker30Day`   "In the past 30 days, have you smoked a cigarette?" ("No" or "Yes")
- `SmokeEveryday`   "Do you smoke cigarettes every day?" ("No" or "Yes")
- `EcigCurrent`   "Do you currently use any electronic nicotine products, including e-cigarettes, vape pens, personal vaporizers and mods, e-cigars, e-pipes, e-hookahs or hookah pens?" ("No" or "Yes")
- `LatinX`   Does the respondent report being Latinx? Derived from questions about race, racial identity, ethnic origin, and ethnic identity ("No" or "Yes")
- `Gender`   "female" or "male"
- `LGBTQPlus`   Does the respondent report being LGBTQPlus? Derived from questions about self-identity and sexual orientation ("No" or "Yes")

---

[16] Proposed by Guangyong Zhou (2004) ("A Modified Poisson Regression Approach to Prospective Studies with Binary Data," *American Journal of Epidemiology* 159(7): 702–706). This model is estimable using the geeglm function of R's geepack package.

[17] This model was developed by British statistician Sir David R. Cox (1970) (*The Analysis of Binary Data*, New York: Chapman and Hall). Exact logistic regression is useful when the sample size is relatively small. In R, the model is approximated using the elrm package.

[18] Introduced by David Firth (1993) ("Bias Reduction of Maximum Likelihood Estimates," *Biometrika* 80(1): 27–38). The model is available in R's logistf package.

[19] The beta-binomial regression model may have appeared first in a paper by English biometrician J. G. Skellam (1948) ("A Probability Distribution Derived from the Binomial Distribution by Regarding the Probability of Success as Variable between the Sets of Trials," *Journal of the Royal Statistical Society, Series B*, 10(2): 257–261). The beta distribution was introduced by Karl Pearson more than 50 years earlier (Pearson (1894), *op. cit.*). The R package betareg and the betabin function of the aod package estimate beta regression models.

- BMI   Body mass index—derived from questions about height and weight

After importing the dataset into R, complete the following exercises.

1. Request a frequency distribution of current e-cigarette use (EcigCurrent) and use it to compute the probability and odds of current e-cigarette use.

2. Request a cross-tabulation of current e-cigarette use (EcigCurrent) and cigarette smoking in the past 30 days (Smoker30Day). Use the cross-tabulation to complete the following:

   a. Compute the probabilities of e-cigarette use among those who smoked cigarettes in the past 30 days and among those who did not smoke cigarettes in the past 30 days.

   b. Compute the OR of e-cigarette use that compares those who smoked cigarettes in the past 30 day to those who did not. Interpret this OR.

3. Estimate a binary logistic regression model with current e-cigarette use as the outcome variable and cigarette smoking in the past 30 days as the explanatory variable. Before requesting the model you'll need to create a numeric binary version of the e-cigarette variable that is coded {0, 1}. Assuming you've called the data frame PATH, the following code does this:

   ```
   PATH$EcigCurrent <-
       as.numeric(as.factor(PATH$EcigCurrent)) - 1
   ```

   Interpret the OR and its 95% CI associated with cigarette smoking in the past 30 days.

4. Extend the logistic regression model estimated in exercise 3 by including LatinX, LGBTQPlus, and BMI as explanatory variables. Interpret the ORs for LatinX and LGBTQPlus.

5. Determine whether the model in exercise 4 (full) fits the data better than model in exercise 3 (nested) using a likelihood ratio chi-square test. What does this test indicate about the relative fit of the two models? Examine the AICs and BICs for the two models. What do these suggest about the relative fit of the two models? Is the evidence from these fit measures consistent? Explain.

6. *Challenge*: two additional measures of fit for logistic regression models are McFadden's R-squared and Tjur's coefficient of discrimination. Determine how to estimate these two fit measures in R and request each one. What do they suggest about how well the model in exercise 4 predicts current e-cigarette use? (Hint: you will need to determine what each one measures about the model.)

17

Conclusions

LRMs offer a powerful tool for analyzing the association between one or more explanatory variables and a single outcome variable. Some novice researchers wish to move quickly beyond this model and learn to use what are deemed more sophisticated methods because they get discouraged about the LRM's limitations and believe that other regression models are more appropriate for their analysis needs. Situations certainly exist that demand alternative regression models (see Chapter 16), yet the linear model should not be overlooked or underutilized. Even when one or more of the assumptions described in the previous chapters are not satisfied, linear regression frequently provides accurate results and good evidence with which to judge conceptual models and hypotheses. Although calling the LRM robust is ill-advised, because robustness has a very specific statistical meaning, claiming that the model works well under a variety of circumstances is not disingenuous. LRMs provide myriad practical uses in social, behavioral, and health research. In fact, these models continue to hold a conspicuous place in quantitative fields such as data mining, machine learning, and analytics, as well as in mathematically rigorous disciplines such as chemometrics, econometrics, engineering, and physics.[1]

The key to using LRMs appropriately is to understand their strengths and weaknesses. As we've learned, for instance, the model is sensitive to nonlinear associations, which, if they exist and are not included, lead to specification bias. But, as shown in Chapter 11, including nonlinear associations in LRMs is feasible. When errors of prediction are not homoscedastic and induce inefficient coefficients, several adjustments are readily available in R and other software. Some of these adjustments, as well as others, can also help guard against the untoward effects of influential observations (see Chapters 9 and 14). Perhaps the most nettlesome problem in social and behavioral sciences is measurement error. But measurement is an issue that is challenging across most studies of social and behavioral phenomena and affects all the methods used in this research. Any problems incurred due to imprecise or incorrect measurement procedures are not unique to LRMs.

---

[1] See, for example, Ethem Alpaydin (2018), *Introduction to Machine Learning*, 3rd Ed., Cambridge, MA: MIT Press; Luca Lista (2017), *Statistical Methods for Data Analysis in Particle Physics*, New York: Springer; Matthias Otto (2016), *Chemometrics: Statistics and Computer Application in Analytical Chemistry*, Weinheim, DE: Wiley-VCH; Richard J. Roiger (2017), *Data Mining: A Tutorial-Based Primer*, 2nd Ed., Boca Raton, FL: CRC Press; and Verbeek (2017), *op. cit.*

Of course, LRMs should not be the first choice in some situations, such as when the outcome variable is binary or takes on only a few values. How many times is "a few" has been a point of contention among researchers. Some claim less than nine, others less than seven. The key issue, though, is whether the assumptions are met, not the precise number of categories. Given the availability of the many regression models designed for categorical outcome variables, however, using LRMs in this situation is rarely necessary or suitable. The binary logistic regression model, for example, provides a solid alternative when the outcome variable includes two categories. Models for three-or-more category outcomes variables, as well as other types, are described in the many books on categorical data analysis and generalized linear models.[2]

We don't have the time to devote to several other key statistical topics that involve LRMs. Given space limitations, we'll briefly discuss only two important issues: sampling weights and establishing causal associations.

Sampling Weights

In several places we discussed the role of survey sampling in data collection and some of its implications for estimating LRMs. Large surveys, whether in-person, via telephone, or web-based, that don't use some form of stratified or clustered sampling are rare.[3] When using most national datasets, considering the effects of stratification and clustering on regression estimates is therefore important. As mentioned in Chapter 8 and elsewhere, the main effect is on the standard errors, so they should be adjusted.

But surveys also utilize *sampling weights*, which can affect all the LRM coefficients. Sampling weights are employed to designate the number of people (or other units) in the population each sample member represents. Recall that we often want the sample to represent some well-understood target population. Each observation in the sample should therefore denote a group in the population. For instance, in a nationally representative sample a 40-year-old White woman may represent several thousand 40-year-old White women from some part of the U.S. Survey data typically include sample weights that may be used in analyses to reflect this representation (e.g., the 40-year-old woman has a weight of 12,340 since she represents an estimated 12,430 40-year-old women).

---

[2] See, for example, Alan Agresti (2018), *Introduction to Categorical Data Analysis*, 3rd Ed., Hoboken, NJ: Wiley; Fox (2016), *op. cit.*; Friendly and Meyer (2016), *op. cit.*; and Hoffmann (2016), *op. cit.*

[3] Web-based surveys offer other sampling challenges we won't discuss. But see Don A. Dillman et al. (2014), *Internet, Phone, Mail, and Mixed-Mode Surveys: The Tailored Design Method*, Hoboken, NJ: Wiley.

Yet using weights in regression models can be problematic. In many programs, the analyst includes the weights in the regression code (for instance, in R: lm(y ~ x, data=data.frame, weight = sample.weight)). In some software, however, if these weights are whole numbers designating the actual number of people each observation represents, the presumed sample size can be inflated considerably. What effect do larger samples have on standard errors? All else being equal, they make them smaller. So, experience shows that almost any regression coefficient is "statistically significant"—whether important or not—if the sample is weighted to represent a large population. This may be seen as an advantage, but it is not and can also lead to misleading and dubious interpretations and inferences. Fortunately, most R functions recognize this issue and adjust the sampling weights accordingly, but, to be safe, rely on the software's survey sampling tools—such as R's survey package—to make sure that sampling weights and survey stage information are treated correctly.[4]

Establishing Causal Associations

Another key issue mentioned in Chapters 2, 6, and 12, but not fully elaborated, involves causal inference: establishing that one variable produces or *causes* another. Causality is a challenging topic because, to ascertain its existence with observational (nonexperimental) data, we should consider all potential confounding variables that might explain why two variables are associated. Yet, few studies have the resources necessary to measure the myriad factors that might account for an association (remember Chapter 12's discussion of omitted variable bias?). Nevertheless, statisticians have developed some promising tools that move us closer to the ability to make causal claims, even with observational data. Two related methods that are growing in popularity are called *propensity score matching* and *weighting*. These methods are simpler to understand if we imagine two groups: the treatment group that is exposed to the presumed cause and the control group that is not exposed to the presumed cause. In a true experiment, researchers may control who or what is and is not exposed to the cause and then observe

---

[4] Whether to use or not use sampling weights in LRMs is a topic of debate and often depends on discipline-specific normative practices. Bollen et al. (2016) describe diagnostic tests that help analysts determine whether weights are appropriate in regression models ("Are Survey Weights Needed? A Review of Diagnostic Tests in Regression Analysis," *Annual Review of Statistics and Its Application* 3: 375–392). Additional information about survey data and sampling weights is available in Paul S. Levy and Stanley Lemeshow (2008), *Sampling of Populations: Methods and Applications*, 4th Ed., Hoboken, NJ: Wiley. For material particular to R, see Thomas Lumley (2011), *Complex Surveys: A Guide to Analysis Using R*, Hoboken, NJ: Wiley.

the putative effect. For example, an experimental vaccine designed to target the coronavirus is given to a randomly chosen sample of 100 people and a placebo to another randomly chosen sample of 100 people. Researchers then compare who does and does not experience COVID-19 symptoms during follow-up, which might be based on a regression model, a simple test of proportions, or some other statistical approach. Because of randomization of participants into the experimental and control groups, in all likelihood the systematic differences between the two groups are controlled or partialled out, so, if the treatment group does not experience symptoms (or they are greatly reduced), the vaccine is, with high probability, the cause of COVID-19 inhibition.

But in observational research projects—such as those represented by the datasets used in the previous chapters—researchers rarely have this level of control over the assignment of the presumed cause, so they must take a different approach and attempt to account for differences between groups in the sample. In an LRM, analysts statistically adjust for differences by including potential confounding variables, which might account for all differences across individuals and allow them to isolate the "cause" of some outcome. However, a more likely scenario is that important variables are left out of the model because they are not available in the dataset or perhaps the conceptual model guiding the research is not developed enough. Using LRMs to make causal inferences also requires other stringent assumptions.[5] Of course, all of this assumes that there is a single "cause" of an outcome and the goal is to identify it. This is often a fair assumption in the medical sciences or program evaluation when one wishes to know whether, say, marijuana vaping causes mouth cancer or a school nutrition program triggers more vegetable consumption. But when studying social and behavioral phenomena many outcomes are multi-causal or may involve complex interactions and nonlinear associations that require a highly developed conceptual model to identify. Thus, the focus on identifying single cause-and-effect relationships is often myopic. Rather than using LRMs in an attempt to identify a single causal factor, it might be preferable in many instances to be guided by a conceptual model that proposes that multiple factors affect an outcome.[6]

Yet the quest to identify uni-causal factors continues and has motivated the development of several statistical approaches. One of the most frequently used approach relies on propensity scores to adjust for differences across groups in observational studies. A propensity score is the probability that an observed unit, such as a person, with a particular set of characteristics is "assigned" to the "treatment" group rather than the "control" group. The scores are estimated by constructing a model, such as a logistic regression

---

[5] Gelman, Hill, and Vehtari (2020), *op. cit.* (chapters 19–20), furnish a discussion of these assumptions.

[6] The philosopher John Mackie (1974) argues that an outcome may be caused not by a single causal factor but rather by a constellation of conditions and still satisfy the criteria for causality (*The Cement of the Universe*, Oxford: Clarendon Press).

model, that includes many variables that might account for differences between those who experience the "treatment" and those who do not (not unlike the selection models discussed in Chapter 12).

For example, suppose we hypothesize that cohabitation has a causal impact on smoking marijuana. Although we might imagine cohabitation as the "treatment," we cannot randomly assign people to cohabitating or non-cohabitating living arrangements and determine who subsequently uses marijuana. We can, though, study the many factors that might lead some people to cohabit or not, such as demographic characteristics, personality traits, religious beliefs or practices, family background, and so forth. If we develop a model that does a good job of predicting the propensity to cohabit, then we may use this information to statistically balance the characteristics that might differ across cohabiters and non-cohabiters, in essence adjusting for potential confounding variables. This then allows us to compare similar groups based on the probability of cohabiting and determine whether, to continue the example, those with a high propensity to cohabit are more likely to use marijuana than those with a low propensity to cohabit.[7]

Though promising, propensity score methods make strong assumptions about unobserved characteristics, in particular that factors that are not measured in the study do not have a large effect on "treatment" assignment. For instance, suppose that the probability of cohabiting is influenced strongly by a personality dimension such as openness or impulsivity, which is also related to marijuana use.[8] Yet we fail to measure this personality trait in our study and cannot include it in the model that predicts the propensity score. We might therefore attribute a causal effect to cohabitation that is actually due to a personality trait. Nonetheless, propensity score methods are promising and can lead us closer to identifying cause-and-effect associations than can traditional LRMs or similar regression models that simply statistically adjust for potential confounding variables.[9]

---

[7] For an empirical illustration, see John P. Hoffmann (2018), "Cohabitation, Marijuana Use, and Heavy Alcohol Use in Young Adulthood," *Substance Use & Misuse* 53(14): 2394–2404.

[8] See Angela Lee-Winn, Tamar Mendelson, and Renee M. Johnson (2018), "Associations of Personality Traits with Marijuana Use in a Nationally Representative Sample of Adolescents in the United States," *Addictive Behaviors Reports* 8: 51–55.

[9] Information on using propensity scores for studying treatment and causal effects in R is available in Leite (2016), *op. cit.* Other causal modeling approaches are discussed in Angrist and Pischke (2009), *op. cit.*, Imbens and Rubin (2015), *op. cit.*, and Pearl (2009), *op cit.* David Freedman (2007) offers a helpful overview and critique of researchers' attempts to identify causal relationships with statistical models ("Statistical Models for Causation," in *The Sage Handbook of Social Science Methodology*, W. Outhwaite and S. P. Turner (Eds.), 127–146, Los Angeles, CA: Sage).

Final Words

We could discuss other important issues, but space is limited. The information provided in the previous chapters should provide the groundwork for further coursework on and self-study of regression models. Perhaps the best advice at this stage is to suggest that interested readers pursue other topics on their own through the many excellent books, tutorials, and courses on linear regression and related statistical techniques. A variety of modeling frameworks, such as Bayesian regression, nonlinear models, generalized linear models (GLMs), simultaneous equations, methods of moments (MM) estimators, generalized additive models (GAMs), marginal models, event history (survival) analysis, log-linear models (including transition models), neural networks, regression and decision trees, text mining, and support vector machines (SVM), have all been used to complement, supplement, or replace LRMs in one way or another.

Linear Regression Modeling: A Summary

To summarize what we've learned, the next few pages present a list of steps to consider for those who utilize LRMs in their research. The list is not meant to be comprehensive since each research project has its own idiosyncrasies that make it unique. Nonetheless, it prescribes the most common steps that should normally be taken if LRMs are part of a quantitative study.

Keep a record of each of the following steps. You should be able to reproduce all of them. Careful recordkeeping is particularly important because the principles of reproducing and replicating research are at the core of the scientific endeavor.[10] In R or any statistical software, create program files (R script files, for instance) that include all your code and comments about what the code is designed to do. The files should include data management and all of the analysis steps.[11]

Know the background literature on your topic. Use your imagination, intuition, reasoning skills, and knowledge of the research literature to evaluate the social or behavioral processes of interest. What theoretical frameworks or conceptual models have you developed—or been used in past studies—to understand the specific phenomenon? What guidance do these provide regarding the choice and treatment of explanatory and outcome

---

[10] See Garret Christensen, Jeremy Freese, and Edward Miguel (2019), *Transparent and Reproducible Social Science Research: How to Do Open Science*, Oakland, CA: University of California Press.

[11] For additional information, see John P. Hoffmann (2017), *Principles of Data Management and Presentation*, Oakland, CA: University of California Press.

variables (e.g., why or how are the variables associated)? Is the ordering of the variables appropriate and justifiable? Do previous studies indicate that the outcome variable is not distributed normally or is not continuous? Has previous research found interactions between explanatory variables? What about nonlinear associations? What control or confounding variables are important for the model? Does measurement error need to be addressed? Lack of independence? Heteroscedastic errors? Endogeneity? Selection issues? What steps are available to assess and remedy these issues? You should establish which variables are needed and the presumed features of their relationships before collecting data or beginning to use a relevant secondary dataset.

Establish hypotheses, decision rules, and evidentiary standards. Decide whether you plan to use hypothesis tests. If not, specify an alternative.[12] Assuming a hypothesis-driven study, will you use directional or non-directional hypotheses? Will you rely on traditional significance testing as evidence regarding the hypotheses? If so, what other evidence will you use? Or will you avoid significance testing? Do you prefer Bayesian inference and analysis? If yes, what implications does this have for the way the model is set up, executed, and interpreted? Will you compare nested models, such as by beginning with a simpler model and adding variables to yield the full model? If yes, how will you evaluate the models (e.g., multiple partial (nested) F-tests, AICs, or BICs)? A good approach is to develop a well-thought-out conceptual model to establish clear hypotheses and outline the evidence you will use to evaluate them.

Know the dataset and understand the variables. You should become deeply familiar with the data. How many observations are in the dataset? Do you need to do a power analysis to estimate whether the sample size offers sufficient power to test hypotheses?[13] Are the data longitudinal? Multilevel? Are there other potential sources of nonindependence among the observations? Do the data come from a survey? What type of survey? Have you read the codebook and documentation for information about how the data were

---

[12] Much of the criticism of traditional hypothesis testing is aimed at using claims about "statistical significance," usually with p-values. As noted in Chapter 2, many critical articles thus use the term null hypothesis (statistical) significance testing (NHST). But most research studies remain beholden to some type of hypothesis testing, at least tacitly, because they presume an association among specific variables. Useful guidance about alternatives to the NHST framework include N. Thompson Hobbs and Ray Hilborn (2006), "Alternatives to Statistical Hypothesis Testing in Ecology: A Guide to Self-Teaching." *Ecological Applications* 16(1): 5–19; Ioannidis (2019), *op. cit.*; and Gelman et al. (2020), *op. cit.* As mentioned in Chapter 2 (fn. 20), a helpful tutorial on different "hypothesis testing theories" (also called "data testing procedures") is available in Perezgonzalez (2015), *op. cit.*

[13] We have not discussed power or power analysis, but they are important issues if hypothesis testing or the Neyman-Pearson approach is used. Statistical power is the probability of rejecting the null hypothesis when it is actually false. Power analysis is a set of techniques that estimates the sample or subsample sizes required to determine if an analysis has sufficient statistical power. Cohen (1988), *op. cit.*, provides a comprehensive review. For R code to conduct power analyses, see Xiaofeng Steven Liu (2014), *Statistical Power Analysis for the Social and Behavioral Sciences: Basic and Advanced Techniques*, New York: Routledge.

collected? What are the survey questions, if applicable, designed to measure and are they appropriate for what you need to measure given your conceptual model or hypotheses? Were scales administered? If yes, and you plan to use them, read the background on each. How are the variables of interest coded? Does the dataset include missing values? How are they coded? How will you handle missing data (see Appendix A)? You will probably need to recode and construct new variables. Does the dataset include sample weights? Establish if and how you will use them.

Check the distributions of the variables. Use q-q plots, kernel density plots, box-and-whisker plots, and similar exploratory statistical tools. Estimate descriptive statistics. Are the explanatory variables normally distributed? If they are not, do you need to consider using a transformation such as the natural logarithm, square root, or quadratic (depending on the direction and degree of skewness)? Or might the results not depend much on the distribution of the explanatory variables? How is the outcome variable distributed? Does it need a transformation? Do you observe outliers or other potential influential observations? Consider their source. If the outcome variable is binary, use logistic regression or an appropriate alternative. Do you need to create indicator variables and include them in the model?

Assess bivariate associations among variables. Use scatter plots for continuous variables. Plot linear and nonlinear lines to determine the bivariate associations (but always use theory as your guide). Compute bivariate correlations for continuous variables (correlation measures for categorical variables are also available). Look for influential observations and potential collinearity problems. Consider t-tests for binary explanatory variables. Plot the outcome variable's means for groups defined by binary explanatory variables or sets of indicator variables (box plots showing the distribution of the outcome variable for each group are useful).

Estimate the LRM. Avoid automated variable selection procedures unless your goal is simply to find the best prediction model and you're not concerned much with external validity. Assess the results. Are there unusual coefficients (overly large or small; negative when they should be positive)? Assess the goodness-of-fit statistics (adjusted R^2; F-value and its accompanying p-value; RMSE). Estimate nested models if appropriate and use multiple partial (nested) F-tests or other tools, such as IC measures, to compare them. If you are concerned with overfit and accurate predictions, employ a cross-validation method or similar procedure. If the data are longitudinal, spatial, or hierarchical, use a suitable regression model, such as a GEE, Prais–Winsten, time-series, or multilevel model.

Check the diagnostics. Construct partial residual plots, a studentized deleted residuals-by-predicted plot, and a normal q-q plot to assess linearity, homoscedasticity, and normality. Do the partial residual plots provide evidence of nonlinear associations? If yes, consider the source and determine if nonlinearities are consistent or inconsistent with the conceptual model or hypotheses.[14] Does the residual-by-predicted plot show evidence of heteroscedastic errors? If the plots are not clear, use Glejser's test or the Breusch–Pagan test to assess homoscedasticity. Try to determine if particular variables or particular observations are inducing this pattern. If heteroscedasticity exists, consider transformations, or using the Huber–White robust (sandwich) estimator or bootstrapping. If the residuals are not normally distributed, consider a transformation. Or, if you've used a large sample size, perhaps normality of residuals is not a substantial concern (but check their distribution anyway).

Save and evaluate the collinearity diagnostics. If they indicate problems, review what solutions are available. If a collinearity issue involves interaction or quadratic terms, use centered values such as z-scores to recompute them. Save the influential observation diagnostics (studentized deleted residuals, leverage values, Cook's D, and DFFITS). Do the diagnostics indicate influential observations? If yes, consider their source and fix coding errors or consider if a transformation is needed. If their source is indeterminate, yet consequential, use a robust regression technique that is affected less by influential observations, or adjust the standard errors using one of the approaches described in Chapter 14.[15]

Interpret and present the results. Present and interpret the model coefficients. What do the goodness-of-fits statistics suggest about the model? Avoid using significance tests, such as p-values, as your primary source of evidence. Confidence intervals are usually better at demonstrating the uncertainty that is a fundamental aspect of statistical models, but also consider several pieces of evidence available in the LRM. Be careful about using language that indicates causality. Without other essential conditions, regression models do not provide causal confirmation and are simply one piece of evidence in a larger research endeavor. Similarly, avoid claiming that you've found the "truth" about an association. Prepare graphs to show the results—such as predictor effects plots and predicted means—on a logical scale (don't shrink the axes of a graph just to support your point). Display graphical illustrations of nonlinear associations and interactions, which can provide intuitive information that is often lost when presenting only slope coefficients and significance tests. Finally, compare the results to the guiding hypotheses or model. Given the decision rules and evidentiary standards adopted early in the study, are the results consistent with the hypotheses or the conceptual model?

Sit back and enjoy your work! ... Then go back and check everything again.

---

[14] We can usually find nonlinear associations if we search long enough. But their mere existence in a set of data does not establish their theoretical importance. Whether nonlinearities are meaningful depends on the guiding conceptual model and hypotheses and not on post-hoc theorizing.

[15] If you've done a careful job of reviewing the literature and exploring the data, some of these corrections may not be necessary after estimating the LRM. For instance, you may have already discovered outliers that might have affected the LRM and taken some ameliorative steps.

Appendix A: Data Management

To become a proficient user of statistical models—whether or not they are regression-based—always remember the key role of *data management*. This term is used to indicate a vital concern with the proper handling and coding of variables in a dataset. Researchers are rarely provided with datasets fully prepared to use in a regression model or in other types of statistical analysis. (The datasets used in the previous chapters notwithstanding.) Rather, variables are often constructed and coded in myriad ways, missing values are included without specific treatment, and categorical variables are not accompanied by indicator variables. This appendix offers some recommendations for managing data and variables.[1] A good idea as you are starting out with R or any other statistical software is to read the documentation carefully and obtain one of the many useful guides to using the software.[2] Datasets are usually accompanied by codebooks that describe the variables and their coding schemes. Studying the codebook and fully understanding the variables of interest is a crucial prerequisite to any statistical exercise.

An elementary, yet important, issue to consider is the nature of datasets. Most datasets appear in spreadsheet format with the variables in columns and the observations in rows. For example, consider Table A.1, which provides a revised and limited portion of the *StateDataMiss.csv* dataset.[3] The variables—RobberyRate, SuicideRate, UnemployRate, and so forth—appear as columns and the states—which constitute the units of observation in the dataset—appear as rows. A good practice is to always scan at least a portion of the dataset (some are huge) to determine if any unusual values or patterns to the data or variables occur (consider the R functions head and tail). Evaluate carefully the variables you plan to analyze. Are indicator variables or binary variables coded and identified properly, whether as characters or numeric values (see Chapter 7)? How are continuous variables coded? (Is income or monetary expenditures measured in dollars or thousands of dollars? What is the base number for rates?) Do the data include unusual values that don't appear tenable (see Chapter 14)? Remember to always conduct exploratory analyses of your variables to check for extreme

---

[1] A more comprehensive treatment of data management principles is found in Hoffmann (2017), *op. cit.* In R, the tidyverse set of packages offers many tools for data management (see www.tidyverse.org; and Ryan Kennedy and Philip D. Waggoner (2021), *Introduction to R for Social Scientists: A Tidy Programming Approach*, Boca Raton, FL: CRC Press).

[2] Some helpful R resources include the following: *Swirl* (https://swirlstats.com); Jared P. Lander (2017), *R for Everyone*, Upper Saddle River, NJ: Pearson Education, and the R tutorials on the UCLA Institution for Digital Research and Education (IDRE) website (https://stats.idre.ucla.edu/r). A few more are listed in Chapter 1.

[3] The dataset is a fabricated version of *StateData2018.csv*.

TABLE A.1

Example of Missing Data in a Data Table

State	RobberyRate	SuicideRate	UnemployRate
Alabama	96.9	15.9	4.1
Alaska	NA	28.1	6.5
Arizona	92.8	19.4	4.7
Arkansas	69.1	21.7	NA
California	125.5	11.5	4.1
Colorado	56.7	21.6	3.1
Connecticut	87.8	10.9	4.2
Delaware	135.6	12.9	4.0
Florida	530.7	6.8	5.7
Georgia	125.2	11.9	3.5
Hawaii	· 78.0	16.2	2.2

Source: StateDataMiss.csv.

values, non-normal distributions, and unusual codes. At the very least, examine the frequencies and the distributional properties of the variables (e.g., means, medians, standard deviations, and skewness) you plan to analyze and evaluate each variable carefully. Use graphs to explore the nature of the variables—don't forget the advice from Chapter 1: plot your data—early and often.

Missing Data

Missing data are ubiquitous in datasets and can make life difficult for analysts. Many books and articles provide information about missing data, so the following provides just a quick taste of what we are faced with when data are missing.[4]

The portion of the *StateDataMiss* dataset that appears in Table A.1 shows the term NA in two of the cells. NA is the most common way to designate missing values in R. But oftentimes missing values are coded with untenable

---

[4] The problem of missing information was probably recognized during the first large data collection effort. Certainly, census takers have always been concerned with this issue. Since at least 1900, statisticians have discussed how to remedy missing data problems (see, for instance, M. O. Heckard (1907), "The Practical Collection of Statistical Data," *Publications of the American Statistical Association* 10(80): 498–502. Although earlier work exists, the 1960s and 1970s saw the development of rigorous and systematic treatments of how to handle missing data (see A. A. Afifi and R. M. Elashoff (1966), "Missing Observations in Multivariate Statistics I. Review of the Literature," *Journal of the American Statistical Association* 61(315): 595–604, and Donald B. Rubin (1976), "Inference and Missing Data," *Biometrika* 63(3): 581–592).

values such as –99 for age or 9 for race/ethnicity when the other choices are 1 = African American, 2 = White, 3 = Latinx, and 4 = other race/ethnicity. Many other possibilities exist; check the codebook accompanying the data for information about how missing values are treated.

Regardless of the coding scheme for missing values, before using a variable you should decide how you are going to handle its missing data. As you do this, though, first consider their source. Why do missing values appear in the dataset? The reasons often depend on the nature of the variable. Data entry staff sometimes forget to place a value in a specific cell of the dataset. Perhaps, in this example, we need to go back to the data source and find the robbery number for Alaska and the unemployment rate for Arkansas. Unfortunately, these data might not exist.

A frequent source of missing information in survey data involves skip patterns. Skip patterns are a convenient way to make responding to survey questions more efficient. For instance, suppose we ask a sample of adolescents to complete a questionnaire about their use of cigarettes. The questions ask details such as how often they smoke, how many cigarettes they smoke in a day or a week, from where they obtain cigarettes, and so forth. Since many adolescents don't smoke at all, our survey includes an initial question about ever using:

Q.1. Have you ever smoked cigarettes? (*select one answer only*)
   (a) No [Go to question Q.7]
   (b) Yes [Go to question Q.2]

Q.2. When was the last time you smoked a cigarette? (*select one answer only*)
   (a) Today
   (b) In the last week
   (c) More than a week ago but less than a month ago
   (d) A month ago or longer

...

Q.7. What do you think about linear regression? (*select one answer only*)
   (a) Love it
   (b) Hate it
   (c) I don't know what it is

In contemporary surveys, these types of questions are often programmed into software so those responding "no" to the first question are automatically skipped past questions Q.2–Q.6. But suppose we wish to analyze information from Q.2 either as an outcome or an explanatory variable (perhaps by creating a categorical variable). Would we only want to analyze data from those who actually answered the question, or should the analysis also include those who never smoked? (Recall the discussion of censoring in Chapter 12.) This is an essential issue to address since those who answered "no" to the first question are "missing" on the second question. If we decide to include those who never smoked in the analysis, then we may create a new variable that includes a code for "never smoked" along with others for those who smoked in the last day, week, etc.

A common occurrence in survey research is to find data missing due to refusals or "don't know" responses (Question: "When was the last time you smoked a cigarette?" Response: "I don't know"). For example, many people don't like to answer questions about personal or family income. A perusal of various survey datasets shows that data on income are frequently missing, with usually more than 10% of respondents refusing to answer. Yet many researchers are interested in the association between income and a host of potential explanatory or outcome variables. Is it worth it to lose 10–20% of your sample so you can include income in the analysis? Are people who refuse to answer income questions (or other types of questions) different from those who do answer? If they are systematically different, then the sample used in the analysis may no longer be a suitable representation of the target population (see Chapter 12's section on selection bias).

Making decisions about missing data is a crucial part of many statistical exercises. However, before deciding what to do about missing data, it is vital to evaluate why data are missing. Data can be missing randomly or non-randomly. If they are *missing completely at random* (MCAR), then one cannot predict their occurrence; they represent a random subsample of the data. If the data are *missing at random* (MAR), then another variable or set of variables in the dataset predicts the missing values and we can use this information to adjust for patterns of missingness. In many situations, however, data are *missing not at random* (MNAR). For example, suppose those who refuse to answer income questions tend to be wealthier people. If we wish to predict income, a regression model will yield biased estimates. Our only recourse when faced with MNAR data is to try to develop a valid model that predicts the "missingness" process and use it to adjust the regression model for these missing data. But this is quite challenging.[5]

Most solutions to missing data issues assume that the data are MCAR or MAR. We shall discuss several of these. First, some researchers simply ignore the missing data by omitting them from the analysis. They argue that a few missing cases do not bias the results of the model too much. If you adopt this approach, be careful not to leave missing observations with seemingly legitimate codes—such as –9 or 999—in the analysis. Imagine, for instance, if

---

[5] See Roderick J. A. Little and Donald B. Rubin (2019), *Statistical Analysis with Missing Data*, 3rd Ed., Hoboken, NJ: Wiley, chapter 15.

you left a missing code of 999 in an analysis of years of education. Now that would be an influential observation!

In any event, omitting all missing data from an analysis is called *listwise deletion*. Also termed *complete case analysis*, this is the R default among those functions that allow missing data.[6] It removes any rows of data with missing information on the variables in the model—the NAs in R. Suppose we analyze the association between robberies and the unemployment rate in the *StateDataMiss* dataset shown earlier with a correlation or an LRM. In this situation, listwise deletion removes both Alaska and Arkansas from the analysis. Now imagine if we estimate a multiple LRM with, say, five explanatory variables. Suppose further that ten different observations are each missing only one value but from five different explanatory variables. If we place all of them in the regression model and use listwise deletion, our sample size goes from 50 to 40; we lose one-fifth of the sample. This can be costly.

A problem that is not infrequent involves the use of multiple nested regression models with different patterns of missing data. Say we estimate a model with violent crimes per 100,000 predicted by the gross state product. Assume neither variable has missing data, so the sample size is 50. But we then add percent poverty, which is missing two observations. A third model adds population density and the unemployment rate, each of which has four missing observations. We thus have up to ten missing observations. A common claim is that the first two models are nested within the third, thus allowing statistical comparisons (see Chapter 6). They are not truly nested, however, because they have different numbers of observations (ns): 50 in the first model, 48 in the second model, and 40–44 in the third model. Not only do model comparison tests, such as Chapter 6's multiple partial F-test, break down when this occurs, but the regression coefficients cannot be compared across models. Missing data must therefore be handled *before* estimating the regression models so that each is based on the same number of observations.

Creating a code for a missing value is another common procedure. Suppose personal income is missing 10% of its observations. We could create a new indicator variable that is coded as 0 = non-missing on income and 1 = missing on income and include this variable in the regression model, along with other income variables. This is only useful mainly if the variable is transformable into a set of indicator variables, though. Thus, income should be categorized into a number of discrete groups (e.g., $0–$10,000; $10,001–$20,000; etc.) that are then the basis for the set of indicator variables.

A variety of techniques are available for handling missing data in addition to listwise deletion or using an indicator variable. Known generally as *imputation methods*, they replace the missing values with other values that

---

[6] Some R functions, such as mean, return an error message when a variable has missing data. The user must tell R to remove the missing data, such as with mean(dataset$variable1, na.rm=TRUE). The latter part of the code requests that the NAs be removed (rm) from the computation.

may then be used in the analysis. Imputation was first used in the decennial U.S. Census in 1940.[7] The most common types include *mean* or *median substitution, regression substitution, raking* techniques, *hot deck* procedures, and *multiple imputation*. Mean substitution replaces missing values of a variable with the mean of that variable. For example, the mean number of robberies per 100,000 for the 46 non-missing observations is 88.5.[8] We can replace the missing value with the mean using the following R code:

R code
```
StateDataMiss$RobberyRate[is.na(StateDataMiss$Robbery
Rate) <- mean(StateDataMiss$RobberyRate, na.rm = TRUE)
```

A problem with mean substitution is that it induces biased and inefficient slope coefficients, especially as a higher proportion of observations is missing. At what point it produces too much bias or inefficiency is not well established, though. In general, avoid this method since more rigorous approaches are available.[9]

Regression substitution replaces missing values with values that are predicted from a regression equation.[10] For instance, suppose a small number of values are missing from the personal income variable that appears in the *GSS2018.csv* dataset. Yet, the variables assessing education, occupational prestige, and age have no missing values. If a regression equation does a good job of predicting the valid income values with these three variables, then the missing values may be predicted by using the coefficients from the LRM shown in Equation A.1.

$$\widehat{\text{income}}_i = \alpha + \beta_1\left(\text{education}_i\right) + \beta_2\left(\text{occupation}_i\right) + \beta_3\left(\text{age}_i\right) \qquad (A.1)$$

In R, after estimating the equation with the lm function, we save the predicted (fitted) values and use them to impute the missing values for personal income. If the variables are not strong predictors of the variable with the missing data problem, however, the slope coefficients and standard errors may manifest substantial bias. Again, more rigorous methods exist.

Raking is an iterative procedure based on using values of adjacent observations to come up with an estimated value of an observation that is missing. Hot deck involves partitioning the data into groups that share similar

---

[7] Fritz Scheuren (2005), "Multiple Imputation: How It Began and Continues," *American Statistician* 59(4): 315–319.

[8] Based on the R code mean(StateDataMiss$RobberyRate, na.rm=TRUE).

[9] Other measures of central tendency, such as the median, could also be used, but the same problems occur.

[10] This method appears to have been introduced by mathematician Samuel S. Wilks in 1932 ("Moments and Distributions of Estimates of Population Parameters from Fragmentary Samples," *Annals of Mathematical Statistics* 3(3): 163–195).

characteristics and then randomly assigning a value based on the non-missing values of this group. Raking and hot deck imputation were used by the U.S. Census Bureau for many years.[11] Both require specialized software to implement (see the R package hot.deck).

Using these various techniques to replace missing values for outcome variables is rarely a good idea. Some analysts argue that raking and hot deck techniques are appropriate if the outcome has only a small number of missing values. In this situation, though, multiple imputation (MI) is a better option.[12] MI is probably the most common corrective for missing data in social, behavioral, and health research.[13] This method involves three steps:

1. Impute, or fill in, the missing values of the dataset, not once, but q times.[14] The imputed values are drawn from a particular probability distribution, which depends on the nature of the variable (for example, continuous or binary). Most MI techniques use regression equations with the other variables in the dataset as predictors to impute the missing values. This step results in q complete datasets.

2. Analyze each of the q datasets, such as with a series of LRMs. This results in q analyses.

3. Combine the results of the q analyses into a final result. Standard algorithms that calculate an average or weighted average of the coefficients and standard errors are used if the model is based on a regression routine.

---

[11] See Patrick J. Cantwell, Howard Hogan, and Kathleen M. Styles (2004), "The Use of Statistical Methods in the U.S. Census," *American Statistician* 58(3): 203–212.

[12] MI was developed by Donald B. Rubin in the mid-1970s to address missing data in U.S. federal government surveys. It was offered as an alternative to raking and hot deck methods used to impute data in most of these surveys (Donald B. Rubin (1977), "The Design of a General and Flexible System for Handling Nonresponse in Sample Surveys," prepared under contract for the U.S. Social Security Administration).

[13] MI's popularity is such that some research journals may mandate its use for regression-based studies. But this does not mean it is flawless. MI works well when the missing data patterns are random, but if there are nonrandom patterns of missing data, it might result in estimates that are often even more biased than those from models that use listwise deletion (Thomas B. Pepinsky (2018), "A Note on Listwise Deletion versus Multiple Imputation," *Political Analysis* 26(4): 480–488).

[14] Relatively few imputations (5–10) are needed to achieve unbiased slope coefficients, but accurate standard errors require more. The R package howManyImputations is designed to help analysts decide, not surprisingly, how many imputations they need for stable estimates (see Paul von Hippel (2020), "How Many Imputations Do You Need? A Two-Stage Calculation Using a Quadratic Rule," *Sociological Methods & Research* 49(3): 699–718; and https://github.com/josherrickson/howManyImputations).

R has several packages suitable for MI, such as mice, mi, Amelia, and Hmisc (e.g., transcan and aregImpute functions).[15] A final method that we don't address, but some experts recommend, is full information maximum likelihood (FIML).[16]

The following furnishes an example of how to use the mice and mi packages to address missing data and estimate an LRM. We'll first create a subset that includes only a few of the variables from the *StateDataMiss* dataset. Next, we'll utilize the mice and mi packages to explore patterns of missing data among the variables.[17]

R code
```
library(mi)
library(mice)
StateMiss <- as.data.frame(StateDataMiss[, 4:8])
  # subset the data
MDF <- missing_data.frame(StateMiss)
  # for convenience
show(MDF)  # supplies information about the variables
hist(MDF)  # histograms of missing data patterns
```

The output (not provided) shows that three of the five variables have missing data that range from two to four observations. According to the md.pattern output, the number of complete cases is 41. We now create the imputed datasets. As noted earlier, the number of imputations varies according to several characteristics, but let's keep it simple by creating ten imputed data sets.

---

[15] A helpful summary of missing data procedures, including imputation methods, is found in Daniel A. Newman (2014), "Missing Data: Five Practical Guidelines," *Organizational Research Methods* 17(4): 372–411. For an in-depth discussion of procedures for missing data, see Little and Rubin (2019), *op. cit.* Madan Lal Yadav and Basav Roychoudhury (2018) provide a review of some methods in R for imputing data ("Handling Missing Values: A Study of Popular Imputation Packages in R," *Knowledge-Based Systems* 160(15): 104–118).

[16] Using ML to address missing data problems predates multiple imputation by several decades (see, for example, Wilks (1932), *op. cit.*). FIML is a technical approach that is difficult to describe but, suffice it to say, it uses MLE to find the most likely data pattern given the observations that are not missing. In most situations, it compares favorably to MI. The psych package includes a function, corFiml, which provides a correlation matrix based on FIML to adjust for missing values. In an LRM context, this adjustment is available in structural equation modeling programs such as R's lavaan or in specialized programs such as MPlus. For more information, see Paul Allison (2012), "Handling Missing Data by Maximum Likelihood," available at http://www.statisticalhorizons.com/wp-content/uploads/MissingDataByML.pdf.

[17] For additional information about the mi package, see Yu-Sung Su et al. (2011), "Multiple Imputation with Diagnostics (mi) in R: Opening Windows into the Black Box," *Journal of Statistical Software* 45(2): 1–31. Examples of using mice and much more are available in Stef van Buuren (2018), *Flexible Imputation of Missing Data*, 2nd Ed., Boca Raton, FL: Chapman & Hall.

R code
```
set.seed(74895) # for replication
imputations <- mice(StateMiss, m=10, print=FALSE)
 # requests 10 imputations
plot(imputations)
```

The plot function provides a set of graphs that show the distributions of the original and imputed variables. If the imputations are reasonable, the lines should be relatively straight (though, again, no set rules about this exist). The lines for suicides per 100,000 demonstrate quite a bit of variability, so additional imputations might be needed.

Finally, we'll use mice's pool function to estimate an LRM using the imputed datasets. This executes MI steps 2 and 3 listed earlier.

R code
```
LRM.original <- lm(RobberyRate ~ SuicideRate +
              UnemployRate + PerPoverty,
              data=StateDataMiss)
 # model from original data with missing values
   (listwise deletion)
LRM.mi <- with(imputations, lm(RobberyRate ~
              SuicideRate + UnemployRate +
              PerPoverty))
 # model using imputations
options(scipen=999)   # turn off scientific notation so
                        it's easier to compare the
                        results
summary(LRM.original) # results based on original data
summary(pool(LRM.mi)) # results based on imputed data
```

R output (abbreviated)

Original data	Estimate	Std. Error	t value	Pr(>|t|)	
(Intercept)	24.750	66.086	0.375	0.710	
SuicideRate	-6.764	1.950	-3.469	0.001	**
UnemployRate	53.895	17.835	3.022	0.004	**
PerPoverty	-1.403	4.651	-0.302	0.765	

Number of observations: 40

Imputed data	Estimate	Std. Error	t value	Pr(>|t|)	
(Intercept)	58.214	54.618	1.066	0.293	
SuicideRate	-7.231	1.593	-4.539	0.001	***
UnemployRate	38.571	12.719	3.033	0.005	**
PerPoverty	0.685	3.469	0.197	0.845	

Number of observations: 50

Several differences result, including distinct intercepts and slope coefficients for the unemployment rate. The standard errors are also smaller in the model based on the imputed data, which is to be expected given the larger average sample size of the imputed datasets. Because of some other peculiarities with the imputations, though, such as the instability shown in the graphs, we might need more imputations to achieve satisfactory results.[18]

Variable Creation, Coding, and Recoding

Researchers often need to recode variables or create new variables from those in a dataset. Recoding consists of changing the codes of a variable, such as when a researcher wishes an indicator variable to have the values 0 and 1 rather than 1 and 2. Another common recoding strategy is to collapse a continuous variable into a categorical variable. These and other recoding steps are not difficult in R. For example, suppose a numeric variable in a dataset called *Descriptives* is labeled gender and includes the following codes: 1 = male and 2 = female. The following demonstrates two simple ways to change this variable so it is coded as 0 = male and 1 = female.

R code
```
Descriptives$new.gender <- Descriptives$gender - 1
# create a new variable called new.gender

library(car)
Descriptives$new.gender = recode(Descriptives$gender,
                  '1=0; 2=1')
 # use the car package's recode function
```

Assume the dataset also includes a numeric variable that measures years of formal education (educate) and has the following range: 6–20 years. We wish to collapse it into an ordered categorical variable that represents 6–11 years, 12 years, 13–15 years, and 16 or more years of formal education. We may conduct this recoding task with the following R code:

R code
```
Descriptives$new.educate <- cut(Descriptives$educate,
                  breaks=c(0,11,12,15,20),
                  labels=c("HSOnly","HSGrad
                  ","SomeColl","CollGrad"))
```

---

[18] As an exercise, try increasing the number of imputations to 50 and compare the results.

The cut function creates a factor variable to which we then apply the labels. The key is to get the breaks or "cut-points" right. The sequence is low through highest value in the category (e.g., 0–11) and so forth. For instance, consider the sequence {11, 12} in the cut function. R interprets this as "create a category that places all of those with years of education *greater than 11* but *less than or equal to 12* into a category." Thus, the categories are defined as {0–11, 12, 13–15, and 16–20}.

Another common task is to create composite variables. Many social and behavioral surveys include sets of questions designed to measure some phenomenon of interest. For example, sets of questions in datasets might measure self-esteem, symptoms of depression, illicit drug use, happiness, anxiety, self-efficacy, job satisfaction, trust, racial prejudice, authoritarianism, religiosity, and many other interesting phenomena. Suppose we're studying adolescent depression and the survey questions are drawn from the Center for Epidemiological Studies-Depression (CES-D) scale, a widely used depression symptom inventory.[19] We plan to estimate an LRM with depressive symptoms as an explanatory variable (along with gender, family closeness, and other measures) and self-efficacy as the outcome variable (we'll ignore possible endogeneity problems; see Chapter 12). Entering each depression question into the LRM separately is rarely a good idea. Rather, we would doubtless attempt to combine the questions in some way to come up with a composite measure of depressive symptoms.

Chapter 13 discusses one way to combine variables: use factor analysis or another latent variable technique. Another method that generally ignores measurement error is to combine variables either by adding up or taking the mean of their values. When combining variables with one of these methods, however, they should be coded in the same direction and in the same way. For instance, if several items are designed to measure symptoms of depression, all of them should be coded so that increasing numbers indicate either more or fewer symptoms. We may thus need to *reverse code* some items so they are all in the same direction. Furthermore, if one variable is coded 1–4 and another is coded 1–10, then summing them or taking the mean of the items will be influenced disproportionately by the variable with more response categories. As an alternative, some researchers recommend first standardizing the variables (taking their z-scores) before adding them up or taking their mean. The reasoning is that even if the number of response categories differs, standardizing will "normalize" them so they are on the same scale.

Simply adding up the variables is risky, however. Suppose, for instance, that we use the following R code to compute a new variable, depress, from five variables in a depression inventory:

---

[19] Introduced by Lenore Sawyer Radloff (1977), "The CES-D Scale: A Self Report Depression Scale for Research in the General Population," *Applied Psychological Measurement* 1: 385–401.

R code

```
depress <- dep1 + dep2 + dep3 + dep4 + dep5
```

An alternative is to use the `rowSums` function to add the values of the items:

R code

```
depress <- rowSums(dataset[, c('dep1', 'dep2', 'dep3',
            'dep4', 'dep5')], na.rm=TRUE)
```

But imagine that the variables have different patterns of missing values. For some observations, the new variable will depend not just on their responses to the questions, but also on whether they are missing one or more items. Respondents with response patterns of {2, 2, 2, 2, 2} will have the same depression score as those with response patterns of {4, 6, NA, NA, NA}, even though their levels of depressive symptoms differ.

Some researchers prefer to take the mean of the items. In R, use the row-Means function to create a variable based on these means:

R code

```
depress <- rowMeans(dataset[, c('dep1', 'dep2', 'dep3',
            'dep4', 'dep5')], na.rm=TRUE)
```

Hence, respondents with the patterns listed earlier will have scores of 2 and (4 + 6)/2 = 5, which reflect more accurately their respective levels of depressive symptoms.

Coding and recoding variables is a crucial part of any data management exercise, whether or not they are needed for statistical analyses.[20] We've already learned about coding indicator variables (see Chapter 7). Indicator variables must be mutually exclusive if they measure aspects of a particular variable (e.g., marital status). Examine cross-tabulations of indicator variables—along with the variable from which they are created—to ensure they are coded correctly.

Make sure that continuous variables are also coded properly. Use coding strategies that are easily understandable and widely accepted, if possible. Measuring income in dollars, education in years, or age in years is commonplace. Yet, for various reasons, coding income into $1,000s is sometimes preferred. We also learned how to transform continuous variables, such as by taking their square roots or natural logarithms (see Chapter 11). Given these myriad approaches to variable creation and modification, keeping track of coding strategies is therefore of utmost importance. Constructing a codebook and understanding the coding and recoding strategies by maintaining

---

[20] A valuable discussion of the principles of good coding for quantitative analysis is in Lee Epstein and Andrew Martin (2005), "Coding Variables," in *Encyclopedia of Social Measurement* (pp.321–327), K. Kempf-Leonard (Ed.), New York: Elsevier. See also Hoffmann (2017), *op. cit.*

copies of R script files are essential and allow researchers to come back later and efficiently recall the steps they took before estimating a model.

R Script Files

We've mentioned R script files a couple of times but have not offered a clear description. A script file is like a recipe for R. It is a text file that lists the code we want R to execute. Here's a short example of a script file that (1) reads the *Nations2018* file into R, (2) creates a scatter plot, (3) estimates an LRM based on the plot, and (4) requests post-LRM diagnostic plots.

```
# R script file example #
# Read the dataset into R's memory
library(readr)
Nations2018 <- read_csv("[path to file goes here]/
                Nations2018.csv")
View(Nations2018) # spreadsheet view of the dataset

# create a scatter plot of labor union by government
  expenditures using the ggplot2 package's ggplot
  function
library(ggplot2)
p <- ggplot(Nations2018, aes(x=perlabor, y=expend)) +
    geom_point(shape=1, color="blue") +
    geom_smooth(method=lm, color="darkred")
p + labs(x="Percent labor union", y="Government
        expenditures")

# Estimate a linear regression model and examine some
  diagnostic plots
LRM.appendix <- lm(expend ~ perlabor, data=Nations2018)
summary(LRM.appendix)
confint(LRM.appendix)
plot(LRM.appendix)
```

Script files may be created and edited in R or RStudio or in a text editor such as Notepad++ or TextEdit.[21] We may then open the files and execute part or all of the code to generate output. (In RStudio, the script window includes a Run button.) Script files can be very large or small, but their main

---

[21] In RStudio, use File – New File – R Script to create a script file. In R, use File – New Document.

purpose is to keep a clear record of all steps, which can include data management, creating graphs, and estimating statistical models.

We will not address the many additional coding issues. Experience is the best teacher. Keeping careful records and backup files of datasets that include before-and-after recodes of key variables should also be a standard practice. Remember that once you save an updated R file using the same name the old file is gone (unless you've made a backup copy). Make sure you are satisfied with the variables created from the original data file and don't overwrite it.[22]

A Vital Step—Exploratory Data Analysis[23]

As emphasized in Chapter 17 and mentioned earlier in this appendix (but worth reemphasizing), after we've completed taking care of missing values, creating new variables, and recoding old variables, a vital step is to examine frequencies (for categorical or indicator variables) and descriptive statistics (for continuous variables) for all of the variables you plan to use in the analysis. Using graphs to examine the distributions of variables is also important. If the preliminary work included recoding variables or missing data imputation, these should also be assessed carefully. R has a host of functions designed to assess the distributions of variables to ensure that the analyst's data manipulations work as expected.

Here's an example that illustrates why carefully checking the variables is critical. Imagine using the *Esteem.csv* dataset to estimate an LRM. First, though, we ask a colleague to transform the variable `stress` using the natural logarithm because previous research indicates that variables measuring stressful life events tend to skew positive. Our colleague uses the following R code to transform the variable:

R code
```
Esteem$newstress <- log(Esteem$stress)
```

After requesting descriptive statistics on the `newstress` variable (e.g., `summary(Esteem$newstress)`) we find that more than half of its cases are missing. Why did this occur? If we didn't pay attention to the original

---

[22] This is why careful researchers *always* write script files from the beginning that work on the original dataset. In other words, they keep their original datasets and write files that work from them. They may create subsequent script files that operate on the files created from earlier script files, but the original datasets are always intact.

[23] This term has a couple of different, though closely related, meanings. John W. Tukey (1977, *op. cit.*) popularized the expression to refer to examining patterns in data as a way to generate hypotheses and develop methods of analysis. Others use the term more generally to describe methods for checking distributions of variables, determining whether recodes work as expected, and as way to assess the assumptions of a statistical model.

variable or failed to look carefully at the data file, we might simply think that most of the adolescents did not respond to this question during survey administration. Perhaps we'd chalk it up to the inexactness of data collection and ignore these cases, though I hope not. Or we might use a substitution or imputation procedure to replace the missing values. But if we looked at the original stress variable—which we should always do—we'd find no missing cases. Something must have happened during the transformation.[24]

Recall that taking the natural logarithm of zero or a negative number is not possible; the result is undefined. But notice that the original stress variable has many negative values because it is mean centered, so the mean is zero and the variable includes many values above and below zero. The code needs to be revised to take into account the negative values in the variable. A simple solution is to revise the code by adding a constant before taking the logarithm:[25]

R code
```
Esteem$newstress <- log(Esteem$stress + 7)
```

Now, those with negative scores on the original variable remain as valid numbers in the new variable. Yet we might not have found this out if we didn't check the descriptive statistics for both the old *and* the new stress variables. Knowing the data and variables, and always checking distributions and frequencies following variable creation and recoding, goes a long way toward preventing modeling headaches later.

---

[24] When transforming the variable in R, the software provides a warning message: NaNs produced, indicating that the function created missing values. But, if our colleague didn't notice the warning, she may not know that a problem exists.

[25] The choice of the constant is based on the minimum value of the original variable, −6.84. The goal is to add a large enough number so the variable no longer has any zero or negative values.

Appendix B: Using Simulations to Examine Assumptions of Linear Regression Models

Books on probability theory and applied statistics provide proofs of the Gauss–Markov theorem, the Central Limit Theorem (CLT), and various postulates, lemmas, and theorems that are at the foundation of regression models. As discussed throughout the chapters, the favorable characteristics of regression models estimated with OLS depend largely on a set of assumptions. This extends to the other models not examined in this book as well; all regression models rest on assumptions. We've now learned in detail about these assumptions for LRMs, how some are critical for estimation, and that a few are related. As noted in several chapters, the assumptions are often not satisfied because we have not thought through the conceptual process that generates the model. For example, even though LRMs are valuable tools for studying relationships among sets of variables, it may be more appropriate to focus—conceptually and empirically—on nonlinear models since variables can be related in a nonlinear manner in more ways than in a linear manner. The main point, though, is that when an assumption is not met, our first goal should be to figure out why rather than to look for a quick analytical fix. Always ask questions such as—what is causing that heteroscedastic pattern in the residuals? What is it about the data that produces a lack of independence among the errors of prediction? Why does a nonlinear pattern appear in a partial residual plot? What's the source of that influential observation?

If we are vigilant in seeking answers to these and related questions, the LRM will continue to be a valuable method for quantitative studies. We should, therefore, have a firm understanding of its assumptions. The previous chapters are designed to help you understand the assumptions on a practical level. If one has sufficient experience with probability and statistical theory, with a good grounding in calculus and linear algebra, then the assumptions make even more sense. If one is proficient with proofs, then all the better. Many students and researchers, however, do not have this type of a background and the role of LRM assumptions can remain opaque, even for those proficient in assessing them. An alternative way to get a better handle on these assumptions, though, is to use statistical simulations to establish important aspects of LRMs and investigate what happens to the estimates when assumptions are not satisfied.

DOI: 10.1201/9781003162230-19

A statistical simulation is an experiment designed to determine the behavior of estimators or statistical properties under various conditions. It usually involves five steps:

1. Choose a specific estimator or assumption to examine, such as the behavior of an estimator from a particular probability distribution (for example, how well does the mean of a sample represent the mean in the population when the distribution is log-normal?).
2. Define a population by setting up a *data generating process* (DGP).[1]
3. Select in a random manner several independent datasets that represent samples from the population.
4. Estimate a model on each dataset.
5. Assemble the results of the model and establish how the estimator—or other aspects of the model—performs.

The DGP is critical since it defines the "true" model in the population and its specific characteristics. In regression simulations the DGP can be a population regression model with known (fixed) slopes and intercept. An advantage of simulations is that researchers may generate different scenarios, so they do not have to wait for datasets that have distinct characteristics, or access population data from which to sample. Why is this useful? Consider the Gauss–Markov theorem mentioned in Chapter 4. This theorem is foundational to the linear model, for it shows that the OLS estimators are BLUE when certain assumptions are met. Its proof is persuasive.[2] Along with the CLT, it gives a firm base to LRMs estimated with OLS but also motivates alternative models, such as robust regression models, that are suitable when certain assumptions are not satisfied. But the OLS assumptions are based on what some claim is an elusive target: we assume what happens when we sample repeatedly from a population. But we rarely have the facility to sample over and over again from a population of, say, all adults in a large geographic area. Simulations provide a valuable alternative: researchers may use them to sample from a specific population that generates a DGP, one that is defined by the researcher.

Several types of simulations are available, but the most common employ Monte Carlo methods, so you will hear the term *Monte Carlo simulations*.[3] How can this help us understand the assumptions of the OLS regression model better? Think about one of its main characteristics: $E\left[\hat{\beta}\right] = \beta$. This

---

[1] A DGP—mentioned briefly in Chapter 3—is a way of creating data that follow a particular probability distribution or joint distributions assuming more than one variable. In most research applications, analysts build models, such as LRMs, that they think (or hope) reflect the actual process that generated the data they observe.

[2] See Lindgren (1993), *op. cit.*, pp.510–511.

[3] See Thomas M. Carsey and Jeffrey J. Harden (2013), *Monte Carlo Simulation and Resampling Methods for Social Science*, Thousand Oaks, CA: Sage Publications.

equality asserts that the expected value of the sample slope equals the slope in the population. But how can we determine this? Some argue we cannot because the definition of a population is so elusive. The major difficulty is that we rarely have sufficient information on the population or even a good idea of what the population is (although this depends on how populations are defined). Others rely on proofs that rest on postulates and are developed using statistical theory. But we may also employ simulations to gather evidence about the presumed equality, $E\left[\hat{\beta}\right] = \beta$. In a simulation, an analyst builds different scenarios or DGPs and then assesses what happens to the sample slopes when, say, the OLS assumptions are met, when one or more is not satisfied, when different sampling distributions are used, and so forth.

Statistical software, including R, Python, Stata, and SAS, provide methods to simulate various situations. Consistent with the rest of this book, we'll use R in the following examples to assess some of the assumptions and characteristics of LRMs estimated with OLS. Although we limit the sizes and other characteristics of the "samples," we could vary these to provide many more examples and tests.[4]

The first simulation is a simple LRM designed to test whether $E\left[\hat{\beta}\right] = \beta$. Some of its characteristics include a sample size of 1,000 and 500 simulations. We could also vary the size of the population, the size of the samples, and the number of samples to conduct additional simulations. In this exercise, though, we define two explanatory variables, X1, which is drawn from a normal distribution, and X2, which is correlated with X1. The DGP defines the "true" LRM in the simulated population as $y = 1 + (0.5 \times X1) + \varepsilon(0,1)$. In other words, the model has parameters $\alpha = 1$ and $\beta = 0.5$, with errors that follow a standard normal distribution (mean = 0; s.d. = 1). We'll also consider a multiple LRM with X1 and X2 to test additional assumptions. The relevant R functions to set up the simulation are

```
set.seed(945793)        # Set a seed for reproducible
                          results
sims <- 500             # Set the number of
                          simulations (try changing
                          the number)
n <- 1000               # Sample size (try varying
                          it)
B.1 <- numeric(sims)    # Create an empty vector for
                          storing simulated slope
                          coefficients, X1
B.2 <- numeric(sims)    # Create an empty vector for
                          storing simulated slope
                          coefficients, X2
```

---

[4] Carsey and Harden (2013), *op. cit.*, provide detailed examples of numerous simulations.

```
a <- 1                         # True value of the
                                 intercept
b1 <- 0.5                      # True value of slope
                                 coefficient 1
b2 <- 0.75                     # True value of slope
                                 coefficient 2
X1 <- rnorm(n,12.45,3.35)      # Create X1, which has
                                 1,000 observations, with
                                 mean=12.45 and sd=3.35
# Note that X1 is treated as fixed in repeated samples.
  It is drawn from a normal distribution, but it could
  also be drawn from others (e.g., uniform)
X2 <- X1*3 + rnorm(n,0,3)      # X2 is constructed to be
                                 based on X1 plus some
                                 normal error
# the correlation between X1 and X2 is about 0.96
```

The next step is to set up the GDP within the actual simulation. This is accomplished with a loop, as shown in the following R functions:

```
# A simple LRM estimated with OLS
for(i in 1:sims){                  # Begin the loop
  Y1 <- a + b1*X1 + rnorm(n,0,1)   # The "true" DGP, with
                                     standard normal
                                     error
  model1 <- lm(Y1 ~ X1)            # Estimate the model
                                     with OLS
  B.1[i] <- model1$coef[2]         # Place the sample
                                     slope coefficients
                                     in the vector B.1
}                                  # End the loop
```

The program then conducts 500 repetitions of the LRM and places each estimated slope coefficient in the data vector B.1. Once this is completed, we may examine the estimated slopes to see how they behave. Since we have met the assumption of normally distributed and constant variance among the errors of prediction, we should find that the mean of the sample slope coefficients equals (with a bit of sampling error) the population slope, 0.5.

```
# Create a kernel density plot of the slope coefficients
plot(density(B.1), main="Distribution of OLS slope
                         coefficients - normal errors",
                   xlab="Slope coefficients")
abline(v=0.5, col="black")
```

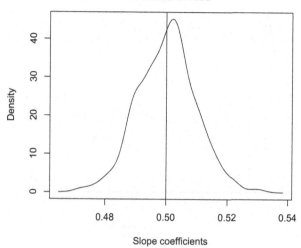

**Distribution of OLS slope coefficients
normal errors**

FIGURE B.1
Distribution of OLS slope coefficients—normal errors.

Figure B.1 is a kernel density plot of the sample slope coefficients. It includes a vertical line at 0.5, the population or "true" slope. The distribution of sample slopes approximates it quite well. The mean (`mean(B.1)`) is 0.500 with a standard deviation of 0.009 and a 95% CI of {0.499, 0.501}. If we increase the number of simulations, which is a proxy for increasing the number of samples from the population, the estimate of the slope becomes more precise.

Let's examine what happens when the errors of prediction do not follow a normal distribution. Recall that the normality assumption calls for errors that follow a normal distribution (although this is not as critical as other assumptions; see Chapter 11). Will the CLT[5] save a model without normally distributed errors? How precise will the estimates be? We'll change the error distribution to log-normal to find out.

```
# Test of the normality assumption
for(i in 1:sims){
Y2 <- a + b1*X1 + rlnorm(n) # DGP with lognormal error
model2 <- lm(Y2 ~ X1)
B.1[i] <- model2$coef[2]
}
```

---

[5] As a refresher, the CLT states: for random variables with finite variance, the sampling distribution of the standardized sample means approaches the standard normal distribution as the sample size approaches infinity. The sample slopes should also approach a normal distribution with repeated sampling even if the underlying distribution is not normal.

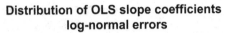

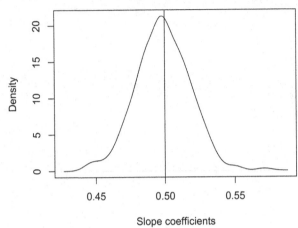

FIGURE B.2
Distribution of OLS slope coefficients—log-normal errors.

Figure B.2 is a kernel density plot of the distribution of the sample slope coefficients. Although the mean of the slopes is approximately the same in both figures, the x-axis scale is wider in Figure B.2 than in Figure B.1 (s.d. = 0.02 and 95% CI = {0.497, 0.502}). This shows the modest loss in efficiency when the errors do not follow a normal distribution. As predicted by the CLT, though, the slope coefficients appear to follow a normal distribution. Try varying the mean and standard deviation of the error distribution in the DGP. The farther away from a normal distribution, the greater the loss in efficiency.

One of the most consequential assumptions of the LRM estimated with OLS is homoscedastic errors (see Chapter 9). The following example displays what happens when this assumption is not satisfied.

```
# Test of the homoscedasticity assumption
for(i in 1:sims){
  Y3 <- a + b1*X1 + rnorm(n,0,sd=1:30)
  # DGP, with heteroscedastic error
  model3 <- lm(Y3 ~ X1)
  B.1[i] <- model3$coef[2]
}
```

The kernel density plot of the slope coefficients suggests a loss of efficiency with heteroscedastic errors, though the estimate of the slope appears largely unbiased (Figure B.3). The mean of the sample slope coefficients is 0.497, with s.d. = 0.164 and 95% CI = {0.482, 0.511}. If we vary the degree of heteroscedasticity, the precision of the estimate also changes.

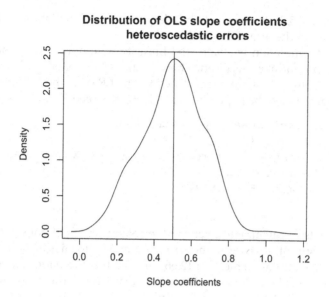

FIGURE B.3
Distribution of OLS slope coefficients—heteroscedastic errors.

The next two simulations illustrate that the slope is biased when certain assumptions of the model are not met. First, recall that one of the derivative assumptions is that the explanatory variables are not systematically related to the errors $(cov(x, \hat{\varepsilon}_i) = 0)$. Omitted variable bias is one way this assumption is not satisfied (see Chapter 12). We designed X2 so it has a positive correlation with X1. What happens when it is included in the DGP, but is left out of the LRM?[6]

```
# Test of omitted variable bias
for(i in 1:sims){
  Y4 <- a + b1*X1 + X2 + rnorm(n,0,1)    # X2 included in
                                           the DGP
  model4 <- lm(Y4 ~ X1)                    # X2 not
                                           included in
                                           the OLS model

  B.1[i] <- model4$coef[2]
}
```

The mean of the sample slopes from this model is 3.47, which is far from the population slope ($\beta_1 = 0.5$). If we include X2 multiplied by its "true" slope ($b2 = 0.75$) in the model, however, the mean of the sample slopes for X1 is

---

[6] This simulation could also be considered a test of measurement error if we assume that the "true score" of X1 is a combination of observed X1 and unobserved X2, which are correlated (see Chapter 13).

very close to 0.5, with only a slightly larger standard deviation and a wider 95% CI than in the original model.

The final simulation evaluates the linearity assumption (see Chapter 11). Recall that one form of specification bias mentioned in Chapter 12 involves non-linear associations that are not captured by the LRM. In the following, notice what happens when the population association between $X1$ and Y is quadratic.

```
# Test of the linearity assumption
for(i in 1:sims){
  Y5 <- a + b1*(X1^2) + rnorm(n,0,1)      # X1-squared in DGP
  model5 <- lm(Y5 ~ X1)
  B.1[i] <- model5$coef[2]
}
```

The results of this model establish the consequences of not satisfying the linearity assumption because the wrong functional form is used. The sample slope coefficients have a mean of 12.66, far from the population slope. A component-plus-residual plot also shows the curvature of the residuals as they are plotted against $X1$ (recall that this plot is built with the crPlots function that is part of the car package).

Other assumptions, such as those involving collinearity and independence of the errors of prediction, may also be examined with simulations. Simulations are also useful for studying the properties of other regression models, such as the logistic model. A limitation of the examples provided here is that they do not investigate what happens when the sample size varies. Examining different sample sizes is not difficult, though. Not surprisingly, larger samples yield more precise estimates.

Appendix C: Selected Formulas

1. Mean (arithmetic average) of a variable

$$E[X] = \mu = \frac{\Sigma X_i}{N} \text{(population)} \quad \text{or} \quad \bar{x} = \frac{\Sigma x_i}{n} \text{(sample)}$$

where N is the population size or n is the sample size

2. Sum of squares of a variable

$$SS[x] = \Sigma(x_i - \bar{x})^2$$

3. Sample variance of a variable

$$\text{var}[x] = s^2 = \frac{\Sigma(x_i - \bar{x})^2}{n-1}$$

4. Sample standard deviation of a variable

$$sd[x] = s = \sqrt{\frac{\Sigma(x_i - \bar{x})^2}{n-1}}$$

5. Coefficient of variation (CV) of a variable

$$CV = \frac{s}{x}$$

6. Standard error of the mean (estimates the standard deviation of the sample means assuming repeated sampling from a population)

$$se(\bar{x}) = \frac{s}{\sqrt{n}}$$

7. General formula for a confidence interval (CI)

$$\text{Point estimate} \pm \left[(\text{confidence level}) \times (\text{standard error}) \right]$$

(Note: the point estimate can be a mean or a linear regression slope coefficient)

DOI: 10.1201/9781003162230-20

8. z-scores computation

$$z\text{-score} = \frac{\left(x_i - \bar{x}\right)}{s}$$

9. Covariance of two variables

$$\text{cov}\left(x, y\right) = \frac{\Sigma\left(x_i - \bar{x}\right)\left(y_i - \bar{y}\right)}{n-1}$$

10. Pearson's correlation coefficient of two variables

$$\text{corr}\left(x, y\right) = r = \frac{\text{cov}\left(x, y\right)}{\sqrt{\text{var}\left(x\right) \times \text{var}\left(y\right)}}$$

$$\text{corr}\left(x, y\right) = r = \frac{\Sigma\left(z_x\right)\left(z_y\right)}{n-1}$$

11. *T*-test comparing two means (equal variances)

$$t = \frac{\bar{x} - \bar{y}}{s_p\sqrt{\dfrac{1}{n_1} + \dfrac{1}{n_2}}}$$

$$\text{where } s_p = \sqrt{\frac{\left(n_1 - 1\right)s_1^2 + \left(n_2 - 1\right)s_2^2}{\left(n_1 + n_2\right) - 2}}$$

12. *T*-test comparing two means (unequal variances) (Welch's test)

$$t' = \frac{\bar{x} - \bar{y}}{\sqrt{\dfrac{\text{var}\left(x\right)}{n_1} + \dfrac{\text{var}\left(y\right)}{n_2}}}$$

13. Sum of squared errors (SSE) for the LRM

$$\text{SSE} = \Sigma\left(y_i - \hat{y}_i\right)^2 = \Sigma\hat{\varepsilon}_i^2 = \Sigma\left(y_i - \left\{\alpha + \beta_1 x_i\right\}\right)^2$$

14. Regression sum of squares (RSS) for the LRM

$$RSS = \Sigma\left(\hat{y}_i - \bar{y}\right)^2$$

15. Residuals (raw or unstandardized) from the LRM

$$\text{resid}_i\left(\hat{\varepsilon}_i\right) = \left(y_i - \widehat{y}_i\right)$$

16. Slope coefficient for the simple LRM

$$\hat{\beta} = \frac{\Sigma(x_i - \bar{x})(y_i - \bar{y})}{\Sigma(x_i - \bar{x})^2}$$

17. Intercept for the simple LRM

$$\hat{\alpha} = \bar{y} - \left\{\hat{\beta} \times \bar{x}\right\}$$

18. Standard error of a simple linear regression slope coefficient

$$se\left(\hat{\beta}\right) = \sqrt{\frac{\Sigma(y_i - \hat{y}_i)^2 / n - 2}{\Sigma(x_i - \bar{x})^2}} = \sqrt{\frac{SSE / n - 2}{SS(x)}}$$

19. *T*-value for a linear regression slope coefficient assuming H_0: $\beta_{pop} = 0$

$$\text{t-value} = \frac{\hat{\beta}}{se\left(\hat{\beta}\right)}$$

20. Standardized linear regression slope coefficient (beta-weight)

$$\left(\hat{\beta}_k^*\right) = \hat{\beta}_k\left(\frac{s_{x_k}}{s_y}\right)$$

21. Multiple linear regression slope coefficients (matrix form)

$$\hat{\beta} = \left(\mathbf{X'X}\right)^{-1}\mathbf{X'Y}$$

22. Variance–covariance matrix for multiple linear regression slope coefficients (used to compute standard errors of the coefficients)

$$V = \left(\mathbf{X'X}\right)^{-1}\hat{\sigma}^2 \text{ where } \hat{\sigma}^2 = \frac{\hat{\varepsilon}'\hat{\varepsilon}}{n-k}$$

23. Standard error of multiple linear regression slope coefficients

$$se\left(\hat{\beta}_i\right) = \sqrt{\frac{\Sigma(y_i - \hat{y}_i)^2}{\Sigma(x_i - \bar{x})^2\left(1 - R_i^2\right)(n-k-1)}}$$

where k = the number of explanatory variables and $\left(1 - R_i^2\right)$ is the tolerance from the auxiliary regression equation: $x_{1i} = \alpha + \beta_1 x_{2i} + \ldots + \beta_k x_{ki}$

24. Coefficient of determination for the LRM

$$R^2 = \frac{RSS}{TSS} = 1 - \frac{SSE}{TSS}$$

where TSS is the (total) sum of squares of the outcome variable y

25. Adjusted coefficient of determination for the LRM

$$\bar{R}^2 = \left(R^2 - \frac{k}{n-1}\right)\left(\frac{n-1}{n-k-1}\right)$$

26. Mean square error and root mean square error for the LRM

$$MSE = \hat{\sigma}^2 = \frac{SSE}{n-k-1} = \frac{\hat{\varepsilon}'\hat{\varepsilon}}{n-k} \quad RMSE(S_E) = \sqrt{MSE}$$

27. F-value/statistic for the LRM

$$F\text{ value} = \frac{RSS/k}{SSE/(n-k-1)} = \frac{\text{Mean square due to regression}(MSR)}{\text{Mean square error}(MSE)}$$

28. Prediction interval for the LRM

$$PI = \hat{y} \pm \left(t_{n-2}\right) \times \left(RMSE\right) \times \sqrt{1 + \frac{1}{n} + \frac{\left(x_0 - \bar{x}\right)^2}{(n-1)s_x}}$$

29. Partial (nested) F-test to compare LRMs

$$F\text{ value} = \frac{RSS(\text{full model}) - RSS(\text{nested model})}{MSE(\text{full model})} \quad df = 1, n-k-1(\text{full model})$$

30. Multiple partial (nested) F-test to compare LRMs

$$F\text{ value} = \frac{\left[RSS(\text{full}) - RSS(\text{nested})\right]/q}{MSE(\text{full})} \quad df = q, \ n-k-1 \text{ (full)}$$

where q is the difference in the number of explanatory variables (*full–nested*)

31. Mallow's C_p to compare LRMs

$$C_p = \frac{SSE(p)}{MSE(k)} - \left[n - 2(p-1)\right]$$

where SSE(p) is from the restricted model, MSE(k) is from the full model, and p = the number of predictors in the restricted model

32. Durbin–Watson statistic to assess serial correlation of residuals

$$d = \frac{\sum_2^n (\hat{\varepsilon}_{ti} - \hat{\varepsilon}_{ti-1})^2}{\sum_1^n \hat{\varepsilon}_{ti}^2}$$

33. Moran's I to assess spatial autocorrelation

$$I = \frac{n \sum_i^n \sum_j^n w_{ij} z_i z_j}{(n-1) \sum_i^n \sum_j^n w_{ij}}$$

w_{ij} is the weight function based on the "distance" between units

34. Variance inflation factors (VIF) in LRMs

$$\text{VIF} = \frac{1}{\left(1 - R_i^2\right)}$$

where $\left\{1 - R_i^2\right\}$ is the tolerance from the auxiliary regression equation (see entry 23)

35. Weighted least squares (WLS) equations

$$\text{SSE}_w = \Sigma \left(\frac{1}{s_i^2}\right)\left(\left(y_i - \bar{y}'\right)^2\right)$$

$$\beta_1 = \frac{\Sigma \frac{1}{s_i^2}\left(\left(x_i - \bar{x}'\right)\left(y_i - \bar{y}'\right)\right)}{\Sigma \frac{1}{s_i^2}\left(x_i - \bar{x}'\right)^2} \quad \text{where } \bar{y}' = \frac{\Sigma \frac{y_i}{s_i^2}}{\Sigma \frac{1}{s_i^2}} \text{ and } \bar{x}' = \frac{\Sigma \frac{x_i}{s_i^2}}{\Sigma \frac{1}{s_i^2}}$$

36. Reliability of a measure

$$r_{xx^*} = \frac{x(\text{true score})}{x\left(\text{true score}\right) + v(\text{error})} = \frac{s_x^2}{s_{x^*}^2}$$

37. Hat matrix to estimate leverage values

$$h_i = \mathbf{H} = \mathbf{X}\left(\mathbf{X'X}\right)^{-1}\mathbf{X'}$$

38. Deleted studentized residuals

$$t_i = \frac{\hat{\varepsilon}_i}{\text{RMSE}_{(-1)}\sqrt{1 - h_i}}$$

39. Cook's D to identify influential observations

$$D_i = \frac{(y_i - \hat{y}_i)^2}{(k+1) \times MSE} \times \frac{h_i}{(1-h_i)^2}$$

40. DFFITS to identify influential observations

$$DFFITS_i = \frac{\hat{y}_i - \hat{y}_{i-1}}{RMSE_i \sqrt{h_i}}$$

41. Intraclass correlation (ICC) to assess dependence

$$ICC(\rho) = \frac{\sigma_u^2}{\sigma_u^2 + \sigma_\varepsilon^2}$$

where σ_u^2 is the variance within units and σ_ε^2 is the variance between units

42. Design effect to assess dependence

$$d^2 = 1 + (\text{ave. group size} - 1) \times ICC$$

43. Logistic function

$$p(y=1) = \frac{1}{1 + \exp\left[-(\alpha + \beta_x x_1 + \cdots + \beta_k x_k)\right]}$$

44. Odds of an event

$$\frac{p}{1-p}$$

where p = probability of an event

45. Odds ratio (OR)

$$\frac{p_1}{(1-p_1)} \bigg/ \frac{p_2}{(1-p_2)}$$

46. Akaike's information criterion (AIC)[1]

$$AIC = Deviance + (2 \times k) \text{ Maximum Likelihood Estimation } (MLE)$$

$$AIC = n \times \ln(SSE/n) + 2(k+2) \text{ OLS estimation}$$

---

[1] Several formulas are used to estimate the AIC and BIC. Those presented here are one version.

47. Bayesian information criterion (BIC)

$$BIC = Deviance + \left(\ln(n) \times k \right) \text{ Maximum Likelihood Estimation (MLE)}$$

$$BIC = n \times \ln(SSE/n) + (k+2)\left(\ln(n) \right) \text{ OLS estimation}$$

Appendix D: User-Written R Packages Employed in the Examples

1. boot
2. car
3. caret
4. DescTools
5. effects
6. gee
7. ggplot2
8. het.test
9. Hmisc
10. interplot
11. ivpack
12. lm.beta
13. lme4
14. lmtest
15. mi
16. mice
17. MuMIn
18. olsrr
19. party
20. performance
21. perturb
22. plotly
23. plyr
24. prais
25. psych
26. quantreg
27. relaimpo
28. robustreg
29. sampleSelection
30. sandwich
31. survey

Note: the list includes only user-written packages that are not part of the native R environment, though some load other packages that are not listed here. R packages that load upon start-up and contain functions used in the modeling exercises are not listed (e.g., methods, stats).

References

Afifi, A. A., and R. M. Elashoff. 1966. "Missing Observations in Multivariate Statistics I. Review of the Literature." *Journal of the American Statistical Association* 61(315): 595–604.

Agresti, Alan. 2018. *Introduction to Categorical Data Analysis*, 3rd Ed. Hoboken, NJ: Wiley.

Aguinis, Herman, Ryan K. Gottfredson, and Harry Joo. 2013. "Best-Practice Recommendations for Defining, Identifying, and Handling Outliers." *Organizational Research Methods* 16(2): 270–301.

Akaike, Hirotugu. 1973. "Information Theory and an Extension of the Maximum Likelihood Principle." In *The 2nd International Symposium on Information Theory*, edited by B. N. Petrov and F. Csáki. Armenia: Tsahkadsor.

Allison, Paul. 2012. "Handling Missing Data by Maximum Likelihood." Retrieved from www.statisticalhorizons.com/wp-content/uploads/MissingDataByML.pdf.

Alpaydin, Etham. 2018. *Introduction to Machine Learning*, 3rd Ed. Cambridge, MA: MIT Press.

Angrist, Joshua D., and Jörn-Steffen Pischke. 2009. *Mostly Harmless Econometrics: An Empiricist's Companion*. Princeton, NJ: Princeton University Press.

Arbuthnot, John. 1711. "An Argument for Divine Providence, Taken from the Constant Regularity Observ'd in the Births of Both Sexes." *Philosophical Transactions of the Royal Society of London* 27(328): 186–190.

Attewell, Paul, and David Monaghan. 2015. *Data Mining for the Social Sciences: An Introduction*. Oakland, CA: University of California Press.

Barnett, Adrian G., Jolieke C. Van Der Pols, and Annette J. Dobson. 2004. "Regression to the Mean: What It Is and How To Deal With It." *International Journal of Epidemiology* 34(1): 215–220.

Bartholomew, David J., Martin Knott, and Irini Moustaki. 2011. *Latent Variable Models and Factor Analysis*, 3rd Ed. New York: Oxford University Press.

Bateson, W. 1908. *The Methods and Scope of Genetics*. Cambridge: Cambridge University Press.

Beaujean, A. Alexander. 2014. *Latent Variable Modeling Using R: A Step-By-Step Guide*. New York: Routledge.

Belsley, David A., Edwin Kuh, and Roy E. Welsch. 2004. *Regression Diagnostics: Identifying Influential Data and Sources of Collinearity*. Hoboken, NJ: Wiley.

Benjamin, Daniel J., and James O. Berger. 2019. "Three Recommendations for Improving the Use of *p*-Values." *American Statistician* 73(S1): 186–191.

Benjamini, Yoav, and Yosef Hochberg. 1995. "Controlling the False Discovery Rate: A Practical and Powerful Approach to Multiple Testing." *Journal of the Royal Statistical Society: Series B* 57(1): 289–300.

Berkson, Joseph. 1944. "Application of the Logistic Function to Bio-Assay." *Journal of the American Statistical Association* 39(227): 357–365.

Berry, William D., and Stanley Feldman. 1985. *Multiple Regression in Practice*. Newbury Park, CA: SAGE.

Bieder, Paul P., Robert M. Groves, Lars E. Lyberg, Nancy A. Mathiowetz, and Seymour Sudman (Eds.). 2004. *Measurement Errors in Surveys*. New York: Wiley.

Bliss, Chester I. 1934. "The Method of Probits." *Science* 79(2037): 38–39.

Bollen, Kenneth A., Paul P. Biemer, Alan F. Karr, Stephen Tueller, and Marcus E. Berzofsky. 2016. "Are Survey Weights Needed? A Review of Diagnostic Tests in Regression Analysis." *Annual Review of Statistics and Its Application* 3: 375–392.

Bollen, Kenneth A., James B. Kirby, Patrick J. Curran, Pamela M. Paxton, and Feinian Chen. 2007. "Latent Variable Models Under Misspecification: Two-Stage Least Squares (2SLS) and Maximum Likelihood (ML) Estimators." *Sociological Methods & Research* 36(1): 48–86.

Box, George P., and David R. Cox. 1964. "An Analysis of Transformations." *Journal of the Royal Statistical Society,: Series B* 26(2): 211–243.

Bradburn, Norman, Seymour Sudman, and Brian Wansink. 2004. *Asking Questions: The Definitive Guide to Questionnaire Design*, Rev. Ed. San Francisco: Jossey-Bass.

Bretz, Frank, Torsten Hothorn, and Peter Westfall. 2016. *Multiple Comparisons Using R*. Boca Raton, FL: CRC Press.

Breusch, T. S., and A. R. Pagan. 1979. "A Simple Test for Heteroskedasticity and Random Coefficient Variation." *Econometrica* 47(5): 1287–1294.

Browne, William J. n.d. "Sample Sizes for Multilevel Models." Retrieved from www .bristol.ac.uk/cmm/learning/multilevel-models/samples.html.

Burger, Scott V. 2018. *Introduction to Machine Learning with R*. Sebastopol, CA: O'Reilly Media.

Cantwell, Patrick J., Howard Hogan, and Kathleen M. Styles. 2004. "The Use of Statistical Methods in the U.S. Census." *American Statistician* 58(3): 203–212.

Carsey, Thomas M., and Jeffrey J. Harden. 2013. *Monte Carlo Simulation and Resampling Methods for Social Science*. Thousand Oaks, CA: SAGE.

Case, Anne, and Angus Deaton. 2020. *Deaths of Despair and the Future of Capitalism*. Princeton, NJ: Princeton University Press.

Chatterjee, Samprit, and Ali S. Hadi. 2012. *Regression Analysis by Example*, 5th Ed. New York: Wiley.

Chernick, Michael R., and Robert A. LaBudde. 2014. *An Introduction to Bootstrap Methods with Applications to R*. Hoboken, NJ: Wiley.

Chihara, Laura M., and Tim C. Hesterberg. 2011. *Mathematical Statistics with Resampling and R*. Hoboken, NJ: Wiley.

Christensen, Garret, Jeremy Freese, and Edward Miguel. 2019. *Transparent and Reproducible Social Science Research: How to Do Open Science*. Oakland, CA: University of California Press.

Cleveland, William S. 1993. *Visualizing Data*. Summit, NJ: Hobart Press.

Cochrane, Donald, and Guy H. Orcutt. 1949. "Application of Least Squares Regression to Relationships Containing Auto-Correlated Error Terms." *Journal of the American Statistical Association* 44(245): 32–61.

Cohen, Jacob. 1988. *Statistical Power Analysis for the Behavioral Sciences*, 2nd Ed. Mahwah, NJ: Lawrence Erlbaum Associates.

Connor, Steven. 2000. *Dumbstruck: A Cultural History of Ventriloquism*. New York: Oxford University Press.

Cook, Dennis. 1977. "Detection of Influential Observation in Linear Regression." *Technometrics* 19(1): 15–18.

Cox, David R. 1970. *The Analysis of Binary Data*. New York: Chapman and Hall.

Cramer, J. S. 2003. *Logit Models from Economics and Other Fields*. New York: Cambridge University Press.

Curto, José Dias, and José Castro Pinto. 2011. "The Corrected VIF (CVIF)." *Journal of Applied Statistics* 38(7): 1499–1507.

Darlington, Richard B., and Andrew F. Hayes. 2016. *Regression Analysis and Linear Models: Concepts, Applications, and Implementation*. New York: Guilford Press.

Darnell, Alfred, and Darren E. Sherkat. 1997. "The Impact of Protestant Fundamentalism on Educational Attainment." *American Sociological Review* 62(2): 306–315.

Daston, Lorraine. 1988. *Classical Probability in the Enlightenment*. Princeton, NJ: Princeton University Press.

David, H. A. 1998. "Early Sample Measures of Variability." *Statistical Science* 13(4): 368–377.

Davies, Neil M., Stephanie von Hinke Kessler Scholder, Helmut Farbmacher, Stephen Burgess, Frank Windmeijer, and George Davey Smith. 2015. "The Many Weak Instruments Problem and Mendelian Randomization." *Statistics in Medicine* 34(3): 454–468.

Dawid, Philip. 2017. "On Individual Risk." *Sythese* 194(9): 3445–3474.

Demidenko, Eugene. 2016. "The *p*-Value You Can't Buy." *American Statistician* 70(1): 33–38.

Deutsch, David. 1997. *The Fabric of Reality*. New York: Penguin Books.

DiCiccio, Thomas J., and Bradley Efron. 1996. "Bootstrap Confidence Intervals." *Statistical Science* 11(3): 189–212.

Dillman, Don A., Jolene D. Smyth, and Leah Melani Christian. 2014. *Internet, Phone, Mail, and Mixed-Mode Surveys: The Tailored Design Method*, 4th Ed. Hoboken, NJ: Wiley.

Dunn, Peter K., and Gordon K. Smyth. 2018. *Generalized Linear Models with Examples in R*. New York: Springer.

Durbin, James. 1954. "Errors in Variables." *Review of the International Statistical Institute* 22(1/3): 23–32.

Durbin, James, and Geoffrey S. Watson. 1971. "Testing for Serial Correlation in Least Squares Regression. III." *Biometrika* 58(1): 1–19.

Edgeworth, Francis Y. 1888. "On a New Method of Reducing Observations Relating to Several Quantities." *London, Edinburgh, and Dublin Philosophical Magazine and Journal of Science* 25(154): 184–191.

Efron, Bradley. 1979. "Bootstrap Methods: Another Look at the Jackknife." *Annals of Statistics* 7(1): 1–26.

Eicker, Friedhelm. 1967. "Limit Theorems for Regression with Unequal and Dependent Errors." In *Proceedings of the Fifth Berkeley Symposium on Mathematical Statistics and Probability*, Vol. 1, edited by Lucien M. Le Cam and Jerzy Neyman, 59–82. Berkeley, CA: University of California Press.

Ekland, Gunner. 1959. *Studies of Selection Bias in Applied Statistics*. Upsalla, SE: Almqvist & Wiksells.

Eliason, Scott R. 1993. *Maximum Likelihood Estimation: Logic and Practice*. Newbury Park, CA: SAGE.

Epstein, Lee, and Andrew Martin. 2005. "Coding Variables." In *Encyclopedia of Social Measurement*, edited by Kimberly Kempf-Leonard, 321–327. New York: Elsevier.

Faraway, Julian J. 2014. *Linear Models with R*, 2nd Ed. Boca Raton, FL: CRC Press.

Fieller, Nick. 2016. *Basics of Matrix Algebra for Statistics with R*. Boca Raton, FL: CRC Press.

Fienberg, Stephen E., and Nicole Lazar. 2001. "William Sealy Gosset." In *Statisticians of the Centuries*, edited by C. C. Heyde E. Seneta, P. Crepel, S. E. Fienberg, and J. Gani, 312–316. New York: Springer.

Finch, W. Holmes, Jocelyn E. Bolin, and Ken Kelley. 2016. *Multilevel Modeling Using R*. Boca Raton, FL: CRC Press.

Finch, W. Holmes, and Brian F. French. 2015. *Latent Variable Modeling with R*. New York: Routledge.

Firth, David. 1993. "Bias Reduction of Maximum Likelihood Estimates." *Biometrika* 80(1): 27–38.

Fisher, Ronald A. 1918. "The Correlation Between Relatives on the Supposition of Mendelian Inheritance." *Earth and Environmental Science Transactions of the Royal Society of Edinburgh* 52(2): 399–433.

Fox, John. 2016. *Applied Regression Analysis and Generalized Linear Models*, 3rd Ed. Los Angeles: SAGE.

Fox, John, Michael Friendly, and Sanford Weisberg. 2013. "Hypothesis Tests for Multivariate Linear Models Using the car Package." *The R Journal* 5(1): 39–52.

Fox, John, and Sanford Weisberg. 2018. *An R Companion to Applied Regression*, 3rd Ed. Thousand Oaks, CA: SAGE.

Franklin, James. 2015. *The Science of Conjecture: Evidence and Probability Before Pascal*. Baltimore, MD: Johns Hopkins University Press.

Franzen, Raymond, and Paul F. Lazarsfeld. 1945. "Mail Questionnaire as a Research Problem." *Journal of Psychology* 20(2): 293–320.

Freedman, David. 2007. "Statistical Models for Causation." In *The Sage Handbook of Social Science Methodology*, edited by William Outhwaite and Stephen P. Turner, 127–146. Los Angeles, CA: SAGE.

Friendly, Michael, and David Meyer. 2016. *Discrete Data Analysis with R: Visualization and Modeling Techniques for Categorical and Count Data*. Boca Raton, FL: CRC Press.

Fu, Vincent Kang, Christopher Winship, and Robert D. Mare. 2004. "Sample Selection ·Bias Models." In *Handbook of Data Analysis*, edited by Melissa A. Hardy and Alan Bryman, 409–430. Newbury Park, CA: SAGE.

Gallo, Amy. 2016. "A Refresher on Statistical Significance." *Harvard Business Review – Analytics*, February 16. Retrieved from https://hbr.org/2016/02/a-refresher-on -statistical-significance.

Galton, Francis. 1886. "Regression Towards Mediocrity in Hereditary Stature." *Journal of the Anthropological Institute of Great Britain and Ireland* 15: 246–263.

Gardiner, Joseph C., Zhehui Luo, and Lee Anne Roman. 2009. "Fixed Effects, Random Effects and GEE: What are the Differences?" *Statistics in Medicine* 28(2): 221–239.

Gauss, Carl Friedrich. 1823. *Theoria Combinationis Observationum Erroribus Minimis Obnoxiae [The Theory of the Combination of Errors in Observations]*. Göttingen, DE: Apud Henricum Dieterich.

Gelman, Andrew. 2008. "Scaling Regression Inputs by Dividing by Two Standard Deviations." *Statistics in Medicine* 27: 2865–2873.

Gelman, Andrew, and Sander Greenland. 2019. "Are Confidence Intervals Better Termed 'Uncertainty Intervals'?" *BMJ* 366: 15381.

Gelman, Andrew, and Jennifer Hill. 2007. *Data Analysis Using Regression and Multilevel/ Hierarchical Models*. New York: Cambridge University Press.

Gelman, Andrew, Jennifer Hill, and Aki Vehtari. 2020. *Regression and Other Stories.* New York: Cambridge University Press.

Gelman, Andrew, and Hal Stern. 2006. "The Difference Between 'Significant' and 'Not Significant' is Not Itself Statistically Significant." *American Statistician* 60(4): 328–331.

Gilbert, Robert P., Michael Shoushani, and Yvonne Ou. 2020. *Multivariable Calculus with Mathematica.* Boca Raton, FL: Chapman and Hall/CRC Press.

Glejser, Herbert. 1969. "A New Test for Heteroskedasticity." *Journal of the American Statistical Association* 64(235): 315–323.

Graybill, Franklin A., and Hariharan K. Iyer. 1994. *Regression Analysis: Concepts and Applications.* Belmont, CA: Duxbury Press.

Gromping, Ulrike. 2006. "Relative Importance for Linear Regression in R: The Package Relaimpo." *Journal of Statistical Software* 17(1): 1–27.

Groves, Robert M., Floyd J. Fowler Jr., Mick P. Couper, James M. Lepkowski, Eleanor Singer, and Roger Tourangeau. 2009. *Survey Methodology,* 2nd Ed. New York: Wiley.

Hahn, Gerry, Necip Doganaksoy, and Bill Meeker. 2019. "Statistical Intervals, Not Statistical Significance." *Significance* 16(4): 20–22.

Hald, Anders. 1999. "On the History of Maximum Likelihood in Relation to Inverse Probability and Least Squares." *Statistical Science* 14(2): 214–222.

Halley, Edmond. 1695. "A Most Compendious and Facile Method for Constructing the Logarithms." *Philosophical Transactions of the Royal Society of London* 19(216): 58–67.

Hanck, Christoph, Martin Arnold, Alexander Gerber, and Martin Schmelzer. 2019. *An Introduction to Econometrics with R.* Retrieved from https://www.econometrics-with-r.org.

Hardin, James, and Joseph Hilbe. 2012. *Generalized Estimating Equations,* 2nd Ed. Boca Raton, FL: CRC Press.

Harris, Randall E. 2019. *Epidemiology of Chronic Disease: Global Perspectives,* 2nd Ed. Burlington, MA: Jones & Bartlett.

Hausman, Jerry A. 1978. "Specification Tests in Econometrics." *Econometrica* 46(6): 1251–1271.

Hay-Jahans, Christopher. 2012. *An R Companion to Linear Statistical Models.* Boca Raton, FL: CRC Press.

Heckard, M. O. 1907. "The Practical Collection of Statistical Data." *Publications of the American Statistical Association* 10(80): 498–502.

Heckman, James J. 1976. "The Common Structure of Statistical Models of Truncation, Sample Selection and Limited Dependent Variables and a Simple Estimator for Such Models." *Annals of Economic and Social Measurement* 5: 475–492.

Held, Leonhard, and Manuela Ott. 2018. "On p-Values and Bayes Factors." *Annual Review of Statistics and Its Applications* 5: 393–419.

Held, Leonhard, Samuel Pawel, and Simon Schwab. 2020. "Replication Power and Regression to the Mean." *Significance* 17(6): 10–11.

Hesterberg, Tim. 2011. "Bootstrap." *Wiley Interdisciplinary Reviews: Computational Statistics* 3(6): 497–526.

Hilbe, Joseph M. 2015. *Practical Guide to Logistic Regression.* Boca Raton, FL: CRC Press.

Hobbs, N. Thompson, and Ray Hilborn. 2006. "Alternatives to Statistical Hypothesis Testing in Ecology: A Guide to Self-Teaching." *Ecological Applications* 16(1): 5–19.

Hoffmann, John P. 2013. "Declining Religious Authority? Confidence in the Leaders of Religious Organizations, 1973–2010." *Review of Religious Research* 55(1): 1–25.

Hoffmann, John P. 2016. *Regression Models for Categorical, Count, and Related Variables.* Oakland, CA: University of California Press.

Hoffmann, John P. 2017. *Principles of Data Management and Presentation.* Oakland, CA: University of California Press.

Hoffmann, John P. 2018. "Cohabitation, Marijuana Use, and Heavy Alcohol Use in Young Adulthood." *Substance Use & Misuse* 53(14): 2394–2404.

Hox, Joop J., Mirjam Moerbeek, and Rens van de Schoot. 2017. *Multilevel Analysis: Techniques and Applications,* 3rd Ed. New York: Routledge.

Hsieh, F. Y., Philip W. Lavori, Harvey J. Cohen, and John R. Feussner. 2003. "An Overview of Variance Inflation Factors for Sample-Size Calculation." *Evaluation & the Health Professions* 26(3): 239–257.

Hubbard, Raymond, and R. Murray Lindsay. 2008. "Why *P* Values are Not a Useful Measure of Evidence in Statistical Significance Testing." *Theory & Psychology* 18(1): 69–88.

Huber, Peter J. 1967. "The Behavior of Maximum Likelihood Estimates under Nonstandard Conditions." In *Proceedings of the Fifth Berkeley Symposium on Mathematical Statistics and Probability,* Vol. 1, edited by Lucien M. Le Cam and Jerzy Neyman, 221–233. Berkeley, CA: University of California Press.

Huff, Darrell. 1993. *How to Lie with Statistics.* New York: W. W. Norton.

Imbens, Guido W., and Donald J. Rubin. 2015. *Causal Inference for Statistics, Social, and Biomedical Sciences: An Introduction.* New York: Cambridge University Press.

Ioannidis, John J. A. 2019. "Options for Publishing Research Without Any *p*-Values." *European Heart Journal* 40(31): 2555–2556.

Johnson, Jeff W., and James M. LeBreton. 2004. "History and Use of Relative Importance Indices in Organizational Research." *Organizational Research Methods* 7(3): 238–257.

Jurečková, Jana, Jan Picek, and Martin Schindler. 2019. *Robust Statistical Methods with R,* 2nd Ed. Boca Raton, FL: CRC Press.

Kabacoff, Robert. 2015. *R in Action: Data Analysis and Graphics with R.* Shelter Island, NY: Manning Publications.

Karlson, Kristian Bernt, Anders Holm, and Richard Breen. 2012. "Comparing Regression Coefficients Between Same-Sample Nested Models Using Logit and Probit: A New Method." *Sociological Methodology* 42(1): 286–313.

Kaufman, Robert L. 2019. *Interaction Effects in Linear and Generalized Linear Models.* Thousand Oaks, CA: SAGE.

Kennedy, Ryan, and Philip D. Waggoner. 2021. *Introduction to R for Social Scientists: A Tidy Programming Approach.* Boca Raton, FL: CRC Press.

Kerr, Margaret, Maarten Van Zalk, and Håkan Stattin. 2012. "Psychopathic Traits Moderate Peer Influence on Adolescent Delinquency." *Journal of Child Psychology and Psychiatry* 53(8): 826–835.

Keynes, John Maynard. 1939. "Official Papers." *The Economic Journal* 49(195): 558–577.

Kleinbaum, David G., Lawrence L. Kupper, Azhar Nizam, and Eli S. Rosenberg. 1998. *Applied Regression Analysis and Other Multivariable Methods,* 3rd Ed. Pacific Grove, CA: Duxbury Press.

Kline, Paul. 1999. *An Easy Guide to Factor Analysis.* London: Routledge.

Koenker, Roger. 2000. "Galton, Edgeworth, Frisch, and Prospects for Quantile Regression in Econometrics." *Journal of Econometrics* 95(2): 347–374.

Koenker, Roger. 2005. *Quantile Regression.* New York: Cambridge University Press.

Krosnick, Jon A., and Leandre R. Fabrigar. 2005. *Questionnaire Design for Attitude Measurement in Social and Psychological Research.* New York: Oxford University Press.

Lam, Raymond W., Erin E. Michalaak, and Richard P. Swinson. 2005. *Assessment Scales in Depression, Mania and Anxiety.* New York: Taylor & Francis.

Lander, Jared P. 2017. *R for Everyone.* Upper Saddle River, NJ: Pearson Education.

Laplace, Pierre-Simon. 1805. *Traité de Mécanique Céleste, Tome IV [Treatise on Celestial Mechanics, Volume IV].* Paris: Chez Courcier.

Lee-Winn, Angela, Tamar Mendelson, and Renee M. Johnson. 2018. "Associations of Personality Traits with Marijuana Use in a Nationally Representative Sample of Adolescents in the United States." *Addictive Behaviors Reports* 8: 51–55.

Lehmann, E. L. 1994. "Jerzy Neyman." *Biographical Memoirs* 34: 395–420.

Leite, Walter. 2016. *Practical Propensity Score Methods Using R.* Los Angeles, CA: SAGE.

Levy, Paul S., and Stanley Lemeshow. 2013. *Sampling of Populations: Methods and Applications,* 4th Ed. New York: Wiley.

Lindgren, Bernard W. 1993. *Statistical Theory,* 4th Ed. Boca Raton, FL: Chapman & Hall.

Lista, Luca. 2017. *Statistical Methods for Data Analysis in Particle Physics.* New York: Springer.

Little, Roderick J. A., and Donald B. Rubin. 2019. *Statistical Analysis with Missing Data,* 3rd Ed. Hoboken, NJ: Wiley.

Liu, Xiaofeng Steven. 2014. *Statistical Power Analysis for the Social and Behavioral Sciences: Basic and Advanced Techniques.* New York: Routledge.

Liu, Ying. 2014. "Big Data and Predictive Business Analytics." *Journal of Business Forecasting* 33(4): 40–42.

Lo, Adeline, Herman Chernoff, Tian Zheng, and Shaw-Hwa Lo. 2015. "Why Significant Variables Aren't Automatically Good Predictors." *Proceedings of the National Academy of Sciences of the United States of America* 112(45): 13892–13897.

Lohr, Sharon L. 2019. *Measuring Crime: Beyond the Statistics.* Boca Raton, FL: CRC Press.

Long, J. Scott, and Laurie H. Ervin. 2000. "Using Heteroscedasticity Consistent Standard Errors in the Linear Regression Model." *American Statistician* 54(3): 217–224.

Long, Jeffrey D. 2011. *Longitudinal Data Analysis for the Behavioral Sciences Using R.* Los Angeles, CA: SAGE.

Lovie, A. D., and P. Lovie. 1993. "Charles Spearman, Cyril Burt, and the Origins of Factor Analysis." *Journal of the History of the Behavioral Sciences* 29(4): 308–321.

Lumley, Thomas. 2011. *Complex Surveys: A Guide to Analysis Using R.* Hoboken, NJ: Wiley.

Lumley, Thomas, Paula Diehr, Scott Emerson, and Lu Chen. 2002. "The Importance of the Normality Assumption in Large Public Health Data Sets." *Annual Review of Public Health* 23(1): 151–169.

Maas, Cora J. M., and Joop J. Hox. 2005. "Sufficient Sample Sizes for Multilevel Modeling." *Methodology* 1(3): 86–92.

Mackie, John. 1974. *The Cement of the Universe.* Oxford: Clarendon Press.

Mallows, C. L. 1973. "Some Comments on CP." *Technometrics* 15(4): 661–675.

Marchenko, Yulia V., and Marc G. Genton. 2012. "A Heckman Selection-*t* Model." *Journal of the American Statistical Association* 107(497): 304–317.

Marin, Jean-Michel, and Christian P. Robert. 2014. *Bayesian Essentials with R*, 2nd Ed. New York: Springer.

McCann, Melinda H., and Joshua D. Habiger. 2020. "The Detection of Nonnegligible Directional Effects with Associated Measures of Statistical Significance." *American Statistician* 74(3): 213–217.

McDonald, James B., and Jeff Sorensen. 2017. "Academic Salary Compression Across Disciplines and Over Time." *Economics of Education Review* 59: 87–104.

McShane, Blakeley B., David Gal, Andrew Gelman, Christian Robert, and Jennifer L. Tackett. 2019. "Abandon Statistical Significance." *American Statistician* 73(supp. 1): 235–245.

Mewes, Jan, Malcolm Fairbrother, Giuseppe Nicola Giordano, Cary Wu, and Rima Wilkes. 2021. "Experiences Matter: A Longitudinal Study of Individual-Level Sources of Declining Social Trust in the United States." *Social Science Research*. Online first. doi:10.1016/j.ssresearch.2021.102537.

Mill, John P. 1926. "Table of the Ratio: Area to Bounding Ordinate, for Any Portion of Normal Curve." *Biometrika* 18(3/4): 395–400.

Moran, P. A. P. 1950. "Notes on Continuous Stochastic Phenomena." *Biometrika* 37(1/2): 17–23.

Mumford, Stephen, and Rani Lill Anjum. 2013. *Causation*. Oxford: Oxford University Press.

Napier, John. 1681. *A Description of the Admirable Table of Logarithmes*. Translated by Edward Wright. London: Simon Waterson.

Nerlove, Marc. 2014. "Individual Heterogeneity and State Dependence: From George Biddell Airy to James Joseph Heckman." *Œconomia* 4(3): 281–320.

Newey, Whitney K., and Kenneth D. West. 1987. "A Simple, Positive Semi-Definite, Heteroskedasticity and Autocorrelation Consistent Covariance Matrix." *Econometrica* 55(3): 703–709.

Newman, Daniel A. 2014. "Missing Data: Five Practical Guidelines." *Organizational Research Methods* 17(4): 372–411.

Neyman, Jerzy. 1934. "On the Two Different Aspects of the Representative Method: The Method of Stratified Sampling and the Method of Purposive Selection." *Journal of the Royal Statistical Society* 97(4): 558–625.

Neyman, Jerzy. 1935. "On the Problem of Confidence Intervals." *Annals of Mathematical Statistics* 6(3): 111–116.

Neyman, Jerzy, and Egon S. Pearson. 1933. "On the Problem of the Most Efficient Tests of Statistical Hypotheses." *Philosophical Transactions of the Royal Society of London, Series A* 231(694–706): 289–337.

Nosek, Brian A., Charles R. Ebersole, Alexander C. DeHaven, and David T. Mellor. 2018. "The Preregistration Revolution." *Proceedings of the National Academy of Sciences of the United States of America* 115(11): 2600–2606.

Nurunnabi, A. A. M., N. Nasser, and A. H. M. R. Imon. 2016. "Identification and Classification of Multiple Outliers, High Leverage Points and Influential Observations in Linear Regression." *Journal of Applied Statistics* 43(3): 509–525.

Oh, Sohee, K. C. Carriere, and Taesung Park. 2008. "Model Diagnostic Plots for Repeated Measures Data Using the Generalized Estimating Equations Approach." *Computational Statistics & Data Analysis* 53(1): 222–232.

Ortiz-Ospina, Esteban, and Max Roser. 2019. "Trust." Retrieved from https://ourworldindata.org/trust.

Otto, Matthias. 2016. *Chemometrics: Statistics and Computer Application in Analytical Chemistry.* Weinheim, DE: Wiley-VCH.

Pan, Wei. 2001. "Akaike's Information Criterion in Generalized Estimating Equations." *Biometrics* 57(1): 120–125.

Paxton, Pamela, John R. Hipp, and Sandra Marquart-Pyatt. 2011. *Nonrecursive Models: Endogeneity, Reciprocal Relationships, and Feedback Loops.* Thousand Oaks, CA: SAGE.

Payton, Mark E., Matthew H. Greenstone, and Nathaniel Schenker. 2003. "Overlapping Confidence Intervals or Standard Error Intervals: What Do They Mean in Terms of Statistical Significance?" *Journal of Insect Science* 3: 34–39.

Pearl, Judea. 2009. *Causality: Modeling, Reasoning, and Inference,* 2nd Ed. New York: Cambridge University Press.

Pearson, Karl. 1894. "Contributions to the Mathematical Theory of Evolution." *Philosophical Transactions of the Royal Society of London Series. Part A,* 185: 71–110.

Pearson, Karl. 1895. "Note on Regression and Inheritance in the Case of Two Parents." *Proceedings of the Royal Society of London Series A* 58(347–352): 240–242.

Pearson, Karl. 1901. "On Lines and Planes of Closest Fit to Systems of Points in Space." *London, Edinburgh, and Dublin Philosophical Magazine and Journal of Science* 2(11): 559–572.

Peng, Richard, and Santosh Vempala. 2021. "Solving Sparse Linear Systems Faster Than Matrix Multiplication." Retrieved from https://arxiv.org/abs/2007.10254.

Pepinsky, Thomas B. 2018. "A Note on Listwise Deletion Versus Multiple Imputation." *Political Analysis* 26(4): 480–488.

Perezgonzalez, Jose D. 2015. "Fisher, Neyman-Pearson or NHST? A Tutorial for Teaching Data Testing." *Frontiers in Psychology* 6(223). Retrieved from https://www.frontiersin.org/articles/10.3389/fpsyg.2015.00223.

Pickover, Clifford A. 2009. *The Math Book: From Pythagoras to the 57th Dimension – 250 Milestones in the History of Mathematics.* New York: Sterling Publishers.

Plackett, R. L. 1958. "Studies in the History of Probability and Statistics. VII. The Principle of the Arithmetic Mean." *Biometrika* 45(1/2): 130–135.

Prais, Sigbert J., and Christopher B. Winsten. 1954. *Trend Estimators and Serial Correlation.* Chicago: Cowles Commission, Discussion Paper.

Pridemore, William A. 2011. "Poverty Matters: A Reassessment of the Inequality-Homicide Relationship in Cross-National Studies." *British Journal of Criminology* 51(5): 739–772.

Radloff, Lenore Sawyer. 1977. "The CES-D Scale: A Self Report Depression Scale for Research in the General Population." *Applied Psychological Measurement* 1: 385–401.

Ramsey, J. B. 1969. "Tests for Specification Error in Classical Linear Least Squares Regression Analysis." *Journal of the Royal Statistical Society: Series B* 31(2): 350–371.

Ritz, Christian, and Jens Carl Streibig. 2008. *Nonlinear Regression with R.* New York: Springer.

Rogerson, Peter A. 2006. *Statistical Methods for Geography,* 2nd Ed. Thousand Oaks, CA: SAGE.

Roiger, Richard J. 2017. *Data Mining: A Tutorial-Based Primer,* 2nd Ed. Boca Raton, FL: CRC Press.

Ross, Sheldon. 1994. *A First Course on Probability.* New York: Macmillan.

Rousseeuw, Peter J., and Annick M. Leroy. 2003. *Robust Regression and Outlier Detection.* New York: Wiley.

Rubin, Donald B. 1976. "Inference and Missing Data." *Biometrika* 63(3): 581–592.

Rubin, Donald B. 1977. "The Design of a General and Flexible System for Handling Nonresponse in Sample Surveys." Prepared under contract for the U.S. Social Security Administration.

Salsburg, David. 2001. *The Lady Tasting Tea*. New York: Owl Books.

Sargan, John D. 1958. "The Estimation of Economic Relationships Using Instrumental Variables." *Econometrica* 26(3): 393–415.

Savitzky, Abraham, and Marcel J. E. Golay. 1964. "Smoothing and Differentiation of Data by Simplified Least Squares Procedures." *Analytical Chemistry* 36(8): 1627–1639.

Scheuren, Fritz. 2005. "Multiple Imputation: How it Began and Continues." *American Statistician* 59(4): 315–319.

Schmidt, Amand F., and Chris Finan. 2018. "Linear Regression and the Normality Assumption." *Journal of Clinical Epidemiology* 98: 146–151.

Schott, James R. 2016. *Matrix Analysis for Statistics*, 3rd Ed. Hoboken, NJ: Wiley.

Schwarz, Gideon E. 1978. "Estimating the Dimension of a Model." *Annals of Statistics* 6(2): 461–464.

Scott, John. 2014. *A Dictionary of Sociology*, 4th Ed. New York: Oxford University Press.

Sheather, Simon. 2009. *A Modern Approach to Regression with R*. New York: Springer.

Shoesmith, Eddie. 1987. "The Continental Controversy over Arbuthnot's Argument for Divine Providence." *Historia Mathematica* 14(2): 133–146.

Shumway, Robert H., and David S. Stoffer. 2010. *Time Series Analysis and Its Applications: With R Examples*. New York: Springer.

Silvey, S. D. 1975. *Statistical Inference*. Boca Raton, FL: CRC Press.

Singer, Judith D., and John B. Willett. 2003. *Applied Longitudinal Data Analysis*. New York: Oxford University Press.

Skellam, J. G. 1948. "A Probability Distribution Derived from the Binomial Distribution by Regarding the Probability of Success as Variable Between the Sets of Trials." *Journal of the Royal Statistical Society, Series B*, 10(2): 257–261.

Spanos, Aris. 2019. *Probability Theory and Statistical Inference*, 2nd Ed. New York: Cambridge University Press.

Spiegelhalter, David. 2019. *The Art of Statistics: Learning from Data*. London: Penguin.

Steel, E. Ashley, Martin Liermann, and Peter Guttorp. 2019. "Beyond Calculations: A Course in Statistical Thinking." *American Statistician* 73(S1): 392–401.

Stevens, S. S. 1946. "On the Theory of Scales of Measurement." *Science* 103(2684): 677–680.

Stigler, Stephen M. 1973. "Simon Newcomb, Percy Daniell, and the History of Robust Estimation, 1885–1920." *Journal of the American Statistical Association* 68(344): 872–879.

Stigler, Stephen M. 1981. "Gauss and the Invention of Least Squares." *Annals of Statistics* 9(3): 465–474.

Stigler, Stephen M. 1986. *The History of Statistics: The Measurement of Uncertainty Before 1900*. Cambridge, MA: Harvard University Press.

Stigler, Stephen M. 1999. *Statistics on the Table*. Cambridge, MA: Harvard University Press.

Stock, James H., and Francesco Trebbi. 2003. "Retrospectives: Who Invented Instrumental Variable Regression?" *Journal of Economic Perspectives* 17(3): 177–194.

Stone, James V. 2013. *Bayes' Rule: A Tutorial Introduction to Bayesian Analysis*. Sheffield: Sebtel Press.

Stone, Mervyn. 1974. "Cross-Validatory Choice and Assessment of Statistical Predictions." *Journal of the Royal Statistical Society,: Series B* 36(2): 111–133.

Student [William S. Gosset]. 1908. "The Probable Error of a Mean." *Biometrika* 6(1): 1–25).

Su, Yu-Sung, Andrew E. Gelman, Jennifer Hill, and Masanao Yajima. 2011. "Multiple Imputation with Diagnostics (mi) in R: Opening Windows into the Black Box." *Journal of Statistical Software* 45(2): 1–31.

Suárez, Erick, Cynthia M. Pérez, Roberto Rivera, and Melissa N. Martínez. 2016. *Applications of Regression Models in Epidemiology*. Hoboken, NJ: Wiley.

Thompson, Bruce. 2004. *Exploratory and Confirmatory Factor Analysis: Understanding Concepts and Applications*. Washington, DC: American Psychological Association.

Tsipursky, Gleb. 2018. "(Dis)trust in Science." *Scientific American*, July 5. Retrieved from https://blogs.scientificamerican.com/observations/dis-trust-in-science.

Tukey, John W. 1977. *Exploratory Data Analysis*. Reading, MA: Addison-Wesley.

Ullah, Muhammad Imdad, Muhammad Aslam, and Saima Altaf. 2018. "lmridge: A Comprehensive R Package for Ridge Regression." *R Journal* 10(2): 326–348.

van Belle, Gerald. 2008. *Statistical Rules of Thumb*. New York: Wiley.

van Buuren, Stef. 2018. *Flexible Imputation of Missing Data*, 2nd Ed. Boca Raton, FL: Chapman & Hall.

Verbeek, Marno. 2017. *A Guide to Modern Econometrics*, 5th Ed. Hoboken, NJ: Wiley.

von Hippel, Paul. 2020. "How Many Imputations Do You Need? A Two-Stage Calculation Using a Quadratic Rule." *Sociological Methods & Research* 49(3): 699–718.

Wachholder, Sholom. 1986. "Binomial Regression in GLIM: Estimating Risk Ratios and Risk Differences." *American Journal of Epidemiology* 123(1): 174–184.

Wallace, Michael. 2020. "Analysis in an Imperfect World." *Significance* 17(1): 14–19.

Weakliem, David L. 2016. *Hypothesis Testing and Model Selection in the Social Sciences*. New York: Guilford.

Weiss, Neil A. 1999. *Introductory Statistics*, 5th Ed. Reading, MA: Addison-Wesley.

Welch, B. L. 1947. "The Generalization of 'Student's' Problem When Several Different Population Variances are Involved." *Biometrika* 34(1/2): 28–35.

Welsch, Roy E., and Edwin Kuh. 1977. "Linear Regression Diagnostics." Working Paper No. 173. Cambridge, MA: National Bureau of Economic Research.

White, Halbert. 1980. "A Heteroskedasticity-Consistent Covariance Matrix Estimator and a Direct Test for Heteroskedasticity." *Econometrica* 48(4): 817–838.

Wickens, Thomas D. 1995. *The Geometry of Multivariate Statistics*. Mahwah, NJ: Lawrence Erlbaum Associates.

Wilcox, Rand. 2014. *Introduction to Robust Estimation and Hypothesis Testing*, 4th Ed. San Diego, CA: Academic Press.

Wilks, Samuel S. 1932. "Moments and Distributions of Estimates of Population Parameters from Fragmentary Samples." *Annals of Mathematical Statistics* 3(3): 163–195.

Williams, Arthur Robin, Julian Santaella-Tenorio, Christine M. Mauro, Frances R. Levin, and Silvia S. Martins. 2017. "Loose Regulation of Medical Marijuana Programs Associated with Higher Rates of Adult Marijuana Use but Not Cannabis Use Disorder." *Addiction* 112(11): 1985–1991.

Williams, Richard. 2020. "Ordinal Independent Variables." Retrieved from https://www3.nd.edu/~rwilliam/xsoc73994/OrdinalIndependent.pdf.

Wood, Simon N. 2006. *Generalized Additive Models: An Introduction with R*. Boca Raton, FL: Chapman & Hall/CRC Press.

Wu, De-Min. 1973. "Alternative Tests of Independence between Stochastic Regressors and Disturbances." *Econometrica* 41(4): 733–750.

Yadav, Madan Lal, and Basav Roychoudhury. 2018. "Handling Missing Values: A Study of Popular Imputation Packages in R." *Knowledge-Based Systems* 160(15): 104–118.

Yah, Bee Wah, and Chiaw Hock Sim. 2011. "Comparisons of Various Types of Normality Tests." *Journal of Statistical Computation and Simulation* 81(12): 2141–2155.

Yong, Ed. 2018. "Psychology's Replication Crisis is Running Out of Excuses." *Atlantic*, November 19. Retrieved from https://www.theatlantic.com/science/archive/2018/11/psychologys-replication-crisis-real/576223.

Yule, George Udny. 1927. "On a Method of Investigating Periodicities Disturbed Series, with Special Reference to Wolfer's Sunspot Numbers." *Philosophical Transactions of the Royal Society of London Series. Part A* 226(636–646): 267–298.

Zeileis, Achim. 2006. "Econometric Computing with HC and HAC Covariance Matrix Estimators." Retrieved from https://cran.r-project.org/web/packages/sandwich/vignettes/sandwich.pdf.

Zhou, Guangyong. 2004. "A Modified Poisson Regression Approach to Prospective Studies with Binary Data." *American Journal of Epidemiology* 159(7): 702–706.

Index

2SLS, *see* Two-stage least squares

A

Absent variables (underfitting)
    confounding, 249
    correlation, 245–246
    illustration of, 248
    omitted variable, 245
Adjusted R^2, 93, 257, 270, 392
Akaike's Information Criterion
    (AIC), 111–112, 394
Alternative hypothesis (H_a), 20–21, 24
    autocorrelation, 149
    Breusch-Pagan test, 175
    F-test and R^2, 94–95
    slope coefficients, 54
American.csv dataset, 35
Analysis of variance (ANOVA), 8, 116,
    316–317, 330–331
    procedures, 32
    table, 89–91
AR(1) pattern, 154–156
Arithmetic mean, 10
Assumptions of LRMs, 42–43, 80–81
Autocorrelation, *see* Serial correlation
Auxiliary regression model, 85, 194

B

Bayesian inference, 16, 17
Bayesian Information Criterion (BIC),
    111–112, 395
Best linear unbiased estimator
    (BLUE), 83
Best statistical practices, 3–4
Beta weights, 76, 77; *see also*
    Standardized slope
    coefficients
BIC, *see* Bayesian Information Criterion
Binary logistic regression model, *see*
    Logistic regression model
Bivariate regression model, 38

BLUE, *see* Best linear unbiased
    estimator
Bootstrapping, 181–182, 308
Breusch–Pagan test for
    homoscedasticity, 175, 177

C

CART, *see* Classification and
    regression trees
Categorical variables, 11, 30, 337, 346,
    356, 374; *see also* Discrete
    variables
    as explanatory variables, 115–116,
    118, 120, 121–123, 125, 129
Causation, 8–9, 48, 357–359; *see also*
    Establishing causal
    associations
Center for Epidemiological Studies-
    Depression (CES-D) scale, 375
Center of gravity, 10
Central Limit Theorem (CLT), 82, 83, 385
Central tendency measures, 9–14
CES-D, *see* Center for Epidemiological
    Studies-Depression scale
Classification and regression trees
    (CART)
    data analytics and machine
    learning, 233
    interactions, 233
    nodes, 235–236
    tree illustration, 234
CLT, *see* Central Limit Theorem
Cluster effect, 318
Clustering adjustment, 141–143
    LRM, no adjustment, 142–143
    LRM adjusted for clustering, 143
    multilevel dataset, 142
Coding, 117, 276, 376
Coefficient of determination, 92, 392;
    see also R^2
Coefficient of variation (CV), 15, 389
Cohen's *d*, 75

411